Blood Traces

Blood Traces

Interpretation of Deposition and Distribution

PETER R. DE FOREST
City University of New York
New York, NY, USA

PETER A. PIZZOLA
Pace University
New York, NY, USA

BROOKE W. KAMMRATH
University of New Haven
West Haven, CT, USA

This edition first published 2021

Registered Offices
John Wiley & Sons, Inc., 111 River Street, Hoboken, NJ 07030, USA
John Wiley & Sons Ltd, The Atrium, Southern Gate, Chichester, West Sussex, PO19 8SQ, UK

Editorial Office
The Atrium, Southern Gate, Chichester, West Sussex, PO19 8SQ, UK

For details of our global editorial offices, customer services, and more information about Wiley products visit us at www.wiley.com.

Wiley also publishes its books in a variety of electronic formats and by print-on-demand. Some content that appears in standard print versions of this book may not be available in other formats.

Library of Congress Cataloging-in-Publication Data

Names: De Forest, Peter R., author. | Pizzola, Peter A., author. | Kammrath, Brooke W., author.
Title: Blood traces : interpretation of deposition and distribution / Peter R. De Forest, Peter A. Pizzola, Brooke W. Kammrath.
Description: Hoboken, NJ : Wiley, 2021. | Includes bibliographical references and index.
Identifiers: LCCN 2021031543 (print) | LCCN 2021031544 (ebook) | ISBN 9781119764533 (cloth) | ISBN 9781119764700 (adobe pdf) | ISBN 9781119764717 (epub)
Subjects: LCSH: Bloodstain pattern analysis. | Forensic hematology. | Crime scene searches. | Evidence, Criminal. | Forensic sciences–Methodology.
Classification: LCC HV8077.5.B56 D45 2021 (print) | LCC HV8077.5.B56 (ebook) | DDC 363.25/62–dc23
LC record available at https://lccn.loc.gov/2021031543
LC ebook record available at https://lccn.loc.gov/2021031544

Cover Design: Wiley
Cover Image: © Courtesy of Jeffrey Buszka and Norman Marin

Set in 10/12pt STIXTwoText by Straive, Pondicherry, India
Printed and bound by CPI Group (UK) Ltd, Croydon, CR0 4YY

C9781119764533_280721

Dedication

We dedicate this book to our students and those scientists striving to improve the science of forensic science and its broader implementation in the service of justice.

For my late wife and life partner Carol, daughter Kimi, son-in-law Jon, son Robb, daughter-in-law Toni, and grandchildren Colby, Wesley, Helena, and Eleanor, who continue to inspire and amaze me.
Peter R. De Forest

To Lori – my wife and best bud – for her support and patience. And for my children Anthony, Paul, Donna, Danielle, along with their spouses and my grandchildren – all of whom keep me on my toes.
Peter A. Pizzola

To my husband Matt, it is a privilege to share this life with you. I love you more. To my children Riley and Grayson, you are my motivation and joy. I love you most.
Brooke W. Kammrath

Epigraph

A poor physical evidence investigation risks punishing the victim, benefits the guilty, and may adversely impact the accused.

Contents

Foreword

From my early childhood memories, I can vividly remember how I became hooked on this black and white American TV series entitled *The Fugitive* that initially ran in the USA between 1963 and 1967 and that I discovered in a French-dubbed version a few years later. In this show, a doctor wrongfully convicted of his wife's murder manages to escape and looks for the real killer. I am not sure how my parents could let a five-year-old watch such a show. However, it planted a seed! Many years later, in the mid-eighties, as an undergraduate student at the University of Lausanne, Switzerland, working under Professor Pierre Margot, and while my passion for forensic science was rapidly growing, I learned that *The Fugitive* might have been loosely inspired by the case of the murder of Marilyn Sheppard in 1954. Like on the TV show (and in the 1993 movie with Harrison Ford), the victim's husband, Dr. Sam Sheppard, was a doctor and was convicted for the murder. However, unlike in the fiction, the husband was acquitted at a retrial in 1967.

Why am I telling you this? First, because blood traces played a significant role in this case, particularly in the reconstruction proposed by the defence, ultimately leading to a non-guilty verdict the second time around. Further, the examination of these blood traces and the reconstruction of the events based on these observations were undertaken by one of the father figures of forensic science, Dr. Paul L. Kirk. At this point, it is worth mentioning that the authors of this book are a direct legacy of Dr. Kirk, particularly Dr. Peter R. De Forest who studied under him and subsequently worked with him for many years. With such a heritage, it is not surprising to discover such a fine book.

Peter De Forest, Brooke Kammrath and Peter Pizzola bring with them over 120 years of combined experience in forensic science, specifically focusing on crime scene reconstructions and blood traces examinations. It is difficult to identify a more authoritative team to cover such a complex and critical topic comprehensively.

In this day and age, forensic science is often criticised for its lack of scientific foundations, and crime laboratories increasingly operate in a mechanistic and self-contained way following what some call a "pill factory paradigm." It is refreshing to see a modern forensic science book calling for improving forensic science through first scientific principles and not mainly normative processes. The need to reframe the crime scene as a scientific problem that has to be addressed using a scientific approach applied by scientifically trained personnel is a prime example of where improvements can occur. A better understanding of fundamental forensic science principles is another area in need, and, unsurprisingly, the authors succeed in anchoring blood traces to these principles. For example, the book includes an enhanced debate about terminology (e.g., stain vs trace; pattern vs configuration) and a discussion about blood traces as a vector of information to develop and answer relevant questions.

I cannot recommend this book strongly enough. After a brief review of the history of blood traces, several key issues are identified in the introduction and ultimately dealt with in more detail later in the text. Basic principles of the interpretation and analysis of blood traces' configuration are well enunciated and then developed in subsequent chapters. Many concepts are illustrated using casework examples. With this book, the authors make a significant contribution to forensic science in general, and the scientific examination and interpretation of crime scene and blood traces in particular.

By reading this book, if you are a student or a trainee, you have the assurance to start your journey in forensic science on the right path. If you are a practitioner, scientist or lawyer, you can benefit from a fresh perspective having its roots in forensic science foundations; it is never too late to learn. If you are a researcher, you will be inspired and develop impactful and relevant research in the future. And honestly, merely finding a forensic science book about blood without focusing on DNA should encourage you to read it straight away...

Distinguished Professor Claude Roux
Director, Centre for Forensic Science, University of Technology Sydney, Australia
President, International Association of Forensic Sciences

Acknowledgements

In addition to the patience, understanding and generosity of time by our families, we have been inspired by generations of past students. It is the role of an educator or mentor to teach critical thinking as well as subject-specific information, but we recognize that students are often the greatest teachers. We have learned more from our students than we have imparted and recognize that one of the most valuable rewards of being an educator and mentor is the opportunity to work with numerous outstanding students.

Much of the early work of this book was based on the concepts of Ralph Ristenbatt, III who also formulated a preliminary outline which was the initial impetus for this project. Throughout the course of this work, he provided many of the references, i.e., textbooks, journal articles, scientific reports, dissertations and theses that we studied and incorporated. Our only regret is that he obstinately refused to become a co-author!

We are grateful to Det. John C. Perkins (retired Yonkers P.D.) who assisted with many of the high-speed studies of blood droplet impacts and the study of backspatter as discussed in this book. He was part of the John Jay College group comprised of Lenore Kodet (retired forensic scientist), Stephen Roth (physicist), Pete Pizzola, and Peter R. De Forest. Stephen Kwechin, retired police officer and member of the Yonkers PD Crime Scene Unit, is thanked for the construction of the device for the controlled projection of blood droplets.

We are also indebted to Detective John Geiss of the Yonkers P.D. cold case unit for providing background information on some of the Yonkers cases discussed in the book and for acting as a liaison with the Department regarding permission to use certain photographs.

There are numerous other colleagues, past and present, who have made indelible impacts on our philosophy and understanding of forensic science. Although there are too many to mention all of them here, in particular, we gratefully acknowledge (*in alphabetical order*) Robert Adamo, Angie Ambers, David Barclay, Peter Barnett, Cliff Brant, Rebecca Bucht, JoAnn Buscaglia, Jeffrey Buszka, Patrick Buzzini, Brian Caddy, Timothy Carron, Donald Christopher, Vincent Crispino, Lisa Dadio, Donald Dahlberg, Josep De Alcaraz-Fossoul, Peter Diaczuk, Fred Drummond, Anna Duggar, Bart Epstein, Charles Hirsch, Brian Gestring, Claire Glynn, Adam Hall, Jack Hietpas, Charles Kingston, Paul Kirk, Koby Kizzire, Thomas Kubic, Philip Langellotti, Pauline Leary, Henry Lee, John Lentini, Colleen Lockhart, Doug Lucas, Pierre Margot, Norman Marin, Kirby Martir, Michelle Miranda, Chuck Morton, Elaine Pagliaro, Chris Palenik, Skip Palenik, Ken Paras, Joseph Peterson, Nicholas D.K. Petraco, Nicholas Petraco (Sr.), Dale Purcell, Lawrence Quarino, Barbara Sampson, Tony Raymond, John Reffner, Katherine Roberts, Linda Rourke, Claude Roux, David San Pietro, George Sensabaugh, Robert Shaler, Francis Sheehan, Jacqueline Speir, Peter Valentin, and Sheila Willis.

Preface to *Blood Traces: Interpretation of Deposition and Distribution*

Why do we need another text on the interpretation of blood traces? Has not the subject of bloodstain pattern analysis been covered adequately in other books? We are of the opinion that there are many aspects of the interpretation of blood traces that are of concern and have not been treated with the thought or caution that should be afforded them. Most important, existing texts do not sufficiently emphasize the need for interpretations to be made by experienced scientists with strict adherence to the scientific method. At one point in the report issued by the National Research Council (NRC) of the National Academy of Sciences (NAS) *Strengthening Forensic Science in the United States: A Path Forward* published in 2009, the question is raised as to whether or not there is a scientific basis behind opinions offered in court in what has been termed bloodstain pattern interpretation.

> *Although there is a professional society of bloodstain pattern analysts, the two organizations that have or recommend qualifications are the IAI and the Scientific Working Group on Bloodstain Pattern Analysis (SWGSTAIN). SWGSTAIN's suggested requirements for practicing bloodstain pattern analysis are outwardly impressive, as are IAI's 240 hours of course instruction. But the IAI has no educational requirements for certification in bloodstain pattern analysis. This emphasis on experience over scientific foundations seems misguided, given the importance of rigorous and objective hypothesis testing and the complex nature of fluid dynamics. In general, the opinions of bloodstain pattern analysts are more subjective than scientific.*
>
> *(NAS Report 2009, page 178)*

As of the writing of this book, there are still no science-based higher education requirements for bloodstain pattern analysts, although efforts are underway in the Organization of Scientific Area Committees (OSACs) and the American Standards Board (ASB) of the American Academy of Forensic Sciences (AAFS). What fails to be appreciated in current practice and publications is that situations involving the interpretation of blood trace configurations are often very complex and the difficulties faced in rendering conclusions in this area are among the most scientifically challenging of those in any area of forensic science.

After a brief survey of the history of the subject of blood traces, several key issues will be identified in the introduction and ultimately dealt with in more detail later in the text. At the end of this chapter, some basic principles of the interpretation and analysis of the configuration of blood traces will be enunciated. These will be discussed at greater length and reiterated in subsequent chapters.

What are blood traces? Why use this as the title for our book rather than the generally accepted term in the English language of "bloodstain patterns?" The early contributors to nascent forensic science, including Edmund Locard, Hans Gross, and Eduard Piotrowski, used the term traces (traces de sang in French) or in the case of the German-speaking authors "Blutspuren" (Piotrowski 1895), the German expression meaning "traces of blood," when referring to the analysis of the physical configurations of blood evidence discovered at the scene of a crime. In addition, not all blood deposits stain the substrate and thus are not truly "stains." Examples would include dried deposits on hydrophobic surfaces, such as polymer coatings (e.g., paint and varnish), plastics, or treated glass, from which they can easily be dislodged. There is the potential for ambiguity with the term blood traces in that it could suggest biochemical genetic testing rather than an interpretation of the three-dimensional geometry of the deposits at a scene. Ultimately, no expression is ideal. However, the authors feel that a return to this original terminology of blood traces more accurately represents all that is encompassed in the interpretation of the deposition and distribution of blood evidence at a crime scene and hope that it will inspire others to recognize that it is more than just pattern classification and analysis.

Why is the term "bloodstain pattern analysis" a poor term for this work? There are several reasons for an objection to this terminology. First, it is not individual patterns that are being analyzed, but rather the totality of the physical aspects of the blood traces. Various blood trace configurations are necessarily examined and contribute to the overall reconstruction. However, the so-called patterns should not be the focus of the forensic examination. Assigning a name to the geometric features of a blood trace, i.e., classifying the pattern, does not necessarily advance the goal of reconstruction and trace interpretation of blood evidence. The interrelation of blood traces with respect to the scene or other objects bearing traces of blood is of much greater importance. Second, there is not a finite number of "bloodstain patterns," thus a crime scene scientist will undoubtedly encounter important geometric configurations that do not fit into a neat classification scheme. For example, blood that has soaked through a fabric or some other porous structure that is no longer present at the time of the scene investigation may represent the vestige of an earlier and more complex configuration, such as blood through bedding that has been disturbed or removed. Third, in the physical evidence context, the term pattern connotes a comparison process, whereby a questioned pattern is compared to a known pattern directed toward source attribution. Although this is the goal when analyzing fingerprints, footwear outsole imprints, tire tread impressions, toolmarks, firearm-produced marks on ammunition components, etc., this is not the central activity of the interpretation of blood traces. Fourth, the common usage of the term pattern indicates a repeating design (such as a textile, wallpaper, or floor pattern), which is certainly not the meaning for blood trace deposits. A recent focus on bloodstain pattern taxonomies in the OSAC Bloodstain Pattern Analysis subcommittee is misdirected, and instead the emphasis should be on providing an understanding of the interrelationship of the totality of blood trace deposits with the goal of reconstructing critical details of the event(s). Despite the expression "bloodstain pattern" having the previously identified weaknesses, we will sometime use this terminology in places in the text because the reader should be familiar with its existing and common usage.

Existing & Commonly Used Terminology	Introduced Terminology	Justification
Bloodstain	Blood traces or deposits	Not all evidence examples of this type are actually stains, thus use of the term bloodstain is inaccurate.
Pattern	Configuration	The word "pattern" is not as general as the term configuration. Configuration is a better term for blood deposits or traces in three-dimensional space.

What are traces? When we are confining our discussing to material evidence, there are two distinct conceptualizations for the term trace in English (De Forest 2001). The first refers to size or amount, where a trace of material indicates a small amount or a low concentration of a component in a larger specimen. An example of the concentration connotation would be trace elemental analysis. Despite the narrower "amount" or "concentration" focus being common in American usage for the term trace, it is not the defining characteristic of the word as it was intended by pioneers Edmund Locard and Paul Kirk. Instead, the term trace in the context used here more appropriately represents evidence of a prior presence or a vestige remaining after a causal object has been removed, as in the phrase "he vanished without a trace." This is consistent with one of the foundational philosophies of forensic science, Locard's exchange principle, which has been commonly, but somewhat inaccurately, reduced to the phrase "every contact leaves a trace." In this context, the term trace refers to the exchange of material or production of a pattern as a result of an interaction from a prior presence or both.

Trace evidence investigations are concerned with the goal of shedding light on an event by taking advantage of the marks that were made and/or material that was transferred or deposited during its occurrence. Unfortunately, blood and other physiological fluids are not commonly considered to be a type of trace evidence, however conceptually it is useful to consider them as such. The reason for the separate consideration of blood traces from trace evidence is because in common practice there exists a self-contained set of laboratory analytical techniques for such biological evidence. Despite this pigeonholing, blood evidence in the context of analysis of the deposition and distribution of blood traces is clearly a form of trace evidence.

Blood trace deposits, often present at scenes of violence, may arguably be one of the most important types of physical evidence in a scene investigation. Despite this, other complementary traces may be present and should be thoroughly considered in a holistic approach to scene investigation. The analysis of blood traces may provide the identification of its origin (species) and subsequent approach to individualization (DNA). Additionally, the analysis of blood may answer questions that extend beyond source attribution to reconstruction. In these cases, the determination of its manner of deposition may prove to be more valuable than the associative evidence aspect.

Historically, the formal introduction of science into criminal investigations has been achieved primarily by physicians and scientists. In the middle of the nineteenth century, physicians first recognized the significance of blood trace configuration evidence. John Swinburne, an American physician, was involved in the investigation of the case of the death of Reverend Henry Budge's wife in 1861. Swinburne testified in the inquest to his analysis of the victim's wounds, blood trace evidence on the body and at the scene, and his experiments performed to illustrate the type of bloodstain patterns expected but not observed around Mrs. Budge (Swinburne 1862). Eduard Piotrowski, a Polish physician, published a study in 1895 in which he bludgeoned and stabbed live rabbits and examined the resultant blood trace geometries (Piotrowski 1895). Balthazard, a French physician, and his colleagues published their research in 1939 involving projected blood droplets (Balthazard et al. 1939). Scientists, many of whom were criminalists including Kirk, MacDonell, and Pizzola, et al., have researched the formation and the configuration of blood traces (Kirk 1953, 1955; MacDonell 1971; Pizzola et al. 1986a, b).

During the last several decades, many groups have reported on various aspects of blood trace formation, blood droplet dynamics, issues involving chronology of two or more blood depositions, and other aspects of blood deposits and their configurations. Unfortunately, some of this work has ignored fundamental aspects of science. In addition, an understanding of the limitations and alternate hypotheses has been absent in the presentation and publication of some studies. In research, these problems usually do not directly affect the outcome of criminal or civil cases; however, when translated into casework, may yield potentially disastrous outcomes. Gross miscarriages of justice due to erroneous, recognizable, and avoidable factors are unacceptable. Put simply, lack of science, poor science, and pseudoscience cannot be tolerated in criminalistics where the goal of the endeavor is to provide objective evidence for the resolution of an issue that is typically critical. Some flawed and dangerously erroneous examples from cases will be discussed later in this volume.

Areas of Concern

There are several aspects of the practice of what is termed "bloodstain pattern analysis" that was the driving force for writing this book, three of which warrant specific discussion here: (i) determining the relevance of blood traces, (ii) the limitations of the value of a classification scheme, and (iii) a lack of appreciation of the complexity of blood trace interpretations and the necessity for a scientific approach, which often manifests itself as deficient educational requirements for those doing this work.

Determining Relevance

There has been inadequate attention directed toward the determination of the relevance of the components of the totality of the blood trace record with respect to the overarching and ultimate goal of reconstructing the event. Not all blood traces at a crime scene are relevant to the inquiry. For example, blood may continue to flow and spread after the event of interest, and post-event activities may alter or eradicate important information. Determining the relevance of specific blood traces to an understanding of the details of the event in question is not a trivial problem, despite the fact that it is both underappreciated and overlooked. When the relevance of blood traces is not critically evaluated, the analysis can lead to both pernicious misinterpretations and mismanagement of resources. The determination of relevance is thus of the utmost importance at the beginning of an investigation to ensure a proper reconstruction as well as economy in the assignment of resources.

Limitations and Pitfalls of a Classification Scheme

Historically and more recently, various individuals and committees of bloodstain pattern analysts have placed considerable emphasis on developing classification schemes (taxonomies and typologies). This focus is misdirected and less important than perceived. One should not expect there to be a finite number of bloodstain patterns. There are in fact an infinite number of possible blood trace configurations. Consequently, no classification scheme of bloodstain patterns can be devised to encompass all possible configurations that can be encountered. Reliance on a classification scheme provides a false sense of confidence for examiners who feel compelled to assign a blood trace configuration to a specific category. Essentially, they may risk trying "*to force a square peg into a round hole,*" which can lead to serious errors. This risk has been demonstrated in research studies and will be discussed further in Chapter 5. When the details of the blood trace geometry are sparse or ambiguous, some analysts may not recognize the uncertainty in identifying its origin and mechanism of deposition. This naïve and forced conformity can contribute to the failure to recognize alternative hypotheses or explanations for the mechanism of deposition of a blood trace configuration and lead to a false interpretation of the overall event.

On many occasions, blood trace configurations are simply too ambiguous for one to have confidence in determining their production mechanism. This issue is an extremely important one as it continues to influence the outcome of some noteworthy cases. A recent and well-publicized example of this is the David Camm triple homicide (Chapter 10). The Camm case was recently adjudicated with a not guilty verdict after the third trial following reversals of the two earlier convictions. The defense and prosecution employed ten different bloodspatter "experts," yet the manner of deposition of the blood traces on Camm's shirt and on the roll bar shroud of the vehicle was the subject of extensive disagreement. Reliance on an oversimplified classification scheme by the initial "expert" propelled the inquiry down a false path. This, combined with inadequate passive and non-existent active photographic documentation, led to a flawed and misdirected investigation resulting in a scientifically unsupported reconstruction.

It is often counterproductive and wasteful to focus on classifying all the blood trace configurations present at an event scene before one has developed insight into their relevance to the entirety of the event and questions raised. The actions of a criminalist when faced with the problem of reconstructing an event at a crime scene do not follow a strictly linear process, whereby one begins with passive documentation followed by classification and lastly interpretation. It should be a recursive process, which is informed by the scientific method involving the development and testing of hypotheses (Chapter 1). Overreliance on a classification scheme can thwart the iterative process that is the hallmark of the scientific method.

The Complexity of Blood Trace Interpretations and the Necessity for a Scientific Approach

The need for a rigorous scientific approach to the physical analysis of blood traces, and thus the necessity of a scientific education as well as specialized training and in-depth experience for crime scene reconstructionists, is discussed throughout this book. The oversimplification of blood trace geometric analysis is a formidable problem in that the complexities of blood trace interpretations are often underappreciated. This is often compounded by the overreliance on classification schemes and simplistic training simulations that do not adequately demonstrate real-world blood trace production and how it can create complex blood trace configurations. By not appropriately challenging trainees, they can develop overconfidence in their abilities and adopt an oversimplified view of this type of analysis. Not understanding the complexities of blood trace deposition interpretations has considerable negative implications that run the risk of inventing flawed and misleading reconstructions.

Cautionary Principles of the Analysis and Interpretation of Blood Trace Configurations

We propose eight fundamental principles that must be remembered and practiced for a comprehensive analysis and interpretation of blood trace configurations at a crime scene. These stem from over a 120years of combined experience by the authors in the field of forensic science, specifically with a focus in crime scene reconstructions. These will be discussed in detail in Chapter 7.

Principle #1: - Properly recognized and understood, blood traces may reveal a great deal of useful information during a crime scene investigation.

Principle #2: - There will be cases where extensive blood traces are present but where a clear understanding of them does not address relevant issues in the case at hand.

Principle #3: - The extent of the blood traces observed at the time of a crime scene investigation is commonly greater than that produced at the time of the initial wounding. This may seem like an obvious statement, but errors are made when this is not given proper consideration. Of course, the time of the initial wounding is often of the most interest in crime scene reconstruction.

This principle has several corollaries. Some of these are principles in their own right.

Principle #4: - The initial wounding may not produce any immediate or useful blood trace configurations.

Principle #5: - Blood traces produced in the course of the initial wounding may be altered or totally obscured by the flow of additional blood from the wound.

Principle #6: - Blood shed by the wound or wounds may be transferred by post-event activities that may alter or obscure the blood trace geometry associated with the initial wounding event.

Principle #7: - Configurations of blood traces consisting of a collection of airborne droplet deposits can be informative with respect to providing an understanding of events that have taken place at a crime scene. Although schemes for assigning such patterns to specific causes or production mechanisms can be helpful, they are often oversimplified. It is naïve to think that patterns encountered in a crime scene can always be assigned to one of a finite number of mechanisms as defined by a typological or taxonomical systems.

Principle #8: - A collection of a few seemingly related dried blood droplets is not necessarily a pattern. It should never be treated as one. The number or extent of blood traces may be inadequate to allow a meaningful interpretation.

References

Balthazard, V., Piédelièvre, R., Desoille, H., and Dérobert, L. (1939). Étude des gouttes de sang projeté. *Congrès de Médecine Légale* 19: 265–323.

De Forest, P.R. (2001). What is trace evidence? In: *Forensic Examination of Glass and Paint: Analysis and Interpretation* (ed. B. Caddy), 1–25. Boca Raton, FL: Taylor & Francis.

Kirk, P.L. (1953). *Crime Investigation: Physical Evidence and the Police Laboratory*. New York: Interscience Publishers, Inc.

Kirk, P.L. (1955). *Affidavit of Paul Leland Kirk*. State of Ohio vs. Samuel H. Sheppard.

MacDonell, H.L. (1971). *Flight Characteristics and Stain Patterns of Human Blood*. Washington, DC: U.S. Government Printing Office.

Piotrowski, E. (1895). *Ueber entstehung, form, richtung, und ausbreitung der blutspuren nach hiebwunden des kopfes*. Wien: Aus dem gerichtsärztlichen Institute der k. k. Universität in Wien.

Pizzola, P.A., Roth, S., and De Forest, P.R. (1986a). Blood droplet dynamics–I. *Journal of Forensic Science* 31 (1): 36–49.

Pizzola, P.A., Roth, S., and De Forest, P.R. (1986b). Blood droplet dynamics–II. *Journal of Forensic Science* 31 (1): 50–64.

Swinburne, J. (1862). *A Review of the Case, the People Agt. Rev. Henry Budge, Indicted for the Murder of his Wife, Priscilla Budge, Tried at the Oneida, New York, Circuit Court, in August and September, 1861*. Albany: C. Van Benthuysen.

CHAPTER 1

Physical Evidence Record

1.1 Generation of Physical Evidence Record

1.1.1 Scene as a Recording Medium

Events leave physical traces which form a physical evidence record of the event. Traces are the elements of the record. Humans may be concerned about such physical evidence records extending over extremely vast timescales. Depending on the timescale, some past events may be the subject of investigations by cosmologists, astrophysicists, geophysicists, geologists, paleontologists, or archeologists (Figure 1.1). The record is created by processes that are subject to physical laws. Carol Cleland has characterized these investigative endeavors as "historical science" (Cleland 2002). All these scientific fields are centered on developing a scientific understanding of the traces that comprise the physical evidence record. Myriad physical evidence records are being created continuously. Some are the result of human activity. The vast majority of more recent events may be inconsequential with respect to a given inquiry. However, some small fraction of much more recent events can be of great concern in accident or criminal investigations.

The physical world is constantly changing. Past changes can best be understood from scientific studies of the physical record that is produced as the result of the processes that create these changes. A scientific understanding of this record in the case of the planet Earth has developed only relatively recently. For example, the acceptance of tectonic plate theory is only about 100 years old. Yet, without the powerful synthesis the tectonic plate theory provides, we would not have valid explanations for myriad observations for the interrelated phenomena of mountain building, earthquakes, sea floor spreading, volcanoes, fossils of marine species in mountainous regions, and fossils of tropical species in Antarctica, to name a few. Focusing on fossils for the moment, these form a record of the development and evolution of life on earth including the emergence of humans. In addition to the fossil record, human activities

Blood Traces: Interpretation of Deposition and Distribution, First Edition. Peter R. De Forest, Peter A. Pizzola, and Brooke W. Kammrath.

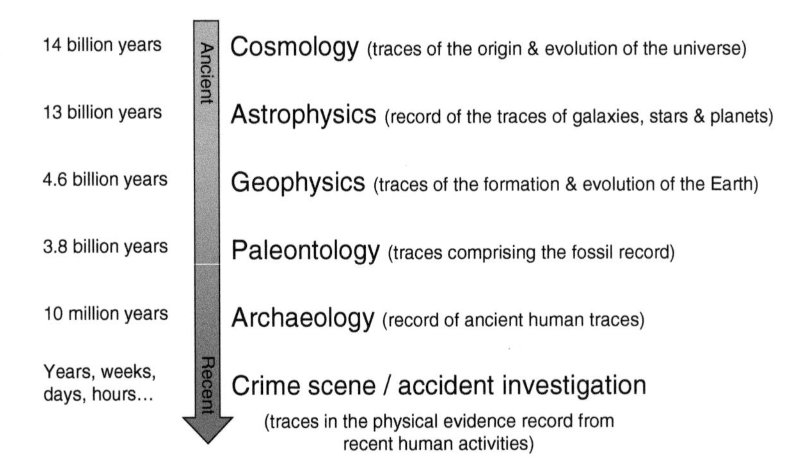

FIGURE 1.1 A timeline of historical sciences that reconstruct the past based on the physical evidence record. This is an illustration of the ideas of Carol Cleland that has been extended by the authors to include crime scene and accident investigation. The arrow refers to how far back in time, or the degree of recentness, an event occurs in which these scientific fields study, i.e., flowing from the ancient past to more recent events.

during the last few hundred thousand years, such as tool making, have left artifacts that have added an important part to the record that archeologists study.

Human activities on a vastly shorter timescale than that confronting archeologists also produce a physical record composed of a variety of traces. Billions of records of the activities of contemporary humans are produced continuously. Most components of these records are of little interest. Additionally, they lack the permanence of the surviving archeological record and are easily degraded, "erased," or "overwritten" by subsequent human activities or natural processes. For example, footprints may be trod upon or be obliterated by wind and rain, and certain material components of a physical record may decay or decompose over time.

The concept of physical interactions producing a physical evidence record of such interactions can be illustrated by the use of Jersey barriers. These are concrete structures that are used to separate lanes of traffic along the highways (Figure 1.2). They demonstrate the somewhat evanescent or complex and fragile nature of the record of traces. If one is reasonably observant while driving or riding in a car along a highway where these barriers are used to divide traffic lanes, many examples of interactions between vehicles and the barrier can be observed, most notably in the form of black rubber abradings and paint smears. The evidence of many interactions can be observed on such a barrier for a distance of many miles unless the barrier has been put in place very recently. Some of these may represent tragic interactions that resulted in the deaths of individuals. Others may have been caused by less violent interactions. One would not expect that the fine details of the interactions could be observed from a moving vehicle because of the lack of proximity and the limited observation time. For example, under these circumstances, it is probably not possible to tell the relative ages of different events unless one is very recent. A very recent interaction would have a richer range of evidence types than simply the black rubber abradings. In such a situation, in addition to varied deposits and marks on the barrier, one could expect that the roadway

FIGURE 1.2 Image of a Jersey barrier, as viewed from a moving vehicle.

at the base of the barrier would have fragments of plastic, glass, paint, and metal. With the passage of time, these would be swept up or otherwise disbursed. Weather could also gradually remove deposits from the barrier itself and wash them away. At some sites, evidence of a newer interaction would be deposited on top of an older one and thus, in effect, "overwrite" it. In crime scene investigations, the overwriting may be unavoidable or it may be the result of avoidable post-event factors arising from poor scene security. One of the problems confronting blood trace configuration reconstructions in particular is that the overwriting can take place on a much more compressed timescale. The timescale can even be shorter than the duration of the event. Blood disbursed late in an event can obscure critical configurations that were created early in the course of the event.

Of the billions of records continuously being produced by human beings, some fraction may be the result of activities society defines as crimes. The natural record that is created during the commission of a crime is the physical evidence record. It is this type of record, and the importance of a study and understanding of the traces which comprise such records, that is integral to crime scene investigation. Although human activity is behind the creation of the physical evidence record, natural phenomena are the mediators, agents, or the means by which the record is produced. During the activities associated with the criminal act, the human actor ultimately responsible is normally creating the record unwittingly or unwillingly. Some component traces of the record may be formed by mechanisms unknown to the criminal leaving them behind.

Conscious attempts to interfere with the production or preservation of the record may fail, because such activities will also create a record, which may be detectable.

At this point, our discussion has been theoretical. In practice, it is probably very rare that the entire physical evidence record of an event at a crime scene is preserved, recognized, understood, and exploited. No crime scene is "pristine" at the outset of an investigation. A significant time delay from the time of the event until it is discovered can result in degradations and loss of evidence due to environmental factors or subsequent unrelated human activity. The discovery of the event and the arrival of the first responders may cause unavoidable alterations to the scene. Hazards such as the presence of an armed criminal, a fire, or chemical contamination may have to be dealt with before the scene is fully secured. If there is an injured party present, the obligation to take measures to save a human life will take precedence over all other activities. Unfortunately, from the physical evidence point of view, the necessary medical intervention may compromise traces that are part of the physical evidence record. The possible adverse consequences of this activity can be minimized by careful documentation of the details of the actions that take place. Once the scene is secured, it is still possible that additional information may be lost, but the onus now lies with those conducting the official investigation. The losses may be the result of errors of commission or omission made by the investigators, the compromising of evidence by the presence of nonessential personnel at the crime scene due to poor scene security, or a failure to recognize the significance of items or features present at the scene. Existing training programs for investigators can address and assist in the avoidance of common errors; however, the skills required for evidence recognition are more subtle and are best gained from a broad scientific education combined with training and extensive experience.

The fact that natural phenomena are directly involved in producing the physical evidence record has important implications. It is necessary to have a sophisticated understanding of the natural phenomena and mechanisms that are the proximal causes of the record being produced. This is essential for the recognition, documentation, and proper preservation of the physical evidence record, which includes developing questions, defining approaches, and determining the problems to be pursued. This calls for the presence of scientists at the scene who possess an extensive and rigorous scientific background, gained through a minimum of a baccalaureate degree in a physical science or its equivalent formal post-secondary science education, practical training, and crime scene experience.

Any item of physical evidence (regardless of its size) represents a trace, vestige, or remnant, of one or more actions that produced it, changed its position, and altered its physical conformation or its composition. Traces do not have to be an article or a physical object, they can be the recognition of an absent material, such as a void in a film of dust or a missing paint chip on a surface. There are properties of evidence that allow us to make inferences about such actions, which may have been directly associated with activities that take place during the commission of a crime. Such properties include the nature of fracture in a glass window, fire patterns, paint smears, and the shape of a bullet hole indicating direction of fire or a destabilized bullet. More closely aligned with the focus of this book are blood traces where the blood itself is not the evidence of interest, but instead the configurations of its deposition are the most important for understanding the event. This will be discussed in great detail throughout this book.

1.1.2 Creation of Blood Traces

Blood traces can be created in a plethora of ways which will be discussed in detail later in this book. In Chapter 5, the authors identify three general types of blood trace configurations: noncontact traces, contact traces, and other conformations. Noncontact traces include blood traces formed from airborne droplets, such as arc ("cast-off"), arterial, trail, expired blood, expectorated blood, sneezed blood, radial ("impact"), and spatter configurations or patterns. The mechanism that produces these airborne droplets contribute to each deposit's characteristic morphology. Contact traces can be formed by either a dynamic or static transfer, with each configuration having its own diagnostic features. There are a number of types of mechanisms for the production of blood trace deposits that cannot be characterized as either contact or noncontact, such as those formed by flowing, pooled, diluted or clotted blood, voids, as well as configurations from splashes, and post-incident events (artifacts). It is important to note that a blood trace configuration can be composed of many individual droplet deposits, and that a single blood drop does not necessarily constitute an interpretable or meaningful pattern; it is the totality of the blood traces that must be taken into account while considering a reconstruction.

In some cases, a full reconstruction solely based on the physical aspects of the blood traces is not possible. However, at times, specific questions may be posed that can be addressed and answered short of attempting a full reconstruction. The utility of blood traces is always context-dependent, as demonstrated in the cases highlighted in Chapters 10–12.

Blood trace configurations can be very informative in reconstructions, but it must be recognized and understood that most blood at scenes and on actors accumulate and may be transferred long after the event of interest. Such traces may be of no help in addressing the critical questions that arise in the investigation. Failure to recognize this can and has led to serious errors and misinterpretations. It must always be borne in mind that the event of interest may not produce blood at the most critical moment (Principle 4). For example, it is widely appreciated that the first blow in a bludgeoning commonly will not produce any airborne spatter because the weapon does not strike any accumulated coalesced blood. This may mean that in a bludgeoning, the site of the initial attack will not be knowable based on blood traces. Similarly, there may be a lack of blood deposits in shooting cases with respect to the first wound.

1.2 Capturing the Physical Evidence Record: Crime Scene Analysis

To maximize the utilization of the information latent in the physical evidence record, forensic science must deal with the physical evidence continuum from recognition of the relevant evidence at the crime scene to its analysis in the laboratory and finally its scientific interpretation for the courts (Figure 1.3 depicts the physical evidence continuum). Traces that are not recognized for their relevance to the event under investigation will not be documented, collected, analyzed, and interpreted. They will not

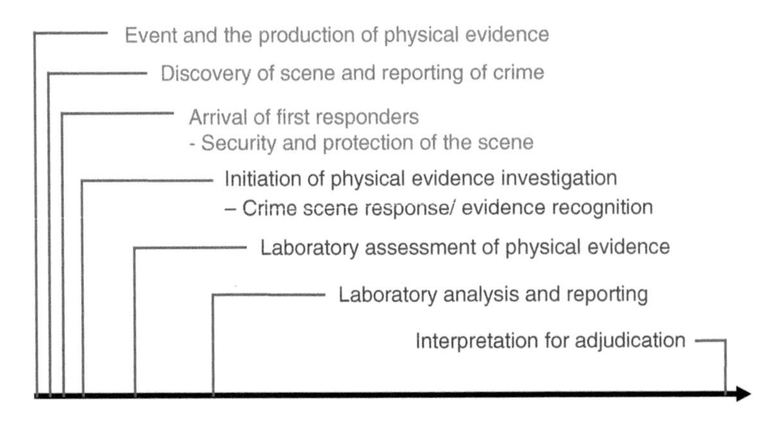

FIGURE 1.3 The physical evidence continuum, where the items written in blue, denote activities that should be overseen by scientists.

form part of the physical evidence continuum and thereby ultimately contribute to the solution and adjudication of the case. Errors and omissions at the crime scene may compromise the physical evidence record to the extent that it is rendered useless. The information that it could have provided may be irretrievably lost. The work at the crime scene is critical to the effectiveness and accuracy of a forensic science system. It cannot be viewed as a separate operation. The crime scene investigation and the laboratory investigation should interface seamlessly.

1.2.1 The Stages of Crime Scene Investigation

The stages of a crime scene investigation can be described as protection, recognition, documentation, recovery, packaging, and transportation. The activities for these stages logically take place in the sequence as listed here, but overlap and some variations are necessary to adapt to specific situations. Following the packaging, the chain of custody (or chain of possession) of the items recovered must be maintained and documented. This is necessary to ensure the scientific and legal integrity of the evidence throughout the physical evidence continuum.

1.2.1.1 Scene Protection and Security Protection of the scene and the integrity of potential traces within it begins once the scene is secured. When immediate dangers and hazards (e.g., armed suspects, fire, explosive, or toxic fumes, etc.) have been dealt with, scene protection and evidence preservation should become an early priority. When critically injured individuals are present at the scene, life-saving measures by emergency medical personnel take precedence over the preservation of physical evidence. However, this does not mean that nothing can be done. Protection of the scene cannot be ignored at this critical stage. The medical personnel can be directed to follow a specific route or path to the victim that is selected to minimize the destruction, compromising, or contamination of evidence. Unavoidable actions taken that alter the scene should be noted and recorded. Patterns or trails of blood that are created in the course of treating or transporting the victim can cause confusion later on, if it is not realized that these are post-event artifacts. These should be documented. This is an example of what can be termed passive documentation. It takes place prior to the recognition of the significance of certain potential evidence and an understanding of its relation to the event under investigation. The scene and the evidence are documented as they are found. Later as the investigation proceeds and following recognition of the significance of certain evidence, active documentation can begin. Although it will be discussed in more detail later, photography provides a clear example of the distinction between passive and active documentation. Close-up photographs would not be taken of everything at the scene. This would be unnecessarily time-consuming and wasteful of resources. After recognition of the significance of certain items, active documentation can begin. There may be overlap. Passive documentation may need to be continued even after active documentation begins with respect to certain items and features. Of course, recovery (collection or sampling) of traces must await their recognition, and obviously packaging must follow recovery. While acknowledging that there will be extensive overlap, each of these stages will be discussed separately below.

If local law does not give the investigators and scientists the necessary authority to exclude nonessential people from the scene, there are steps that can still be taken to reduce the compromising of the scene and potential traces it contains. The most difficult situations arise when high-ranking police personnel, prosecutors, and politicians, who do not understand the damage they could cause, want to enter the scene. If explaining the purpose of protecting the scene does not suffice, they can be asked to submit to the production of fingerprint and footwear outsole pattern exemplars. This request may underscore the point that the scene is fragile and may cause them to rethink their need to enter the scene. This is more than a tactical ploy. If the difficult to exclude individuals still insist on entering, having obtained the exemplars for elimination purposes could prove to be very useful later on. It could save much time and effort pursuing misleading theories. An alternative strategy for excluding nonessential personnel is to inform them that their presence will be documented photographically and/or by the use of a continuously recording video camera. Another deterrent strategy, but one that is also of value if the deterrence aspect fails, is to supply them with protective gear, especially gloves and footwear coverings. They can also be asked to sign in, to carry a pen and clipboard, and to make notes of what they observe and what they do. Even if it does not serve as a deterrent, keeping the hands busy may reduce the touching or handling of evidence. The preferable situation is one where laws exist to protect crime scenes and where those charged with scene protection duties have the clear authority to exclude any and all nonessential individuals no matter who they might be.

The news media are commonly attracted to crime scenes. Of course, they, like the others cited above, would have no useful function at the scene and must be excluded to prevent the degradation of the physical evidence record. This may require considerable effort. Some of the pressure behind the desire of the news media and high-ranking officials to enter a cordoned off scene can be reduced if a video camera can be set up within the cordon to supply a video signal feed to a law enforcement command post set up outside. The video feed can be used to supply case detectives who have a need to know certain case details, but who have no essential function within the scene cordon, with valuable information. This video feed can be managed to provide selected portions to satisfy the curiosity of high-ranking officials and the news media.

The authors continue to be surprised that despite decades of discussing the problem of evidence destruction by the presence of unnecessary personnel, it persists (De Forest et al. 1983). Stringent protocols and regulations need to be in place and enforced in order to protect the physical evidence record.

1.2.1.2 Evidence Recognition The most intellectually demanding of the activities undertaken at the crime scene is evidence recognition. It is also critically important. Potential traces that remain unrecognized are likely to remain at the crime scene at the conclusion of the scene investigation or be destroyed in the process, and as result it cannot contribute to the case solution. Once the scene is released, unrecognized evidence is *lost and gone forever*. Some traces at crime scenes are obvious, such as a bloody fingerprint. One would think that even inexperienced amateurs would recognize its possible significance. Astounding perhaps but some "vaunted" crime scene personnel have rationalized why bloody ridge patterns can be ignored since they lack sufficient characteristics. A dangerous practice indeed! Other potential evidence could be easily overlooked. Shortcomings with respect to evidence recognition may have little adverse impact on the case, if the obvious evidence is the most important. Unfortunately, obviousness and ultimate importance are not correlated. Every crime scene is unique. Objects or evidence types that are the most important in one case may have little or no significance in other cases. The reverse is also true. Objects or potential evidence types that have had no value or have not even been encountered in prior cases may take on critical significance in the case at hand. There is no fixed procedure, protocol, or formula that assures recognition of evidence. At times, perfunctory approaches may result in the recovery of useful evidence, but this will be hit-or-miss in the long run. A flexible but systematic approach that facilitates insights is needed. These seemingly contradictory attributes can be found in an approach based on the scientific method. The scientific method has been described in different ways. A useful way of summarizing it for our purposes is presented in Figure 1.4.

The data-gathering phase in Figure 1.4 corresponds to surveying the overall scene and making observations. No action is taken initially. Physical interaction with the scene is kept to a minimum at this stage to avoid compromising both obvious evidence and yet-to-be recognized evidence. The temptation to do something other than some possible passive documentation at this point needs to be resisted. The transition to

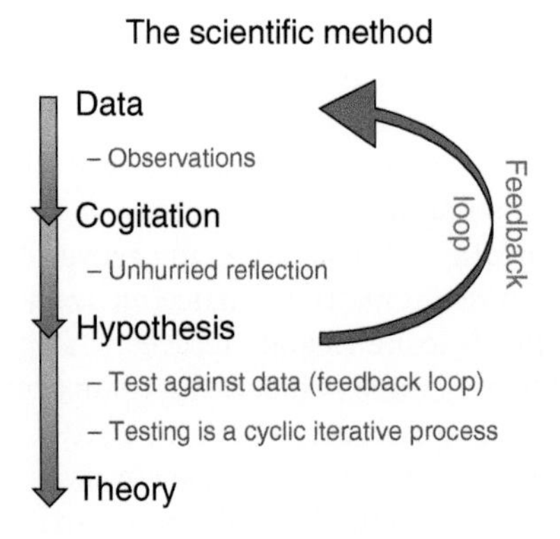

FIGURE 1.4 The scientific method.

Deductive: The Result given the Rule and the Actuality.
Inductive: The Rule given the Actuality and the Result.
Abductive: The Actuality given the Rule and the Result.

FIGURE 1.5 Deductive, inductive, and abductive reasoning summary.

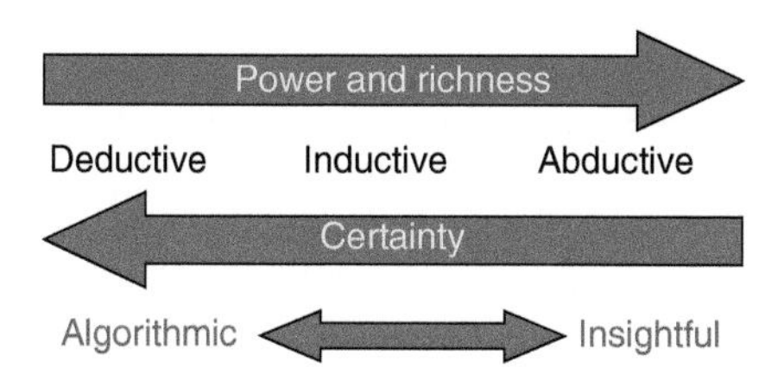

FIGURE 1.6 This image demonstrates the value of deductive, inductive, and abductive reasoning in terms of their power, richness, and certainty. Deductive reasoning is the most algorithmic and can provide absolute certainty to its conclusions, but produces little in the way of predictive insight. Abductive reasoning is the most insightful and imaginative form of reasoning and thus produces the most powerful and rich ideas with the least certainty. Insights gained through the abductive reasoning process need to be verified through hypothesis testing via other components of the scientific method.

the next phase, cogitation, may be gradual and involve some overlap. Here, the scientific investigator engages in unhurried reflection and lets the mind range over the problem and the accumulating observations. As an adjunct to the familiar terms deductive reasoning and inductive reasoning, the philosopher, Charles Sanders Peirce, coined the term abductive reasoning (Figures 1.5 and 1.6). Umberto Eco and others have pointed out that this is the kind of reasoning and mental activity that is useful in the preliminary stages of investigations (Eco and Sebeok 1983). This then leads to the development of an explanatory hypothesis or perhaps several hypotheses. The hypotheses generated in this process need to be tested rigorously. It is the rigorous testing of hypotheses that is the key component of the scientific method. Earnestly attempting to destroy a hypothesis one has created, and may identify with, runs counter to human nature. However, this process is essential to avoid operating on untested assumptions, which can lead an investigation in the wrong direction. It is the essence of science.

After successive tests, hypotheses are rejected and replaced, modified, or supported. Refined hypotheses that fit the data are developed. A hypothesis that is developed this way can then be used as a working hypothesis that informs and guides the investigation. It provides a custom-made framework for conducting the scene investigation that adapted to the unique needs of the particular case. The resulting approach was arrived at in a flexible way and is systematic.

The consequences of errors or omissions at the recognition stage should be clear. They can have profound ramifications for evidence recovery and ultimately case solutions. Unrecognized evidence is lost and gone forever; it remains at the scene never to be submitted to the laboratory or to be subjected to a subsequent analysis of any kind. Without a rational and scientifically based selection process, relevant evidence will not be recovered or, perhaps equally damaging, irrelevant material will be collected and submitted to the laboratory. Overloading the laboratory with irrelevant material is not only wasteful of resources, but it may also impede the physical evidence investigation.

Informed selectivity is essential. This should flow from a scientifically based evidence recognition process. Selectivity should take place at the scene where the full context can be appreciated more readily, not in the laboratory.

1.2.1.3 Evidence Documentation

Several recording media are available for crime scene and evidence documentation. There is no need for all of these to be used at every scene, but it would be a rare situation where several of these documentation tools could not be used profitably in a single investigation. In general, all available media of documentation are to some degree complementary. The use of one does not automatically eliminate the need for the use of another. A list of these is provided in Table 1.1 and they are discussed in more detail below.

Handwritten notes made during the course of a scene investigation are valuable. They can be supplemented and augmented by notes made using a voice-activated electronic recorder. These records should be considered permanent. The handwritten notes are taken contemporaneously with the observations they record and are made in ink. Corrections, where necessary, are made by putting a single line through the error and initialing and dating the correction. The notes are preserved after the report is written. In most circumstances, they are not destroyed until after the final adjudication of the case. Actually, with respect to homicide cases, original handwritten notes should not be destroyed even after final adjudication. Most states have statutes that regulate the preservation of such records and must be carefully followed. If an audio recording is made, it should be preserved even after it is transcribed. Original digital audio recordings can be copied digitally onto a more permanent medium, if the electronic time stamp or other authenticating information is retained.

In addition to documenting observations made at the scene, the handwritten notes may be used to document activities at the scene as well. One of these may be to document what photographs are taken at the scene and the conditions under which they are taken. This record is known as a photographic log or simply a photo log. Maintenance of the photo log should be contemporaneous with the photography. As will be mentioned later, the maintenance of the photo log can be facilitated with the help of a video camera.

Sketches are a useful accompaniment to handwritten notes. Measurements of dimensions can be recorded on them. There are features that can be shown on sketches

TABLE 1.1 Modes of available documentation.

Contemporaneous handwritten notes in ink
Notes made by voice-activated audio recording
Sketching
Measuring (including total station equipment and three-dimensional laser imaging)
Photography
Video recording
Casting

that are more difficult to illustrate or document with photography. Key features can be emphasized. A high degree of drafting skill is not necessary in their preparation. They need not be works of art. The accuracy of the information they convey is more important than their aesthetic appeal. Following the scene investigation, the sketch can be used as the basis for preparing a finished sketch, scale drawing, or computer-generated plot. As was the case with the handwritten notes described above, the original sketch should be retained even after the more refined graphics have been produced.

The preparation of the scale drawing or computer plot would not be possible without measurements. Traditionally measurements are made at a crime scene using a tape measure. For large outdoor scenes, other measuring techniques may be used. For large distances over smooth ground, the simplest of these is a measuring device that operates on a principle similar to the odometer in an automobile. A wheel attached to a handle is pushed ahead of the operator who assures that it is kept in uniform contact with the ground during the measurement operation. A mechanical or electronic mechanism keeps track of the number of rotations of the wheel and taking the wheel's circumference into account displays the distance traveled.

When a measuring tape is employed, the use of a two-person team can facilitate the taking and recording of the measurements. One makes the measurements with the tape, while the other maintains a handwritten record that describes what is being measured and records the associated dimensional datum. This process can be conducted more rapidly with the use of a video camera. Here the one conducting the measurement calls out the value of the measurement, so that it is recorded on the audio track of the recorder. For redundancy, the camera operator repeats the value of the measurement orally. The one conducting the measurement then confirms this. The camera operator pans the camera over the area or object being measured, and where possible, focuses on the numbers on the tape. Skillful use of the camera avoids the ambiguity that sometimes exists in written descriptions of the feature being measured.

Recently developed technology allows dimensional data at crime scenes to be acquired and recorded with greater ease and more rapidly. The main technologies are "total station" surveying equipment and laser ranging and three-dimensional plotting technology. Two- and three-dimensional plots can be generated quite easily. The equipment is relatively expensive. It does save a considerable amount of time, and it does allow more data to be acquired, but it is not essential. A two-person team using a measuring tape can acquire data that are adequate for the vast majority of scene investigations.

Still photography was discussed previously in a limited way in connection with the discussion of passive and active documentation. It is the single-most versatile and generally applicable documentation technique available for use at crime scenes. Whereas scientists and investigators should have wide discretion concerning which documentation techniques to employ at a given crime scene, there is probably no scene encountered where photography could not be used to advantage. The use of sketches, drawings, and modern plotting technologies does not obviate the need for photography. Similarly, the use of photography does not necessarily eliminate the need for sketches. Each serves a different purpose.

Crime scene photography is very demanding of photographic skills. In its goals and operation, it is most like the field of scientific photography. Automation and computerization of camera functions have not lessened the degree of skills required. In some ways, this automation has complicated the situation. Any camera being considered for purchase for crime scene work should have manual override features that are robust and easily implemented. Despite the convenience of camera-mounted flash illumination in casual or amateur photography, there are few situations in crime scene photography where using a camera-mounted flash or a built-in flash is appropriate. In many cases, using such illumination will produce glare and totally unacceptable results. Such illumination is inappropriate for close-up photography. Crime scene investigators should be provided with equipment that allows them to take the flash unit off of the camera and employ it at varied illumination angles. In addition, equipment for alternatives to flash illumination should be readily available.

Original images acquired in the course of conducting a crime scene investigation and in carrying out laboratory work need to be of the highest quality that can be obtained with the available camera equipment. Such images are associated with large digital files that have not been compressed into smaller file formats such as JPEG or JFIF. For scientific and legal reasons, when copies are made, these original raw image files need to be retained unaltered. Any manipulations of the images ranging from a simple rotation to sophisticated enhancements must be performed on copies of the image. The process of maintaining the original raw image files and the manipulations employed with the altered copies must be documented. Images that are distributed for use by investigators, attorneys, and others in the criminal justice system must be traceable to the unaltered original file. This is analogous to the practice of securing and preserving the original negatives when silver emulsion film was used. Some hardware and software packages developed for forensic science use allow unlimited copies of the original unaltered raw file to be made, but they do not allow it to be altered without creating an electronic "trail" or record. No matter what system is adopted, a protocol must be in place that identifies and preserves copies of the original unaltered raw image files. This does not mean that the camera chip cannot be erased after the original image files have been transferred from the camera chip to a computer or stored on an optical disk or alternative solid-state memory. It only means that exact copies of the original raw image file are identifiable and are maintained unaltered. By way of contrast, it would be wrong to erase or destroy the raw image file after it had been used to produce and store a compressed, altered, or enhanced copy. In the absence of an automated authentication system, procedures should be set up to accomplish the same goal. One protocol could be set up to require the burning of raw image files to an archivable memory medium (e.g., next-generation optical disks such as those utilizing gold nanomaterials) at the time they are transferred from the camera chip. This memory medium would be labeled appropriately, and written records and a master file of the disks would be maintained.

Video recording can be a very useful tool for crime scene investigations. One use of video recording was mentioned above in connection with the discussion of measurements. As was described, it can greatly speed up the manual measurement process while simultaneously contributing to its reliability. Another principal use of video recording at a crime scene is overall scene documentation. With proper use of the video camera, the relationships of different parts of the scene to each other can be documented and illustrated. Video recording should not be used as a substitute for the still camera. The video camera and the still camera have distinctly different capabilities and thus they have distinctly different purposes at crime scenes. The smooth continuous flow of images produced by the video camera is superb for overall scene documentation. It is relatively poor for recording fine detail. The recording of fine detail is where the high-quality still camera excels. Both types of cameras have complementary roles to play in documenting crime scenes. These roles should not be confused. The two most common mistakes made with the video camera are to keep turning it on and off and to repeatedly zoom in on specific objects or features at the scene. The resulting "choppy" record is a poor and unsatisfactory hybrid between detailed still photography and the smooth-flowing overall documentation of the scene that constitutes the strength of video recording. The errors may be compounded further, often fatally, if fewer still photographs are taken because it is reasoned that the video recording took care of documenting many details. This kind of thinking must be avoided. The single-most important image recording at crime scenes is accomplished with the use of the high-resolution still camera. The video camera is no substitute. If one is limited to having only one type of camera to bring to a crime scene, it should be a high-quality still camera.

As mentioned in the earlier discussion concerning the importance of the photo log, the video camera can also be used to assist with its maintenance. Data beyond that automatically preserved in the metadata accompanying digital camera images can be important. The sequence of photographs, the location and orientation of the still camera, and other photographic data for each photographic frame can be readily be documented by taking advantage of both the video and audio capabilities of the video camera. Investigators are often reluctant to have an open microphone at a crime scene. As a result, the audio recording capability of the video camera is commonly disabled. This is a mistake. The preferable alternative is to enable the audio channel and caution those present to avoid potentially embarrassing oral utterances. Two of the more specialized uses of the video camera at crime scenes discussed above, assisting with recording measurements and maintaining the photo log, take full advantage of the audio channel. The audio channel is also very useful for recording commentary and descriptions during general video documentation. Video recording knits together salient features of the scene and the photographic images of it. The original video recording should be retained.

The time and date should be recorded on both digital and video images. This is a useful documentation feature available on most digital and video cameras as well as other equipment. However, confusion can be created if the accuracy of the camera clock is not verified just prior to beginning of each recording session. This is a habit that should be developed when using all equipment that time stamps records.

The best means of documenting three-dimensional markings or impressions, such as toolmarks or footwear outsole patterns, is often casting. However, caution and careful evaluation is necessary before casting is employed. With marks in some soft or easily deformable materials (e.g., sand, soil, snow, etc.), there is the risk of damaging or destroying the mark when the casting medium is poured. In these situations, photographing the mark prior to casting is an important requirement. Normally, several photographs should be taken employing different angles of illumination. Even if there is no risk of damage, such as would be the case with a toolmark in metal, this precaution should still be observed. Additionally, it would still be important to consider the possible impact of casting on traces that might be present.

Forensic scientists have used many casting media over the years. Most of these are mainly of historic interest. There are two that are used most frequently at present. These are dental stone (a harder version of Plaster of Paris) and pigment-filled silicone rubber. The silicone rubber is best for replicating and documenting extremely fine details in nonporous materials (e.g., toolmarks in metal). The pigment filling renders the silicone rubber opaque and makes observations using the high-intensity illumination under the comparison microscope easier. For larger volume impressions where the detail is not extremely fine (e.g., footwear outsole and tire tread impressions), dental stone is commonly used.

1.2.1.4 Evidence Recovery, Packaging, and Transportation

Following the recognition of particular items of evidence, additional documentation may be necessary. Again, as noted earlier, documentation prior to recognition is passive documentation and documentation that follows recognition is largely active documentation. Documentation records the traces in the context in which they are found. Once this is complete, recovery and packaging can commence, although in many situations there is no need for recovery to immediately follow the recognition of each individual bit of evidence. As long as each is at no risk of being lost or otherwise compromised, recovery and packaging can be delayed until the scene investigation is nearly concluded. In this way, the focus can remain on the recognition and documentation aspects. Leaving items in situ as long as possible can be advantageous. In cases of homicide where it is clearly too late to attempt to save the victim's life, it is important for the body to remain at the scene until late in the scene investigation. In this way, scene disturbance is minimized and insights with respect to evidence recognition may be developed. It is useful to view the body as an integral part of the scene, although practice in some jurisdictions and cultures can make this difficult.

Details of specialized recovery procedures appropriate for certain kinds of evidence are beyond the scope of this book. There exist numerous references on crime scene investigation that contain this information, although some general comments are appropriate here. With some kinds of evidence and in certain circumstances, recovery should be initiated immediately following recognition. These are situations where the evidence is likely to be compromised or lost due to circumstances such as weather at outdoor scenes (e.g., wind or rain) or due to the nature of the evidence itself (e.g., biologically labile, volatile, chemically unstable, etc.). Leaking containers may need to be secured, volatile samples may need to be placed in airtight containers, reacting chemicals may need to be knowledgeably separated, and biological evidence may need to be dried and/or refrigerated. Expertise and judgment are needed in these situations.

There are two broad types of evidence recovery. These are collection and sampling. With collection, the entire item of evidence or the object on which the evidence is present is collected and packaged. With sampling, the evidence or some portion of it is removed from a large or immovable object and packaged. In general, where the situation allows, collection should be favored over sampling. However, considerable judgment is called for here. Collecting large items may create evidence storage problems.

The securing and packaging of evidence should follow collection or sampling without delay. Securing without final packaging may be necessary where biological evidence needs to be dried prior to final packaging and shipping. Special secure drying cabinets for this purpose can be fabricated or purchased. Small amounts of moisture contained within biological evidence, such as that retained in dried blood traces on garments, can facilitate microbial activity that may lead to the degradation of genetic markers in the deposits. Vapor permeable containers such as kraft paper bags are best for packaging clothing items containing biological traces. Plastic bags should never be used for this purpose.

With fire debris samples suspected of containing residues of a volatile hydrocarbons or other flammable liquids, non-permeable primary containers are necessary. Polyethylene film bags are not suitable for this purpose because they are permeable to hydrocarbon vapors. Special nylon film bags have been developed and marketed for this purpose. Large, newly purchased, unused paint cans are also used for packaging and hermetically sealing fire debris samples. Liquid samples need to be collected in appropriate leak-proof primary containers.

The most suitable primary packaging for microtraces, including hairs and fibers, is often a "druggist fold" or similar paper fold. These can be prepared onsite from a pad of paper carried for this purpose. Such paper folds are superior to paper envelopes or plastic bags for holding microtraces. Paper envelopes have small openings at the corners through which small particles or fibers can escape. The paper folds containing the collected microtraces can be sealed within paper envelopes, if necessary. For very small paper folds placing them in envelopes can facilitate sealing, labeling, and maintaining the chain of custody. Electrostatic attraction phenomena limit the utility of small plastic bags for the packaging of small amounts of particulates.

There are several critical aspects to the packaging of dried blood traces. These are necessary to preserve the configuration and to limit decomposition or degradation. In the case of blood traces on a victim's body or clothing, body coverings and bags can seriously compromise or even destroy the blood trace configuration. This is demonstrated in several cases discussed later in this book, including the "Murder of an Off-Duty Police Officer" (Figure 11.17), the "Dew Theory Case" (Chapter 11), and also in the O.J. Simpson murder case (Chapter 10) where valuable blood trace deposits on Nicole Brown Simpson's body were initially captured in passive photographs but lost due to covering the body with a blanket taken from the house. This practice also compromised the trace evidence, specifically microtraces found on the victim were carpet fibers consistent with the defendant's Bronco and hair consistent with those taken from O.J. Simpson's head, but at trial these were attributed to the blanket covering by the defense. A viable alternative to ensuring the victim's privacy and modesty is the use of screens that do not contact the body but prevent it from being seen by curious onlookers (Figure 1.7). When clothing with blood traces is collected, it should be inspected for possible trace evidence, then knowledgeably and carefully removed from the victim, and after drying, packaged to prevent the transfer of blood to unstained areas.

These general guidelines and comments on packaging cannot come close to addressing every situation that might be encountered. More details can be found in other references. Consultation with the laboratory where the evidence will ultimately be analyzed is important in decisions about developing evidence recovery and packaging protocols. For the sampling of biological traces, scientists in the field may prefer selective sampling techniques such as cutting, scraping, or tape lifting, but many analysts and laboratory protocols prefer swabbing, which can be less selective depending on the sizes of the deposit. Swabbing is the most common method of blood sampling; however, it is not always the best as it can alter or destroy the blood trace configuration (Figure 1.8). Consultation should take place and consensus should be reached on such matters in advance. It is often advisable to have laboratory personnel directly involved in the evidence recognition and recovery process.

Testing at the scene is rarely advisable or appropriate. Even with sophisticated testing kits and advanced portable instrumentation, testing is best carried out by scientists in the laboratory.

FIGURE 1.7 The photograph shows the use of protective screens to shield the body from view. These screens can serve as nondestructive replacements for body coverings without compromising the physical evidence. Unfortunately, in this case, a body covering was unnecessarily used in addition to the screens.

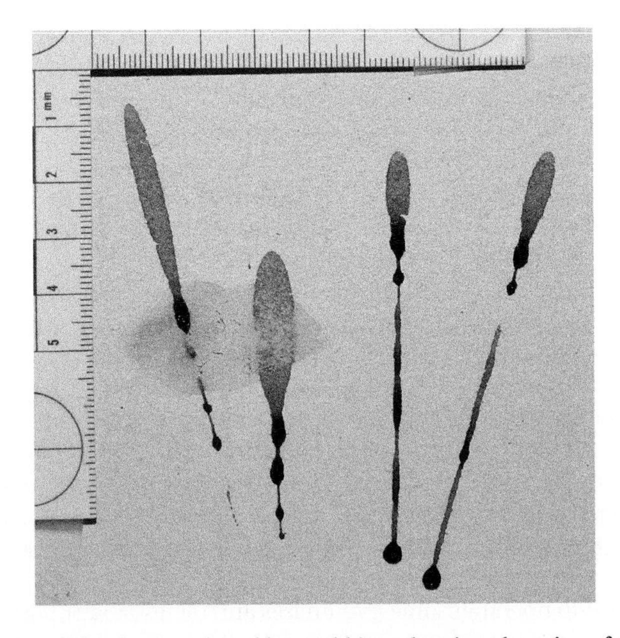

FIGURE 1.8 Image of blood traces altered by swabbing, showing alteration of the deposits and nonselective sampling (commingled samples from two traces).

1.3 Reconstruction of Past Incidents from the Physical Evidence Record

1.3.1 Definition

Reconstruction is the process of recognizing, assembling, and analyzing information derived from the physical evidence record in order to determine the sequence and details of past events. The physical evidence record is mediated by physical laws, thus an understanding of the details of these laws is necessary for a knowledgeable interpretation of the physical evidence record. Subjects (victims and perpetrators) may engage in various activities that are of interest in a criminal investigation; it is the physical laws that enable the relevant actions to be recorded as naturally produced traces. Experienced forensic scientists are needed in order to recognize and interpret relevant traces which enables an understanding of this physical evidence record to reconstruct the activities that comprise an event of the past.

Earlier, we discussed historical science. We pointed out that historical science deals with the problem of reconstructing the past. It involves developing knowledge of past events from a scientific study of the physical evidence that the event itself creates. These are the traces of the event. These traces are the physical changes that are left behind after the event or series of events of interest take place. It is the role of reconstruction to explicate these elements of the physical evidence record.

1.3.2 Art or Science, or Both?

Cleland (2002) points out that historical science is every bit as valid as experimental science. Reconstructing the past depends on the researchers carrying out this activity having an in-depth understanding of the physical world. Stated somewhat differently, they need a broad-based scientific background. However, as important as this is, it is not enough. Scientists at the scene also need experience in developing hypotheses from observations and in evaluating and testing these hypotheses. There is a need for creativity or what might be called art. Hypothesis development often involves the need for an active imagination informed by extensive experience allowing for some "educated" guessing. In this process, the scientist's mind ranges over a large number of possibilities to be evaluated. Here, we may recognize many of the attributes of an art. This is what Charles Sanders Peirce termed abductive reasoning. The process does not end here. Each hypothesis developed this way must then be subjected to rigorous scientific testing in an effort to disprove and to reject it. The cycle of observations, hypothesis development followed by hypothesis testing, epitomizes the scientific method. Hypotheses are never proved. They can be supported, if successive efforts to reject them fail. Rejected hypotheses are eliminated because they cannot be supported by traces comprising the physical evidence record. They cannot contribute to an explanation for the past event.

It is important to note that some hypotheses are not testable or falsifiable and thus cannot be the subject of a scientific inquiry. This Popperian philosophy (Stufflebeam 2017) has important implications for forensic reconstructions, where investigators' pet theories allow them to ignore or overlook contradictory physical evidence completely. Conclusions based on untestable or untested hypotheses have been the cause of several wrongful convictions, such as in the Jeffrey Deskovic case (Innocence Project 2020). Hypothesis testability (or falsifiability) is the hallmark of a scientific endeavor; it provides the cleanest demarcation between science and nonscience.

1.3.3 Importance of the Scientific Method

Most books provide a few words about the scientific method but do not then develop a coherent intellectual structure for criminalistics. We propose such a structure (Figure 1.4) and, in the context of cases, examine its profound implications for scientific investigators. Science and the scientific method provide the best answers possible at a given moment in time, but these answers or explications are always provisional. This provisional nature is due to the fact that if additional information were to become available, the conclusions may need to be modified. The scientist must recognize and acknowledge this explicitly.

Crime scene reconstruction, in which a blood trace configuration interpretation component would be an integral part, operates at two levels, an informal level and a formal level. At the informal level, hypothesis development and testing may take place as a solely mental process. This is where scientific investigators pose questions to themselves, almost subliminally, and look for support or contradiction from the physical evidence. Hypotheses that fail to be eliminated at this level can help guide the evidence recognition process and provide a flexible framework for guiding the scene investigation as discussed earlier in this chapter. The more formal process is used to develop the refined expert opinions that are used in reports and testimony.

Another important advantage of employing the scientific method is that it helps to guard against the possible introduction of unconscious bias relative to conclusions drawn by a crime scene scientist. By continuously and rigorously testing hypotheses, we strive to ensure that all conclusions formed are supported by the physical evidence and are not swayed by personal preconceptions. This is discussed further in Chapter 7.

1.3.4 Reconstruction vs. Reenactment

As previously described, reconstructions must be based on a scientific interpretation of the traces comprising the physical evidence record. Too often, courtroom presentations that fall far short of a scientific reconstruction are accepted as evidence at trials. An example would be video animations. A scientifically based reconstruction can often provide details of events at specific points in time over the course of events in question. The problem with most video animations is that they, of necessity, must fill in the gaps in scientifically based knowledge with speculative assertions and illustrations in order to have a smooth-flowing presentation. In some cases, audio recordings of a 911 call have allowed gunshots to be fixed at various points in time. Since this provides a timeline, it is very tempting to use these points in time as part of a continuous video; however, this is objectionable because there is no valid information about geometric details to allow these unknowable details of the voids to be filled in. Fortunately, it is possible to employ animation technology in ways that avoid this serious objection. For example, the details of an event at a single point in time can often be visualized by changing the viewer perspective continuously such as a flyover or rotation of the scene. These types of conservative video animations can be very valuable in illustrating a scientifically based reconstruction without having to invent unsupported fictional activities. Some commonly fabricated actions are the position of limbs (e.g., a raised arm in the case of a shooting) and body orientations of either the victim or the perpetrator. The use of video animations in court should deliberately avoid the use of fiction and only be used to responsibly represent the scientific record to the triers of fact.

Some court presentations may violate the principle illustrated above without using video technology in attempting to present speculative and scientifically unsupportable evidence by employing reenactments. Reenactments are essentially a non-video illustration of a critical event by human actors. This can be acceptable as long as it is supported by a scientific interpretation of the physical evidence record. However, too often reenactments may suffer the same shortcomings as video animation demonstrations with their tendency to fill in the voids with scientifically foundationless conclusions.

1.3.5 Holistic Philosophy: Blood Trace Configuration Interpretation Is Only One Aspect of Reconstruction

A holistic approach recognizes that the whole is often worth more than the mere sum of its parts, and this is true with respect to crime scene reconstruction. While carrying out a blood trace configuration-based reconstruction, it is important to bear in mind that analysis of blood traces is an integral part of an overall scene investigation and that other types of physical evidence will likely contribute to the reconstruction. One should not expect to encounter crime scenes where the only evidence present consists of blood traces. This would be unusual. In fact, the opposite may be the case. In a given case, although there may be blood traces present, they may not be very informative about resolving important questions in understanding the specific event. Other evidence may address these questions much more effectively. Such evidence should not be overlooked due to the fact that the initial focus or request is on bloodstain patterns. Several such case examples are detailed in Chapter 12. Experienced crime scene scientists who are called upon to interpret blood trace configurations at a scene must maintain a holistic perspective and avoid focusing on blood traces to the exclusion of other physical evidence.

The holistic approach is facilitated by having a case manager, also called a "conductor" or "Maestro," with a broad range of education and experience (De Forest 2018; De Forest et al. 2015). It is this generalist-scientist who is responsible for the coordination and scientific direction of a case investigation. This idea needs to be more broadly promoted to greatly improve the quality of information obtained from traces during a scene investigation.

The holistic approach is difficult to implement in police departments. In 2016, it was reported that there were in excess of 15000 law enforcement, police, and sheriff departments with armed officers (Greenberg 2016). This is a tremendous number of departments, and it would be unreasonable and unnecessary to expect each to possess the requisite expertise to scientifically investigate crime scenes. This unfortunately contributes to the poor scene documentation and investigations that the authors have observed in our roles as laboratory scientists and consulting criminalists (i.e., the case of the Contested Fratricide in Chapter 10). A potential solution to this would be the availability of scientific investigative centers with criminalists available to serve as a resource to smaller departments. Although this exists in varied form in several U.S. states, it is not always formalized, may not be staffed with scientists in lieu of nonscientist law enforcement investigators, or utilized by local departments.

References

Cleland, C.E. (2002). Methodological and epistemic differences between historical science and experimental science. Philosophy of Science 69 (3): 474–496.

De Forest, P. (2018). Physical aspects of blood traces as a tool in crime scene investigation. *The CAC News*, Third Quarter, 30–33.

De Forest, P.R., Gaensslen, R.E., and Lee, H.C. (1983). Forensic Science: An Introduction to Criminalistics. McGraw-Hill.

De Forest, P., Bucht, R., Buzzini, P. et al. (2015). The making of the criminalistics maestro: on the knowledge, skills, and abilities to oversee and coordinate the work on non-routine and complex cases. Proceedings of the 67th Meeting of the American Academy of Forensic Sciences, Orlando, FL (February 2015).

Eco, U. and Sebeok, T. (eds.) (1983). The Sign of Three. Indiana University Press.

Greenberg, J. (2016). How many police departments are in the United States? *Politifact.* https://www.politifact.com/punditfact/statements/2016/jul/10/charles-ramsey/how-many-police-departments-are-us (accessed 1 October 2020).

Innocence Project. (2020). Jeff Deskovic. https://www.innocenceproject.org/cases/jeff-deskovic (accessed 1 October 2020).

Stufflebeam, R. (2017). *Popper: Conjectures and Refutations.* Department of Philosophy, University of New Orleans. [Video]. YouTube. https://www.youtube.com/watch?v=q3qUzLvOKJo (accessed 1 October 2020).

CHAPTER 2

Historical Perspective

The history of the use of information derived from a scientific interpretation of the physical aspects of blood traces at crime scenes, and on other items of potential evidence, has been checkered and punctuated with significant gaps in foundational research. Clearly, even though there has been recognition that the physical configurations of blood traces can be utilized effectively to understand and shed light on the details of an event for the last 50 years or more, there are problems with its implementation especially with respect to the allocation of scientific resources. This and the negative fallout stemming from the inept performances of unqualified investigators in well-publicized cases has led to questioning the scientific validity of work in this area. The National Academy of Sciences raises this issue in its report *Strengthening the Forensic Sciences – The Path Forward* published in 2009. Part of the unfortunate situation described in this report was caused by a failure of forensic scientists and investigators to fully appreciate and understand the value of the physical aspects of blood traces and instead placed more emphasis on routine laboratory analysis (often required by law). This is not a new problem, as demonstrated by Paul Kirk in his 1967 chapter "*Blood – A Neglected Criminalistics Research Area,*" where he recognized the potential of blood trace configurations and its limitations. Dr. Kirk was clear that this was a neglected area of criminalistics.

2.1 Edgar Allen Poe and Sir Arthur Conan Doyle: History in Fiction

In 1841, Edgar Allen Poe published his short story "The Murders in the Rue Morgue" in Graham's Magazine. In the locked room with the bodies of the two victims, among other traces, were a razor and tufts of gray hair, both covered with blood (Poe 1927).

In Sir Arthur Conan Doyle's "A Study In Scarlet" (1887), Sherlock Holmes observed blood traces surrounding a body which had no physical wounds and deduced that the blood originated from the assailant as opposed to the victim. The idea of knowledge gained from the examination of blood traces was not fully developed.

Blood Traces: Interpretation of Deposition and Distribution, First Edition. Peter R. De Forest, Peter A. Pizzola, and Brooke W. Kammrath.

2.2 Hans Gross

Hans Gross, an Austrian investigating magistrate, was a pioneer of forensic science for recognizing the potential of physical evidence. He is credited with coining the term "criminalistics." In his 1893 "Handbuch für Untersuchungsrichter als System der Kriminalistik" (Handbook for Examining Magistrates as a System of Criminalistics), he includes a chapter on blood marks and their preservation. Among other things, he details how the morphology of a blood trace deposit can provide information about the motion of the source and presents relevant illustrations (Gross 1924).

2.3 History of Research in Blood Traces

One of the more prominent early research works was that of Dr. Eduard Piotrowski involving living rabbits as test subjects (1895). Piotrowksi whose background was in medicine studied the production of blood spatter generated from wounds he had administered to the rabbits under a variety of conditions. This work was essentially of a descriptive nature illustrating the distribution of bloodstains that were obtained from a series of experiments that were recorded by drawings/paintings. No photographs were published in this book, although there were several detailed color drawings (Figure 2.1).

In 1914, Ziempke, E. authored "Die Untersuchnung von Blutspuren" translated as "The Investigation of Blood Tracks" in "Gerichtsärztliche und polizeiärztliche

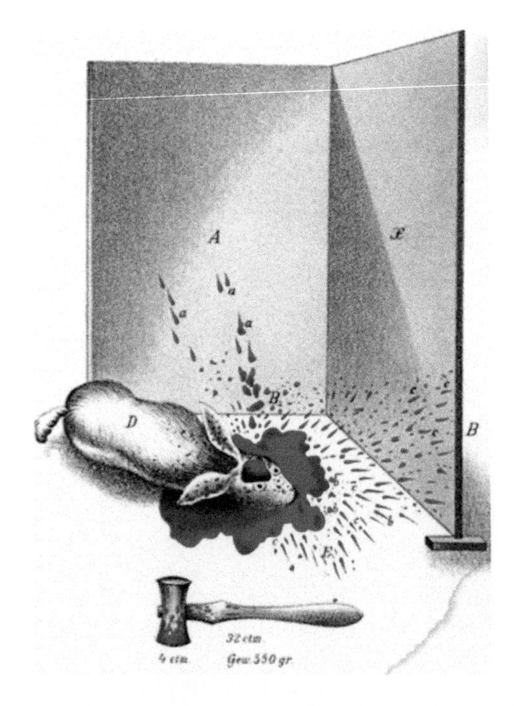

FIGURE 2.1 An image from Piotrowski's book (1895) depicting the blood traces produced from the experimental bludgeoning of a rabbit. *Source: © Eduard Piotrowski.*

Technik: eine Handbuch für Studierende, Ärzte, Medizinalbeamte und Juristen." This early work systematically investigated blood traces produced by droplets under a range of conditions, including vertical falling from different heights, and projection on both horizontal and vertical surfaces (Ziempke 1914).

In 1931, Von Dr. W. F. Hesselink published research on blood traces entitled "Blutspuren in der kriminalistiscen Praxis" in the German Journal "Zeitschrift für angewandte Chemie." Hesselink concludes the paper, saying "Bei Bluntuntersuchunger ist also der eigentliche Blutnachweis oft bei weitem nicht so wichtig wie die Aufklärung von Nebenumstanden," which translated means "In the case of blood tests, the actual proof of blood is often far less important than the explanation of secondary circumstances." The authors of this book wholeheartedly agree with this insightful observation made almost a century ago.

In 1939, V. Balthazard, R. Piédelviévre, H, Desoille, and L, Dérobert published the results of their scientific research, "Etude des Gouttes de Sang Projete" (Study of Projected Drops of Blood). Their research included the study of the relationship between the impact angle and the ratio of the minor and major diameters of the elliptical portion of individual droplet traces, the area of convergence based on an examination of radial impact configurations, the influence of different substrates on trace morphology, showed that the height of fall could not be estimated from the trace diameter (since it is a function of drop volume and impact velocity), and high-speed cine photography to examine the dynamics of droplet impacts with solid surfaces.

2.4 Detective Charlie Chan: History in Film

A significant demonstration on of early layperson awareness of the potential use of blood traces for reconstruction was illustrated by a 1936 Charlie Chan movie entitled "Charlie Chan at the Race Track." In this film, the detective used blood trace configurations to disprove an accidental claim of death by the kicking of a horse in a homicide investigation (Figure 2.2). Although this was not yet widely supported by forensic scientific research, it indicated the reconstruction capabilities of blood traces.

2.5 Paul Kirk

In his classic 1953 textbook entitled "Crime Investigation," Dr. Kirk briefly introduces the concept of reconstructions based on physical blood traces. As one example, he described a case with a relatively complex interpretation derived from physical traces of blood and serum on the victim's person and on her clothing. It is interesting to speculate on how such a blood trace situation would be recognized and interpreted now, some seven decades later. What background scientific knowledge would the investigator have to have to recognize such a situation? The passage from the textbook is quoted below.

> *In one instance in which a female victim was stabbed through the heart, a considerable amount of the blood had flowed between the breasts and below and around the right breast. In addition to this, a much larger quantity of blood*

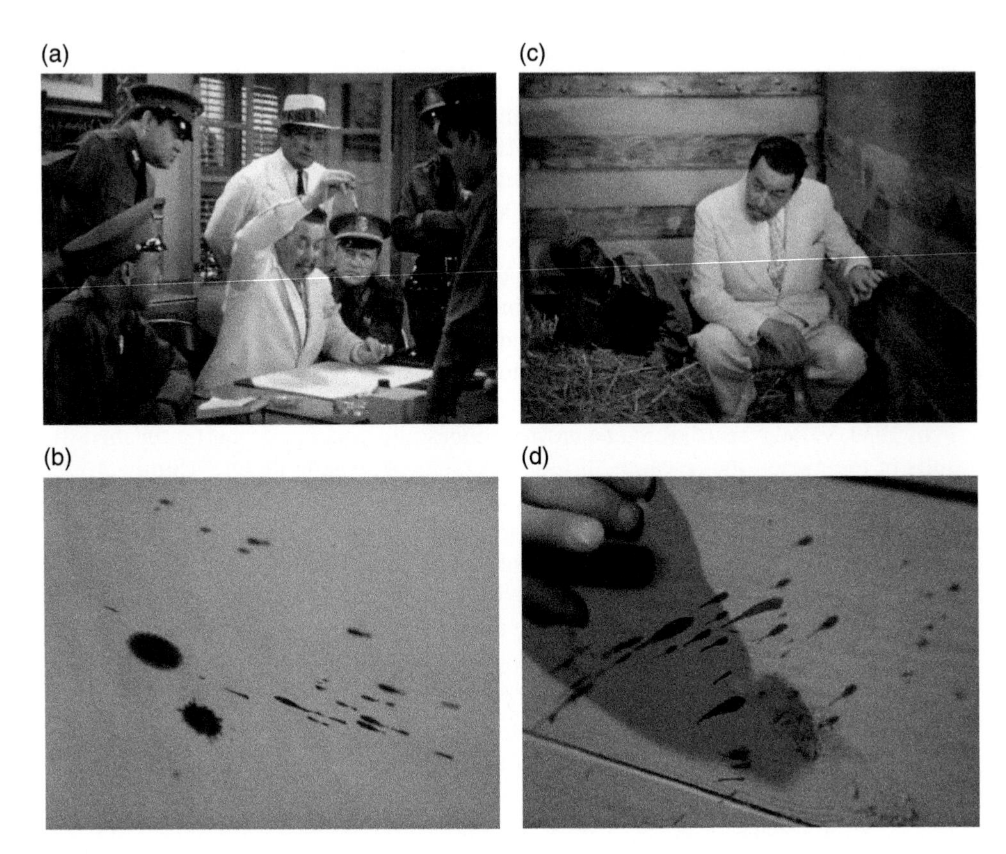

FIGURE 2.2 Images from the 1936 film "Charlie Chan at the Race Track," showing the character Charlie Chan demonstrating principles of blood trace formation to law enforcement (a and b) and investigating the murder in the horse stall (c and d). *Source: © John Stone.*

> *serum had flowed around the left side and to the back of her clothing which was soaked with serum. From these facts, it was clear that she was in a sitting or partly reclining position tilted to the right at the time the knife was withdrawn. The body was not moved for sometime, as indicated by the extensive clotting of the blood shed internally. The body was then moved to a lying position, being tilted somewhat to the left either while being carried or as placed, and during the carriage, the head was low, allowing serum to flow over the left shoulder and down around the left side of the back. Since the crime occurred in the culprit's automobile, the sequence of events was of importance in reconstructing the actions of the murder.*

Dr. Samuel Sheppard was convicted of the murder of his wife Marilyn in December 1954 (see Chapter 10 for a detailed discussion of this case). The homicide had taken place in the couple's home in a suburb of Cleveland, Ohio on 4 July 1954. The trial attracted a huge amount of nationwide public attention. Rapidly on the heels of the conviction, the defense attorney William Corrigan initiated the appeals process. His quest for a defense expert led him to contact Professor Paul L. Kirk of the University of California at Berkeley to retain him as an expert. Dr. Kirk accepted the case.

He was told that the crime scene had been preserved by the defendant's family following the completion of the state's investigation in July 1954. It was decided that he would examine and investigate the crime scene as soon as possible. He traveled to Cleveland in January 1955. He spent two days at the scene and then another day examining evidence retained by prosecutor's office. Additional evidence was taken from the scene and sent to his laboratory in Berkeley. On concluding his work, which encompassed the analysis of several types of traces including blood deposits, Dr. Kirk was asked to prepare an affidavit for use by the defense in an appeal.

The 1955 affidavit of Dr. Paul Kirk has been widely distributed among forensic scientists. Although it is not a research article, it has been generally appreciated as a seminal document in the field of crime scene investigation. The value and significance of the affidavit stemmed from its illustration of the wealth of important information that could be derived from traces at a crime scene.

2.6 Herbert MacDonell

In 1971, MacDonell and Bialousz, with the financial support of the precursor to the National Institute of Justice (NIJ), studied the production of blood trace configurations. In ostensibly the same year, MacDonell published the results in the booklet "Flight Characteristics and Stain Patterns of Human Blood" as both a sole author and as a co-author with Bialousz. This document was widely distributed to law enforcement and forensic laboratories throughout the country in the early to mid 70s. In 1974, it was reprinted including L. Bialousz as a co-author. Also in 1971, MacDonell, authored a chapter entitled "Interpretation of Bloodstains: Physical Considerations" which is essentially a reworking of the former reference. Some interesting observations are made in these publications, such as the relationship between the ellipticity of a stain and the angle of impact and the critical importance of a substrate's surface characteristics in the formation of stains. However, there are also unsupportable erroneous claims made, which include the determination of a "normal" drop volume (MacDonell and Bialousz 1971, pp. 3–4), the classification of bloodstain patterns as being medium velocity or high velocity, and the use of the ratio of the height to what is referred to as the "dense blood height" for estimating the impact angle on nonporous substrates (MacDonell and Bialousz 1971, p. 12).

In a second and third edition of his 1971 publication, MacDonell presented a model of a single droplet of blood impacting a horizontal surface at an oblique angle and demonstrated that the sine function could be used to estimate the impact angle. Interestingly, Rizer had previously introduced the use of the sine function for the estimate of the impact angle (1955). Concomitantly, MacDonell initiated a myth that the stain's minor diameter (d) or width, as it is commonly referred to, corresponds directly or is identical to the drop diameter (1982, p. 52; 1993, p. 35).

MacDonell initiated short courses or 40-hour courses in bloodstain pattern analysis, with the first being taught in 1973 in Jackson, Mississippi. There have been a huge number of these "instant experts," with thousands of attendees having been students in these workshops. These have had both positive and negative repercussions. There has been a huge growth in the awareness of the utility of blood trace configurations for event reconstructions that would have otherwise been overlooked. MacDonell has acknowledged the negative consequences of short courses in his 1993 self-published book "Bloodstain Patterns":

> *Although the author is the originator and director of the majority of these specialized courses in the geometric interpretation of bloodstain pattern evidence, I completely agree with Dr. Peter DeForest in his concern that there are those who feel they are qualified beyond their actual ability. In a recent journal review Dr. DeForest wrote "Numbers of individuals without the scientific backgrounds have been trained in these courses. With this type of training, these individuals have stepped beyond this important investigative role to offer scientific evidence as expert witnesses in court. The danger inherent in this development cannot be overemphasized. No amount of experience can supplement [sic supplant] scientific knowledge and a thought process based on careful adherence to the scientific method."*

The above mentioned book review by De Forest was published (1990) in the Journal of Forensic Science.

2.7 Bloodstain Pattern Analysis Committees and Organizations

In the early 1970s, the first American-based bloodstain pattern analysis groups were formed. This has developed from the earliest of committees forming in 1970 under the American Society of Testing and Materials (ASTM E-30), to the formation of the International Association of Bloodstain Pattern Analysts (IABPA) in 1983, to the current and relatively newly formed Organization of Scientific Area Committees (OSAC) Bloodstain Pattern Analysis group. Details of these groups and efforts are discussed in Chapter 14: Resources.

References

Balthazard, V., Piédelièvre, R., Desoille, H., and Dérobert, L. (1939). Étude des Gouttes de Sang Projeté. *Annales de Médecine Légale de Criminologie, Police Scientifique, Médecine Sociale, et Toxicologie* 19: 265–323.

De Forest, P.R. (1990). A review of Interpretation of bloodstain evidence at crime scenes. *Journal of Forensic Science* 35 (6): 1491–1495.

Gross, H. (1924). *Criminal Investigation: A Practical Textbook for Magistrates, Police Officers and Lawyers. Adapted from the System Der Kriminalistik*. Sweet & Maxwell, Limited.

Hesselink, W.F. (1931). Blutspuren in der kriminalistischen Praxis. *Angewandte Chemie* 44 (31): 653–655.

Kirk, P.L. (1953). *Crime Investigation: Physical Evidence and the Police Laboratory*. Interscience Publishers, Inc.

Kirk, P. L. (1955). *Affidavit of Paul Leland Kirk in state of Ohio vs. Samuel H. Sheppard.* State of Ohio, Cuyahoga County, Court of Common Pleas, Criminal Branch, No. 64571, April 26, pp. 1–27.

Kirk, P.L. (1967). Blood-a neglected criminalistics research area. *Law Enforcement Science and Technology* 1: 267–272.

MacDonell, H.L. (1971). Interpretation of bloodstains: physical considerations. In: *Legal Medicine Annual* (ed. C. Wecht), 91–136. *Appleton-Century Crofts.*

MacDonell, H.L. (1982). *Bloodstain Pattern Interpretation.* Corning, NY: Laboratory of Forensic Science.

MacDonell, H.L. (1993). *Bloodstain Patterns*, 35–36. Corning, NY: Laboratory of Forensic Science.

MacDonell, H.L. and Bialousz, L.F. (1971). *Flight Characteristics and Stain Patterns of Human Blood.* PR 71-4, November 1971. Washington, DC: U.S. Department of Justice, Law Enforcement Administration, National Institute of Law Enforcement and Criminal Justice, U.S. Government Printing Office.

Piotrowski, E. (1895). *Über Entstehung, Form, Richtung u. Ausbreitung der Blutspuren nach Hiebwunden des Kopfes.* Slomski.

Poe, E.A. (1927). *Collected Works of Edgar Allan Poe.* New York: Walter J. Black.

Rizer, C. (1955). *Police Mathematics: A Textbook in Applied Mathematics for Police.* Police Science Series. Charles C. Thomas.

Ziempke, E. (1914). Die Untersuchnung von Blutspuren. Translated: "The Investigation of Blood Tracks." In: T. Lochte and J.F. Bergmann (eds.), *Gerichtsarziliche und polizieartzliche Technik.* Wiesbaden, pp. 152–166.

CHAPTER 3

Characteristics of Liquids Including Blood

3.1 Physical Properties and Fluid Mechanics of Liquids

Liquids are one of the three commonly recognized states of matter, in between solids and gases. Gases and liquids are classified as fluids because they can be made to flow and cannot resist any sheer force applied to them. Liquids take the shape of their container, like a gas, but exhibit cohesion among their constituent components (e.g., atoms or molecules) and at a given temperature have a constant volume, density and are incompressible, like a solid. Supercritical fluids are a fourth state of matter that exists when gases subjected to temperatures and pressures above the critical point exhibit properties (i.e., density, viscosity, and diffusivity) intermediate between a liquid and a gas.

Blood is a liquid, and thus attention is given to important physical properties of liquids in this chapter.

3.1.1 Surface Tension and Weber Number

Surface tension is a property that is the result of an imbalance of forces at an air–liquid interface (Figure 3.1). At the surface of a liquid, the cohesive forces in the liquid far exceed the adhesive forces with the air, resulting in a net inward force that acts as a skin-like membrane. It is this property that allows insects and other materials to walk or float on the surface of water. It is also what causes liquid droplets to form a spherical shape due to the minimization of the surface area to volume ratio (Table 3.1).

The Weber number encapsulates the competitive relationship between surface tension and the forces which tend to disrupt the surface, e.g., inertia and impact. It is used in a number of different fields of study when investigating the flow of liquids in thin films and heat pipes, as well as the formation of bubbles and droplets. It is calculated through the equation:

Blood Traces: Interpretation of Deposition and Distribution, First Edition. Peter R. De Forest, Peter A. Pizzola, and Brooke W. Kammrath.

FIGURE 3.1 Diagram showing the imbalance of forces between molecules at the surface of a fluid, resulting in surface tension, when compared to the balanced forces on the interior molecules. The arrows represent the attractive intermolecular forces.

TABLE 3.1 **Approximate surface tension of common liquids at approximately 20 °C**

Liquid	Surface tension (mN/m)
Ethyl alcohol	22
Soapy water	25–45
Chloroform	27.1
Ethylene glycol	47.7
Blood	56
Glycerin	64
Water	73
Saturated salt (NaCl) water	82
Mercury	425

$$W_e = \frac{\rho v^2 l}{\sigma}, \tag{3.1}$$

where W_e is the Weber number, ρ is the density of the liquid, v is the velocity, l is the characteristic length, and σ is the surface tension. The Weber number is commonly used when studying flowing liquids with the v term referring to the fluid's velocity. However, this is not directly applicable to crime scene reconstruction work. Impact velocity and fluid velocity are not synonymous; however, Hulse-Smith et al. (2005) equate the v in the equation to the impact velocity of the blood droplet.

3.1.2 Density

Density (mass density) is an important property of matter that is equal to the mass of a given amount of a material divided by its volume. For pure substances, density has a fixed value. However for mixtures, such as blood, it can vary with proportions of the components. The concept of specific gravity is similar to that of density, but it is a unitless term that relates the density of a material to that of a selected liquid, commonly water, at a given temperature (Table 3.2).

It should be noted that the above table results report the density measurement at approximately room temperature. It can be expected that at physiological temperature (37 °C), the density of blood would be somewhat lower, where the density of blood in animals ranged from 1.044 to 1.050 g/cm^3) (Kenner et al. 1977). These same authors report that transient changes in density can occur with the infusion of hypertonic or hypotonic solutions, but they return to a normal value in a matter of several seconds.

3.1.3 Newtonian and Non-Newtonian Fluids

A Newtonian fluid is characterized by the viscosity (to be discussed below) of a fluid remaining constant at a given temperature regardless of the shear force that is applied, such as stirring, flowing, or having an object travel through it. Shear can be viewed as the opposing parallel planar forces tending to cause hypothetical layers of a deformable material or object to slide over one another, such as when the top and bottom of a stack of papers are pushed in opposite direction causing the papers to slide and the stack to deform (Figure 3.2). A simple theoretical model of shearing in a liquid is represented by the telescopic sliding of concentric cylindrical layers of fluid flowing inside a straight inflexible tube at a constant velocity. For a Newtonian fluid, its stress is linearly proportional to the strain put on it. Water is the most commonly encountered example of a Newtonian fluid.

TABLE 3.2 Approximate densities of common liquids at roughly room temperature

Liquid	Density (g/cm^3)
Ethyl alcohol	0.79
Olive oil	0.80–0.92
Water	0.997
Sea water	1.02
Blood	1.06
Saturated salt (NaCl) water	1.20
Honey	1.38–1.45
Chloroform	1.46
Mercury	13.6

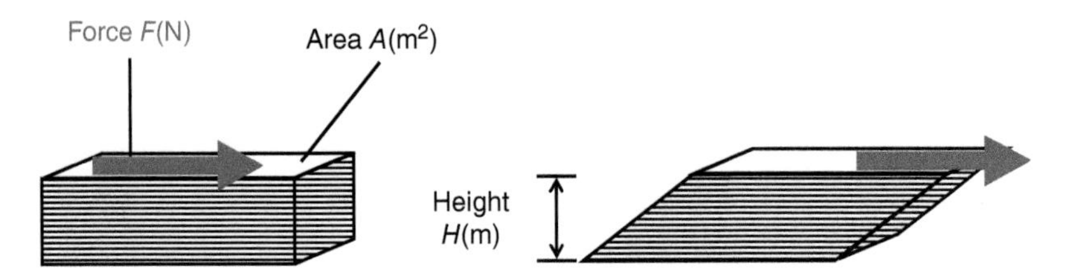

FIGURE 3.2 A diagram of showing the physics of sheer as applied to a stack of papers.

The viscosity of non-Newtonian fluids varies with shear force. Examples of non-Newtonian fluids are blood, non-drip paint, and even oobleck (a term invented by Dr. Seuss, the pseudonym of Theodor Geisel, in one of his stories, and subsequently given to the child's toy made of a thick suspension of cornstarch grains in water). With oobleck, the viscosity increases with increased shear or force, thus making the material feel like a solid when poked or when a finger is drawn through it.

3.1.4 Viscosity and Poiseuille's Equation

Viscosity, or a fluid's resistance to flow, originates from the internal friction amid particulate components and molecules in a bulk fluid. It was first studied in detail by the French scientist Jean Léonard Marie Poiseuille in the mid-eighteenth century. He derived an equation for the laminar flow of incompressible Newtonian fluids in tubes with uniform circular cross sections along their lengths, known as Poiseuille's equation. It is sometimes called the Hagen–Poiseuille equation, which credits Gotthilf Heinrich Ludwig Hagen with independently deriving the equation at around the same time, although Poiseuille was the first to publish. Poiseulle's equation, expressed in terms of the pressure drop across the ends of a tubular pipe, is

$$\Delta P = \frac{8\eta LQ}{\pi R^4}, \tag{3.2}$$

where ΔP is the pressure difference between the two ends, L is the length of pipe, η is the dynamic viscosity, Q is the volumetric flow rate, and R is the radius of the pipe. For determining the viscosity of a liquid using an appropriate tubular apparatus, the equation can be rearranged as follows:

$$\eta = \frac{\Delta P \pi R^4}{8LQ}. \tag{3.3}$$

Newton is credited with the equation that relates viscosity (η), shear rate (τ), and shear force (γ):

$$\eta = \tau / \gamma \tag{3.4}$$

where the shear force is applied to a layer of fluid which causes its motion in relation to an adjacent layer of fluid. The velocity gradient that exists between two adjacent layers of fluid (divided by the distance between the layers) is the shear rate (Table 3.3).

TABLE 3.3 Approximate viscosities of common liquids at roughly room temperature

Liquid	Viscosity (centipoise, cp)
Ethyl ether	0.22
Water	0.89
Saturated salt (NaCl) water	1.99
Milk	3
Blood	3–4
Ethylene glycol	16.2
Motor oil (SAE 10W-40)	120
Glycerin	950
Honey[a]	2000–40 000

[a] Depending on water content.

3.1.5 Flow Stability, Reynolds Number, and Rayleigh Number

Poiseuille's equation is not applicable when the flow of fluids is unsteady or turbulent (Figure 3.3). Turbulent flows are characterized by eddy currents which are observed in both liquids and gases (Figure 3.4). Based on extensive experimentation, it has been found that the Reynolds number can be used to determine if the flow is turbulent, laminar, or unstable. The equation for the calculation of the Reynolds number (dimensionless) in a pipe or tube is

$$N_r = \frac{2\rho\bar{v}R}{\eta}, \tag{3.5}$$

where ρ is the fluid density, $\bar{v}$ is the average velocity, R is the radius of the tube, and η is the viscosity.

The prediction of flow characteristics is quantified as a function of Reynolds number values. The Reynolds number predicting a laminar flow is less than 2000, while that predicting a turbulent flow is greater than 3000. A Reynolds number between 2000 and 3000 predicts an unstable flow.

The Rayleigh number is another dimensionless number that characterizes the motion of liquids within liquids, or buoyancy-driven flow, that is known as free or natural convection. It can also be understood as the ratio of buoyancy and viscosity forces multiplied by the ratio of the momentum and thermal diffusivities. Similar to the Reynold's number, it is used to characterize a fluid's flow, where laminar flow or turbulent flow are denoted by low and high ranges of Rayleigh numbers.

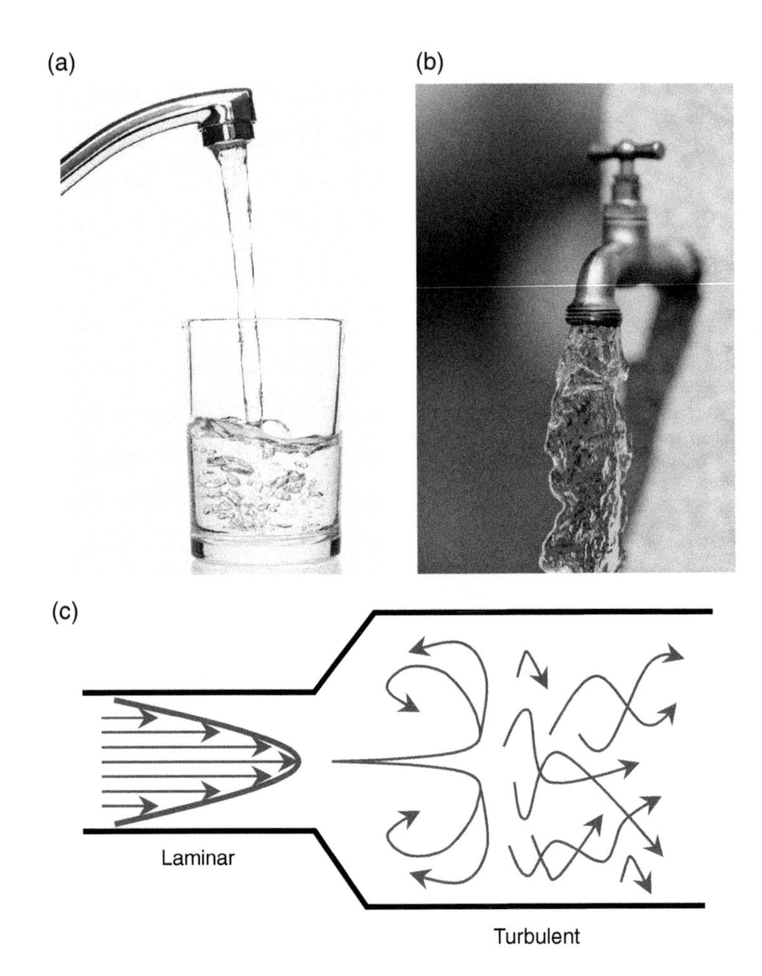

FIGURE 3.3 Laminar (a) and turbulent (b) flows of water from a faucet. Figure (c) shows these two types of flows diagrammatically. *Sources: (a) Jurisam/iStock/Getty Images © Getty Images John Wiley & Sons and (b) Bigandt_Photography/Getty Images. © Getty Images John Wiley & Sons.*

3.1.6 Viscoelasticity

Viscoelasticity is a property that takes both viscosity and elasticity into account when undergoing a deformation. It has been previously used to study blood in circulation of the body (Thurston and Henderson 2006) and may have applicability to blood traces. More research is needed to understand its relevance to blood trace deposit interpretation.

3.1.7 Caveats

Most of the research in fluid dynamics has been focused on the practical problem of understanding the flow of fluids in tubes or pipes. Little of this relates to the impact of droplets on solid surfaces. It is the forces involved with impacts and the deposit of blood on a target substrate that has the most relevance for the forensic interpretation of blood traces. The situation is very complex, and the influence of these properties on the impact of liquid blood droplets encountered in practice has not been comprehensively studied or evaluated.

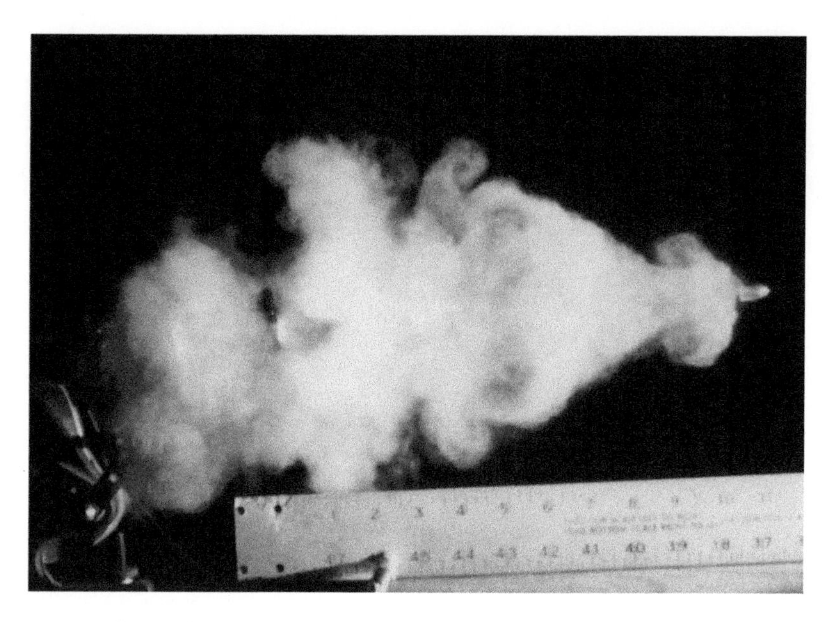

FIGURE 3.4 High-speed photograph of a bullet from a gunshot followed by particulate and vaporous muzzle residues exhibiting eddy currents. *Source: Pizzola, Doctoral Dissertation (1998) – photography by Kirby Martir & Pizzola.*

The numerous fluid dynamics equations may offer useful insights but cannot be applied directly to case situations involving blood trace deposits. As discussed by Adam (2012), it is important that there be "...a proper understanding on how the fundamental equations need to be modified to account for the surface properties of a wider range of substrates such as those that might be found routinely in casework. Such surfaces may exhibit different characteristics of roughness, elasticity, wettability and porosity, and systematic studies of the effect of such properties on the spreading and splashing of blood remain to be done."

In terms of relevance to blood trace deposit interpretation, droplet dynamics are more important. This is discussed in Chapter 6.

3.2 Physical Characteristics of Blood

It is important to understand some of the physical properties and characteristics of blood in order to be able to understand deposit formation. This science-based knowledge is clearly necessary for the interpretation of blood trace configurations.

3.2.1 Definition and Description of Blood

Blood is a complex mixture and would best be described as a *suspension* or *multi-phase heterogeneous mixture* that is composed of two phases – plasma and cellular. The latter phase is comprised of thrombocytes (platelets), leukocytes (white blood cells), and erythrocytes (red blood cells or RBCs). Plasma contains numerous dissolved substances such as proteins, antibodies, enzymes, and nutrients. Blood is occasionally referred to as a colloidal fluid by some authors (Bevel and Gardner, 2008, 111) which is incorrect according to Ristenbatt (2008). The dimensions of the cellular components in blood exceed the range used to denote *colloidal* by an order or two in magnitude (Everett 1972; Petrucci 1982). Particles in a colloidal suspension stay suspended for a very long time, if not perpetually, in contrast to the fast settling characteristic of blood's large solid components (Petrucci 1982). The term "colloidal" describes some medium that contains molecular particles (or particles of polymolecules) that are dispersed in at least in one direction possessing a dimension of roughly speaking between 1 nm and 1 μm; or discontinuities exist that possess distances of that order (IUPAC, 2014, Gold Book, 295).

Blood is a non-Newtonian fluid, as noted above, because of the property of viscosity varying with certain conditions, such as at higher stress (e.g., as it flows in capillaries). While erythrocytes might be thought of as solids because of their cellular composition, they are not very rigid. According to Merrill, they are readily "deformed and distorted" – not nearly as rigid as perhaps thought (870–871). The ability to bend or deform contributes to their ability to pass through vessels that rigid disks could not. Blood exhibits large increases in viscosity at low-shear rates because of red cell aggregation (Lowe 1987, 600). In plasma, assuming static or a low flow rate, erythrocytes will form linear as well as secondary aggregates which are elastic. A certain level of shear stress is necessary to break up the networks and promote flow. "The non-Newtonian rheology of blood is dominated by the interaction of fibrinogen with red cells." (Merrill 1969, 876). Aggregation among RBCs during flow is due to the presence of large proteins in the plasma. Fibrinogen is the protein most responsible for aggregation of RBCs, whereas albumin does not promote aggregation because of its minimal molecular length (Lowe 1987, 614). The much larger fibrinogen, in contrast to albumin (much shorter), allows the repulsive force between the cells to be mitigated (Lowe 1987, 614). Additionally, red cell aggregates are dispersed at high-shear rates and "red cells are deformed into ellipsoids in parallel with the flow streamlines and participate in flow. This shear-induced deformation of red cells renders blood a Newtonian fluid of minimal viscosity under the high-shear conditions which it encounters in large vessels (arteries and veins) in the normal circulation" (Lowe 1987, 606). The viscosity of blood is minimized in large blood vessels by red cell deformation. The deformation in large vessels is affected by a number of extrinsic factors including plasma, the hematocrit, and shear conditions. Narrow vessels also influence RBC deformation. Erythrocytes, or RBCs, of healthy mammals readily travel through capillaries having mean diameters of 4 μm and are very deformable. Intrinsic factors also have an influence on RBC deformation; flexibility of the membrane, internal viscosity, and geometry of the cell (Lowe 1987, 622).

It is interesting that although blood predominately is considered a non-Newtonian fluid, it can also behave under certain conditions as a Newtonian fluid (Lowe 1987, 606) as mentioned above. The cellular components of blood have a controlling influence on its overall behavior as a non-Newtonian fluid. Without the cellular components, the individual behavior of plasma (in vivo) or serum (in vitro) would dominate and therefore act as Newtonian fluids.

3.2.2 Factors that Influence Droplet Deposit Periphery

The Weber number, described above, has importance when considering liquid droplet impacts with surfaces and the formation of blood droplet deposits. The equation for Weber number (Eq. 3.1) essentially deals with the competition between inertia of the droplet (numerator) and its surface tension (denominator). Higher Weber numbers are associated with the kinetic energy dominating and overcoming the surface tension; this favors spreading of the impacting drop, on a solid or liquid surface, or even the occurrence of splashing where coronas (also known as crowning) can form depending on the nature of the surface.

There have been a number of other terms used in publications to describe the disturbances around the periphery of the collapsing blood droplet and resulting traces. Some authors also distinguish between the kinds of protrusions (i.e., scallops, spines, or fingers) based on a normal impact versus an oblique one (Adam 2012). It is noteworthy that the same author (Adam 2012) referenced a much earlier semiquantitative research effort by two of the present authors (Pizzola et al. 1986b). The 1986 research incorporated high-speed photography of the formation of blood traces (1986b) and demonstrated the equivalence of deposit features created from an impact with an inclined surface to those formed from a perpendicular impact on a moving horizontal substrate (1986b; see Chapter 6). More recent fluid dynamics studies have focused on the impact of droplets with inclined surfaces where the velocity (tangential) has been controlled via moving belts and rotating cylinders. Adam stated that the formation of the features of the splashes, also in a non-perpendicular context, are determined by the impact velocity's normal component which can be evaluated by the "corresponding" Weber and Reynolds numbers. These crowns may be accompanied by scallops or spines. Smaller droplets may also form on the tip of the spines.

Arthur Mason Worthington (1852–1916) was among the first scientists to study splashes involving liquids as well as solids. Much of his record of splashes was captured via drawings, in addition to photography, because he felt that the photographic images were blurred during his early experiments. He attributed the lack of clarity in the photographs to the source of illumination (electric spark) rather than the photographic plates (Worthington 1963, xi). Ostensibly, the problem was due to the flash duration, with it being on for too long. Subsequently, after communication with Sir Charles Vernon Boyd, and using the same electrical system as Lord Rayleigh, two pioneers of high-speed studies during that era, he was able to capture photographic images that ". . .had come out well" (Worthington 1963). Approximately a half-century later, Edgerton pioneered the high-speed electronic flashes that captured photographic images of drop collapses which have become well known (see Figure 3.5). At lower Weber numbers, the surface tension is able to resist the inertial motion (or kinetic energy) and hold the impacting drop together preventing splashing or spreading. When splashing is obviated microdroplets or satellites (the latter term used by Mehdizadeh et al. 2004) will not be able to form around the periphery of the impacting drop (see hi-speed montage of blood drop impacting sessile drop – Figure 6.26). It is reported that the development of fingers in a drop impact is influenced by irregularities on the surface of the substrate at a broad range of frequencies (Mehdizadeh et al. 2004). Recently, Stotesbury et al. (2016) have reported that crowning in blood deposits is influenced mainly by three factors during the mechanism of impact of a droplet with a solid surface. Chapter 6 provides more details regarding this phenomenon and blood trace formation.

FIGURE 3.5 Rayleigh spout. *Source: Cranberry Juice into Milk by Harold Eugene Edgerton.*

3.2.3 Factors that Influence Droplet and Deposit Size

Although there has been a persistent myth of a "normal drop" (MacDonell and Bialousz 1971; Bevel 1983; see discussion in Chapter 13), the size of a blood droplet does vary based on the mechanism of its formation and separation. One of the simplest mechanisms of blood droplet formation is the result of separation from an orifice or surface due to acceleration. This may be due to gravitational acceleration (i.e., dripping) or centripetal acceleration (i.e., cast-off), among others. Another mechanism, which is more complex, is the product of impacts to accumulated liquid blood (i.e., primary impact from moving objects and secondary impacts from blood droplets). Additional mechanisms could include the rupture of bubbles, interactions with energetic air streams (i.e., atomization).

The physical properties of liquids discussed previously affect the droplet size. The size of a droplet is influenced by a competition between the inherent surface tension of the liquid and the disruptive force of the interacting object. In order to have a droplet separate from a larger mass of blood, the surface tension must be overcome by the applied force.

Among the factors that can affect drop size is the size of the orifice from which the droplet is forming (assuming of course that one is forming from some sort of orifice). While it is true that pipettes are ordinarily not the source of blood droplets during a real-world incident, the fact that the orifice diameter can influence drop size is important. First, the inner diameter of a blood vessel can be considered similar to the orifice of a pipette. Second, since different drop sizes can be readily produced under laboratory conditions, the same can be expected in a real-world event such as during a violent struggle. Laber and Epstein used special pipettes to generate different drop volumes in order to demonstrate the effect of different drop volume on deposit diameter (1983, p 7). Kinnel (1972), working with aqueous droplets (not blood), produced different drop sizes by systematically varying tube orifices.

The mechanism responsible for blood droplets formed from the separation from blood bearing surfaces under the influence of acceleration (e.g., gravitational or centripetal) is similar to that discussed above for blood droplets forming from an orifice. Droplet size is related to the contact area and nature of the surface from which it is separating. For example, a blood droplet separating from the tip of an icepick would be expected to be smaller than one forming from a larger object like a baseball bat (Pizzola et al. 1986a).

Another factor that has an influence on the size a blood droplet is the wettability of the surface from which it is separating. Wettability is the ability of a liquid to adhere to a surface, which is due to a balance between intermolecular forces. In the case of more wettable releasing surfaces, the stronger adherence to the surface from which the droplet is formed would result in the formation of larger blood droplets. Conversely, smaller droplets are formed from less wettable releasing surfaces. Wettability of the receiving surface (i.e., target or substrate) is also a major influence on the size of a blood trace deposit. For a given liquid drop size impacting on a target, a more wettable receiving surface would result in larger diameter deposit than for a less wettable surface (Figure 3.6). Wettability is commonly measured by a concept known as the contact (or wetting) angle. This angle is measured tangent to the droplet at the triple-point interface between the droplet, the receiving surface, and the air. As seen in Figures 3.6 and 3.7, the contact angle is greater for the droplet on the right which has a more wettable surface than that on the less wettable or more repellant surface on the left.

FIGURE 3.6 Photographs of sessile (or nonmoving) liquid blood droplets on a glass substrate with perfluorooctane sulfonate, e.g., Scotchguard® (left) and on uncoated glass (right).

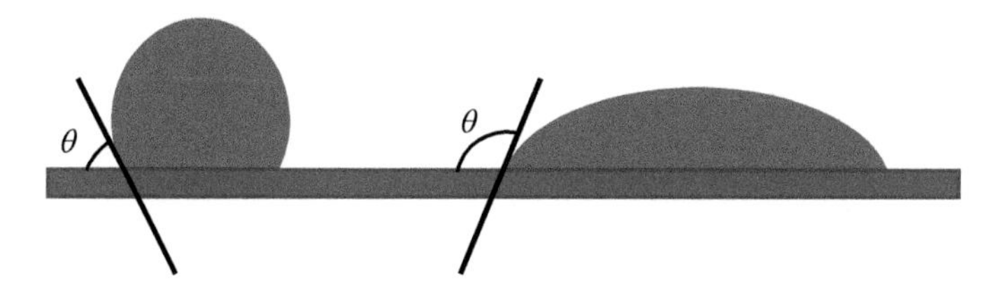

FIGURE 3.7 Figure showing the effect of the wettability of the surface on the contact or wetting angle (θ) of a liquid droplet. The figure on the right shows a surface with a greater wettability and thus a larger resulting contact angle than the figure on the right.

In addition to droplet size, a major factor that affects the deposit size is the impact velocity. The impact velocity is dependent on factors such as the dropping height and projection velocity. Large impact velocities produce larger deposits.

Lastly, the impact substrate texture will not only affect the size of the deposit but also its morphology. Recently, there has been significant research on the role of textile structures and fiber composition to the formation of blood deposits, which demonstrate the extreme complexity of droplet–substrate interactions (Reynolds and Silenieks 2016).

3.2.4 Sedimentation and Hematocrit

RBCs in liquid blood that is left to stand will gradually settle out of suspension due to their slightly greater density. This process of sedimentation can be accelerated by centrifugation. The extent of the volume of sedimentation is known as hematocrit. Quantitatively, hematocrit is expressed as the proportion by volume of RBCs in blood and is represented as a percentage. Hematocrit levels, which roughly range between 35 and 55% in a healthy adult, will affect the blood's viscosity. Rogers (2009) and Dubey (2019) studied the effect of hematocrit on blood trace configurations resulting from the impact of blood droplets. Rogers (2009) observed that changes in hematocrit levels had a significant effect on the length and width of a blood trace deposit, but their ratio (which is used to calculate the impact angle) remained relatively constant. Dubey (2019) demonstrated, not unexpectedly, that the hematocrit has a direct influence on the density of fluid blood. It also has a significant influence on the viscosity. Details on the effect of hematocrit on surface tension are unclear, and its effect on blood trace deposit morphology is in need of further study.

In 2019, Aplin et al. examined the value of hematocrit for its influence on estimating the area of origin of blood spatter. The authors varied the hematocrit values, ranging from 8 to 90%, of a single individual's blood by systematically altering the concentration between the liquid and cellular fractions. Ultimately, they did not find a "...statistically significant difference between the calculated..." and actual area of origin corresponding to the six hematocrit values.

3.3 Optical Properties of Blood Deposits

Matter generally has a number of optical properties, only some of which are of direct relevance to the interpretation of blood traces. Those optical properties that are encountered in absorption spectroscopy (i.e., the Beer–Lambert law), have implications in the understanding of the appearance of blood deposits. Simply stated, those optical properties of importance are color, opacity, scattering, absorption, and transmission.

Color of blood deposits may be useful for rough estimates of the age since the deposition for relatively recent traces, but it has the serious potential to be misleading. The perceived color can depend greatly on the nature and color of the underlying substrate along with the thickness or path length through the deposit and the illuminating light intensity, directionality, and color temperature.

3.4 Physiological Characteristics of Blood

3.4.1 Hemostasis and Clotting

Human beings, and mammals generally, have evolved a very complex system for maintaining homeostasis by preventing or limiting excessive blood loss from wounds. When a blood vessel is damaged, several different mechanisms that are part of the overall process of hemostasis take place. These are vascular constriction, production of a plug with platelets, generation of a clot, and the development of fibrous tissue which grows into the clot (Hall 2015, 483). It should be noted that there is a nomenclature problem: clotting is very commonly referred to as coagulation although there are substantial differences. Strictly speaking, coagulation is a very general term for the denaturation and subsequent precipitation of proteins from solution as a result of a number of causes including heating and exposure to a range of chemical compounds. The common observation of the changes that take place in cooking an egg or exposing it to alcohol are good illustrations of coagulation. By way of contrast, clotting is the result of a very specific and sophisticated process. The complex process that takes place in the body (in vivo) is detailed in Figure 3.8. An important medical concern relates to the problem of clotting taking place inappropriately within organs or blood vessels rather than as a result of wounding. For this reason, in vivo clotting has been studied extensively. On the other hand, in vitro clotting, such as clotting in blood collection devices and in pools of blood at a crime scene, is far from being well understood. In medical

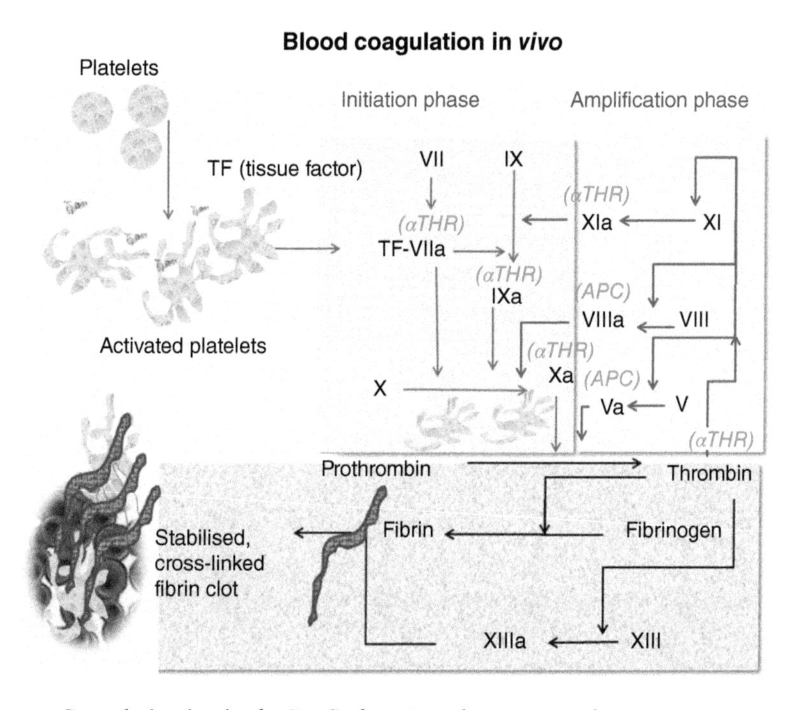

FIGURE 3.8 Coagulation in vivo by Dr. Graham Beards – Own work. *Source: Dr Graham Beards, Blood coagulation pathways in vivo, Own work. Licensed under CC-BY-SA-3.0.*

situations, there are a number of methods which have been developed for preventing in vitro clotting from taking place. The common method is the use of an anticoagulant (a commonly used term, although anticlotting agent would be more accurate), such as heparin or calcium-chelating agents (i.e., ethylenediamine tetraacetate [EDTA] or calcium-precipitating agents such as citrate and oxalate ions). The clot will be composed of the clotting proteins along with the cellular components, while the liquid portion (minus the clotting proteins) is called serum.

As mentioned above, knowledge of details of the process of clotting taking place outside the body is lacking due to a large number of unexplored variables. Thus, from the forensic science point of view in dealing with blood deposits, extreme caution is necessary in attempting to establish time estimates for clotting. Whereas, the authors encourage the timely input at the investigative stage, it is strongly recommended that the scientific scene investigator refrain from offering premature opinions about the time since the event in official communications without explicit caveats dealing with the issues of uncertainty.

3.4.1.1 Postmortem Clotting In 2006, Jackowski et al. studied changes to blood in cadavers, including postmortem clotting, internal livores, and sedimentation by magnetic resonance imaging (MRI) and multi-slice computed topography (MSCT) imaging; one purpose of this project was to provide some recommendations on how to distinguish between the above factors and other forensic findings. According to the researchers, who studied 44 cadavers by MSCT and MRI prior to traditional autopsy, the "cessation of the circulation induces clotting processes. . .within the cardiac cavities. . . because the fibrinogen within the postmortem serum as well as the platelets are still capable of functioning for a variable period of time." They asserted that while sedimentation occurs because of gravitation, postmortem fibrinogen and other blood proteins are still active. Out of 44 cases, postmortem clotting was observed in 14 cases. Some of these cases are quite interesting in that, for example, in a case where the cause of death was a gunshot wound to the head, clotting was observed in the right atrium, left atrium, left ventricle, and the aorta. In another case where the cause of death was a gunshot wound to the head, clotting was observed only in the right atrium.

3.4.1.2 Lack of Clotting Clotting may not take place under certain conditions (Mole 1948; Takeichi et al. 1984, 1985, 1986). Denison et al. (2011) provided interesting historical insights regarding the fluidic state of blood postmortem and that it can stay as such for days. Denison et al. stated that the founders of pathology, i.e., Giovanni Battista Morgagni (1682–1771), John Hunter (1728–1793), and Rudolf Virchow (1821–1902), were well aware that violent deaths were often accompanied by unclotted blood and this was briefly discussed in the research by Takeichi. As cited by Mole (1948), Morawitz (1906) showed that fluid cadaver blood lacked fibrinogen and attributed it to the presence of a fibrinolysin activity. Furthermore, as cited by Mole, Virchow had much earlier (1871) found that capillaries in cadavers contained fluid blood without evidence of clotting. Mole studied cadavers in 61 random cases and observed that entirely fluid blood was more likely to be encountered with sudden deaths. Mole's observations were subsequently verified by Takeichi et al. in 1984 and further examined in subsequent publications (1985, 1986). These latter researchers showed that both physical exercise and vasoactive agents could activate fibrinolysis, and this process could be initiated very quickly. This undoubtedly would benefit from further study as it has implications for crime scene reconstructions.

3.5 Use of Blood Substitutes in Training and Simulations

Numerous liquids have been used for blood substitutes in blood trace deposition experiments throughout the years. Initial studies used human or animal blood; however, to avoid concerns with disease and animal or human cruelty, there is a desire to create artificial blood substitutes. Some liquid formulations have been poorly thought out, with one of the more questionable recent substitutes being a mixture of an unknown ratio of human blood, barium sulfate suspension, and acrylic paint (Schyma et al. 2015). The physical properties of this mixture, such as viscosity, surface tension, and density, were never compared to those of blood, which is critical to know prior to using it for blood trace deposition experimentation. The question arises as to whether different liquids are adequate substitutes for human blood in both the research context or for ad hoc experimentation for casework and education. There also remain questions as to whether there are meaningful differences that would influence their deposition and configurations, which would cause issues in interpretations. This would depend on the type of simulation that is being conducted. We are not aware of any proposed blood substitute that can be used for all anticipated simulations involving blood trace deposition. For example, training demonstrations that involve temporal variables (i.e., drying time or post-deposition modifications) are not adequately replicated by any proposed blood substitute. However, these substitutes may be satisfactorily used for non-casework demonstration simulations of other components of blood trace analyses, such as for area of origin determinations.

In 2015, Stotesbury et al. commented on the need for creating a synthetic blood substitute for forensic science. The authors state that there is a need in the bloodstain pattern analysis community to possess a blood substitute with "precise fluid dynamics equivalent to whole blood" delineating viscosity, surface tension, and density. The authors point out that there are some blood substitutes that are commercially available, but there is a dearth of published literature on the validation of these products as blood substitutes. They asserted that various animal sources are being used ". . .and their use is well justified" since the physical characteristics mimic that of human blood. While the use of animal sources is widely accepted throughout the forensic community, these cannot possibly be of equal quality. These authors propose using a sol–gel chemistry-based system; three of the authors are currently conducting experiments with sol–gel materials under simulated impact conditions. We agree with the authors that a blood substitute that behaves like blood in terms of its physical, chemical, and biological properties would be of significant value; however, it is doubtful that a universal blood substitute for use in casework is even possible.

There are several commercially available blood substitutes that are commonly used in the field which lack a demonstrated rigorous scientific foundation. Some of these claim to have been tested but fail to provide supporting documentation to prove this assertion (e.g., no certificate of analysis) and there are no refereed publications. Two approaches can be discerned for the creation of these blood substitutes. The first involves modifying nonhuman blood to prevent clotting. A major issue with this is that there is no information provided on what physical properties were measured or compared for these blood substitute products. The suppliers state that the fibrinogen has been removed and thus will not clot. It seems clear that if the fibrinogen has been removed, the lack of this relatively large molecule will have some influence on the blood such as viscosity. The inability to clot is evidence that there is some significant difference from whole blood. One such product claims that it has been defibrinated. But to remove fibrin, the fibrinogen would have to be converted to fibrin first, meaning that the blood was allowed to clot. Albeit the source of this type of blood refers to it as "defibrinated" in all likelihood, it has been defibrogenated. But, if the interaction between erythrocytes and fibrinogen is the dominant factor that makes blood rheologoically a non-Newtonian fluid, it is reasonable to assume that it does not have the same physical properties as its "defibrinated" counterpart, and thus is no longer a non-Newtonian fluid. It is unknown how the relevant physical properties of this fluid have been changed, and consequently how these would affect their depositions and configurations. The second approach is a blood substitute created specifically for blood spatter experiments, and according to this source, it was developed to have the same physical properties as blood. These include surface tension, color, and viscosity, and the suppliers state that it can be used for training or for analysis but do not provide adequate documentation of this.

Other practices described in the literature include freezing and storing whole blood for lengthy periods of time and then thawing it out prior to use. Routinely, the blood is allowed to reach room temperature for experimentation. Freezing of blood will lyse the RBCs and since they are responsible in large part (hematocrit) for the viscosity of whole blood, it seems likely that there would be a considerable effect on the relevant physical properties of the blood. Furthermore, the temperature dependence on some of the physical properties of blood (i.e., density and surface tension) could cause meaningful differences in observed characteristics of blood trace deposits when simulations are conducted at room temperature (20–25 °C) rather than at physiological temperature (37 °C).

In conclusion, we are skeptical about the use of any blood substitute for casework. Although they have their use for some training demonstrations, even this practice should be done with caution and explicit recognition of their limitations.

References

Adam, C.D. (2012). Fundamental studies of bloodstain formation and characteristics. *Forensic Science International* 219: 76–87.

Aplin, S., Reynolds, M., Mead, R.J., and Speers, S.J. (2019). The influence of hematocrit value on area of origin estimations for blood source in bloodstain pattern analysis. *Journal of Forensic Identification* 69: 163–173.

Bevel, T. (1983). Geometric bloodstain interpretation. *FBI Law Enforcement Bulletin* 52: 7.

Bevel, T. and Gardner, R. (2008). *Bloodstain Pattern Analysis*, 3e. CRC Press, Taylor & Francis Group.

Denison, D., Porter, A., Mills, M., and Schroter, R.C. (2011). Forensic implications of respiratory derived blood spatter distributions. *Forensic Science International* 204 (1–3): 154.

DuBey, I.S. (2019). *A Study of the Impact of the Physical Properties of Blood on the Interpretation of Bloodstain Patterns in Forensic Investigations.* CUNY Academic Works.

Everett, D.H. (1972). Manual of symbols and terminology for physicochemical quantities and units, appendix II: definitions, terminology and symbols in colloid and surface chemistry. *Pure and Applied Chemistry* 31 (4): 577–638.

Hall, J.E. (2015). Hemostasis and blood coagulation. In J.E. Hall (Ed.) Guyton and Hall Textbook of Medical Physiology. 13. (pp. 483–494). Elsevier. https://vitalebooks.store/product/guyton-and-hall-textbook-of-medical-physiology-e-book-13th-edition/?msclkid=24e1ebbf430e16afa1c5b4c874aa22fc (accessed 1 October 2020).

Hulse-Smith, L., Mehdizadeh, N., and Chandra, S. (2005). Deducing drop size and impact velocity from circular bloodstains. *Journal of Forensic Sciences* 50 (1): 54–63. JFS2003224-10, https://doi.org/10.1520/JFS2003224. ISSN 0022-1198.

International Union of Pure and Applied Chemistry. (2014, February 24). Compendium of Chemical Terminology: Gold Book Version 2.3.3.

Jackowski, C., Thali, M., Aghayev, E. et al. (2006). Postmortem imaging of blood and its characteristics using MSCT and MRI. *International Journal of Legal Medicine* 120: 233–240.

Kenner, T., Leopold, H., and Hinghofer-Szalkay, H. (1977). The continuous high pressure measurement of the density of flowing blood. *European Journal of Physiology Pflugers-Archiv* 370: 25–29.

Kinnel, P.I.A. (1972). The acoustic measurement of water drop impacts. *Journal of Applied Meteorology* 11 (4): 691–694.

Laber, T.L. and Epstein, B.P. (1983). *Experiments and Practical Exercises in Bloodstain Pattern Analysis*, Callen Publishing, Inc.

Lowe, G.D.O. (1987). Blood rheology in vitro and in vivo. *Baillière's Clinical Haematology* 1: 597–636.

MacDonell, H.L. and Bialousz, L.F. (1971). *Flight Characteristics and Stain Patterns of Human Blood*, 3. U.S. Department of Justice, Law Enforcement Assistance Administration.

Mehdizadeh, N.Z., Chandra, S., and Mostaghim, J. (2004). Formation of fingers around the edges of a drop hitting a metal plate with high velocity. *Journal of Fluid Mechanics* 510: 353–373.

Merrill, E.W. (1969). Rheology of blood. *Physiological Reviews* 49: 863–888.

Mole, R.H. (1948). Fibrinolysin and the fluidity of the blood post-mortem. *The Journal of Pathology and Bacteriology* 60: 413–427.

Petrucci, R.H. (1982). *General Chemistry: Principles and Modern Applications*, 3e. Macmillan Publishing Co., Inc.

Pizzola, P.A. (1998). *Improvements in the Detection of Gunshot Residue and Considerations Affecting Its Interpretation*. UMI Company.

Pizzola, P.A., Roth, S., and De Forest, P.R. (1986a). Blood droplet dynamics – I. *Journal of Forensic Sciences* 31 (1): 36–49.

Pizzola, P.A., Roth, S., and De Forest, P.R. (1986b). Blood droplet dynamics – II. *Journal of Forensic Sciences* 31 (1): 50–64.

Reynolds, M. and Silenieks, E. (2016). Considerations for the assessment of bloodstains on fabric. *Journal of Bloodstain Pattern Analysis* 32: 15–20.

Ristenbatt, R.R. (2008). *Review of: Bloodstain Pattern Analysis with an Introduction to Crime Scene Reconstruction*, 3e. Boca Raton, FL: CRC Press (Taylor & Francis Group), 402 p. *Journal of Forest Science* 54: 234. doi: 10.1111/j.1556-4029.2008.00932.x.

Rogers, N. (2009). Hematocrit implications for bloodstain pattern analysis. MS thesis, University of Western Australia, Perth.

Schyma, C., Lux, C., Madea, B., and Courts, C. (2015). The 'triple contrast' method in experimental wound ballistics and backspatter analysis. *International Journal of Legal Medicine* 129: 1027–1033.

Stotesbury, T., Illes, M., Wilson, P., and Vreugdenhil, A. (2015). A commentary on synthetic blood substitute research and development. *Journal of Bloodstain Pattern Analysis* 31: 3–6.

Stotesbury, T., Illes, M., Jermy, M. et al. (2016). Three physical factors that affect crown growth of the impact mechanism and its implication for bloodstain pattern analysis. *Forensic Science International* 266: 254–262.

Takeichi, S., Wakasugi, C., and Shikata, I. (1984). Fluidity of cadaveric blood after sudden death. Part I. Post mortem fibrinolysis and plasma catecholamine level. *American Journal of Forensic Medicine and Pathology* 5: 223–228.

Takeichi, S., Wakasugi, C., and Shikata, I. (1985). Fluidity of cadaveric blood after sudden death. Part II. Mechanism of release of plasminogen activator from blood vessels. *American Journal of Forensic Medicine and Pathology* 6: 25–29.

Takeichi, S., Tokunaga, I., Hayakumo, K., and Maeiwa, M. (1986). Fluidity of cadaveric blood after sudden death: part III. Acid-Base balance and fibrinolysis. *American Journal of Forensic Medicine and Pathology* 7: 35–38.

Thurston, G.B. and Henderson, N.M. (2006). Effects of flow geometry on blood viscoelasticity. *Biorheology* 43 (6): 729–746.

Worthington, A.M. (1963). *A Study of Splashes.* With an Introduction and Notes by K.G. Irwin, xi. The McMillan Co.

CHAPTER 4

Detection, Visual Enhancement, Identification, and Source Attribution of Blood Deposits and Configurations

For both field and laboratory testing of suspected blood deposits, it is beneficial to review the concepts for their detection, enhancement, and identification. For the interpretation of blood trace configurations, it is inadequate to assume that an observed trace is blood as well as to presume to know the identity of the source. There are numerous catalytic and protein tests for the presumptive identification of blood in the field and in the laboratory, as well as confirmatory laboratory-based tests. The science of these will be briefly discussed here, however, not with the intention of providing a complete guide to blood testing, but to inform the reader of their availability. There are numerous books and references on blood identification available that can be consulted.

Blood Traces: Interpretation of Deposition and Distribution, First Edition. Peter R. De Forest, Peter A. Pizzola, and Brooke W. Kammrath.

4.1 Optical Visualization of Blood Trace Deposits

Before applying chemical reagents or stains to a suspected blood trace deposit, one should give careful consideration to the use of nondestructive visualization techniques. The perils of not doing so are detailed below and in the Stomping Case detailed in Chapter 11. Perhaps surprisingly, one straight-forward visualization technique is to illuminate deposits for visual examination using an intense polychromatic (white) light over a range of illumination and visualization angles. Oblique illumination can reveal much with certain substrates. Intense white light sources have been usefully applied for many years. White light illumination employing various source and viewing angles should be a first step for visualizing blood trace deposits. To augment the white light, tunable light sources are used to provide illumination consisting of a narrow band of wavelengths or a single color. These light sources use monochromators, interference filters, lasers, or LED technologies to provide the selected wavelengths (i.e., colors). For example, use of a blue light source in the range of 415 nm, paired with clear or yellow filters or goggles, has been demonstrated to be particularly useful for the visualization of blood traces on many substrates and backgrounds. There are several commercially available specialized light sources for use in crime scene and evidence examinations.

Infrared (IR) photography is an excellent method for blood trace visualization (Raymond and Hall 1986; Perkins 2005; Miskelly and Wagner 2005), as well as the documentation of other numerous traces including, but not limited to, body fluids, gunshot residue, and questioned documents. There are two methods for accomplishing this in the near-IR range. The first technique, IR luminescence, uses visible light as an excitation source and captures the IR fluorescence produced by some materials. The second is to illuminate the sample with IR radiation, which takes advantage of the differential IR absorption. For this method, it may be advantageous to block the UV–visible light from the source to prevent this radiation from reaching the film or image sensor. Both of these described methods require cameras sensitive in the IR and lens-mounted IR transmitting filters which block visible radiation. For guidance on choosing appropriate IR filters, the reader is referred to the book "Kodak Filters for Scientific and Technical Uses (B-3)" (Kodak Company 1985). Digital imaging has made possible a convenient and increased use of IR photography for forensic documentation and examinations. With film cameras, there was no method to preview the end result before the film was chemically developed; however, digital IR photography has greatly simplified the process by enabling the image to be easily previewed and assessed in real time. Brown and Watkins (2016) describe how digital cameras can be made into UV/IR cameras: "By simply removing the interior filter and replacing it with clear glass, camera manufacturers have created UV/IR cameras that 'see' in the spectrum of approximately 350 nm to 1000 nm, thus visualizing 'invisible' evidence." Ultimately, photography at a crime scene and in the laboratory is best accomplished by having both a high quality DSLR camera and an IR camera available for the documentation of traces, as demonstrated in Figure 4.1. This was particularly effective because both the red and black dyes of the garment are transparent in the IR, thus enabling the impressive contrast between the printed pattern and the bloody imprint.

With some substrates the scientist can take advantage of a highly fluorescent background since blood traces will absorb (quench) short wavelength radiation

(a)

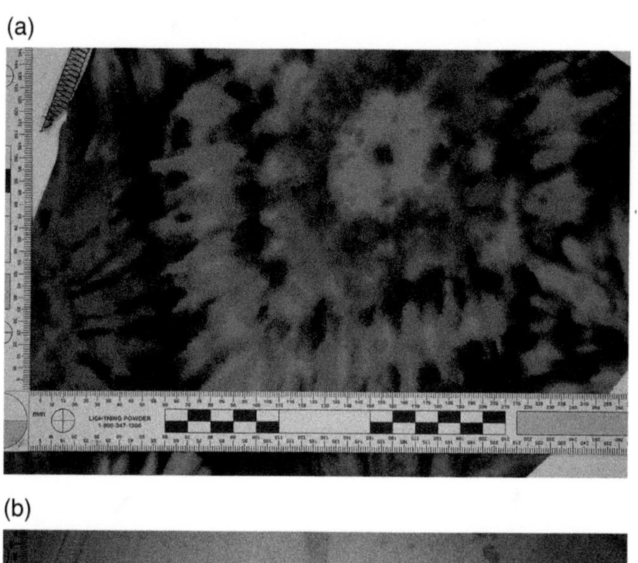

(b)

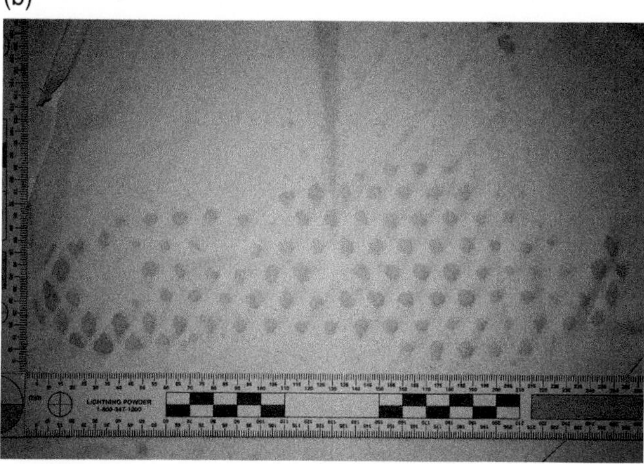

FIGURE 4.1 Visible (overhead fluorescent) light (a) and IR photographs (b) of a multi-colored garment. The IR image reveals a bloody footwear outsole pattern, which was captured using a 35 mm digital camera, Fuji X-T1, using a Schneider Kreuznach 098 longpass IR filter (transmits >695 nm) illuminated with a quartz-halogen lamp. IR image adjusted for desaturation, brightness, and contrast in Photoshop. *Source: Photographs Courtesy of Jessica Hovingh, Forensic Science Graduate Program, Pennsylvania State University.*

allowing visualization of the traces (See Figure 4.2). In this figure, the bloodstained white cloth was washed making the traces latent. The traces became easy to detect and document when irradiated with short wavelength (430 nm) visible light. This can also be accomplished with ultra-violet radiation.

Another method for the nondestructive photographic documentation of blood traces on dark-colored substrates is the use of photography with crossed polarizers (Figure 4.3). This method uses a polarizing filter in front of the light source as well as a second polarizer over the camera lens. A detailed explanation of this enhanced visualization method was published by the National Institute of Justice (De Forest et al. 2009).

The use of photography with crossed polarizers was critical in a case involving the murder of a seven-year-old girl. A black leather jacket belonging to the victim's

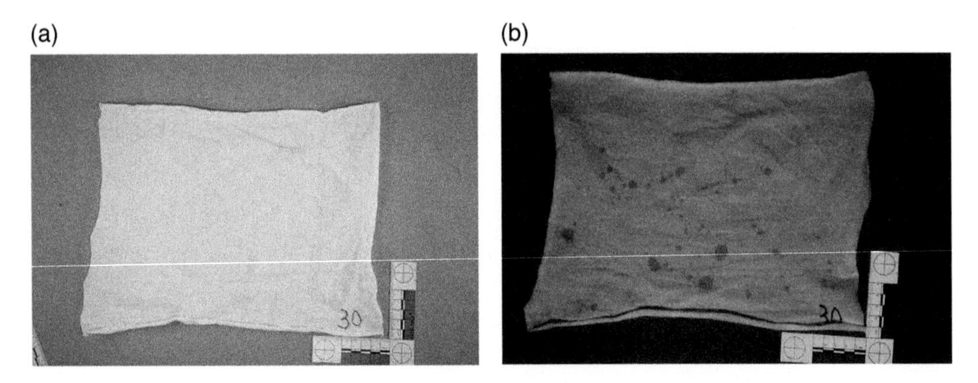

FIGURE 4.2 No visible blood traces are observed with normal white light (a), however stains are evident when illuminated with 430 nm excitation with a yellow barrier filter (b). Photographs taken with a Nikon D200 DSLR camera, f/4.5, 20sec ss, ISO 100. *Source: Photographs courtesy of Norman Marin & Jeffrey Buszka.*

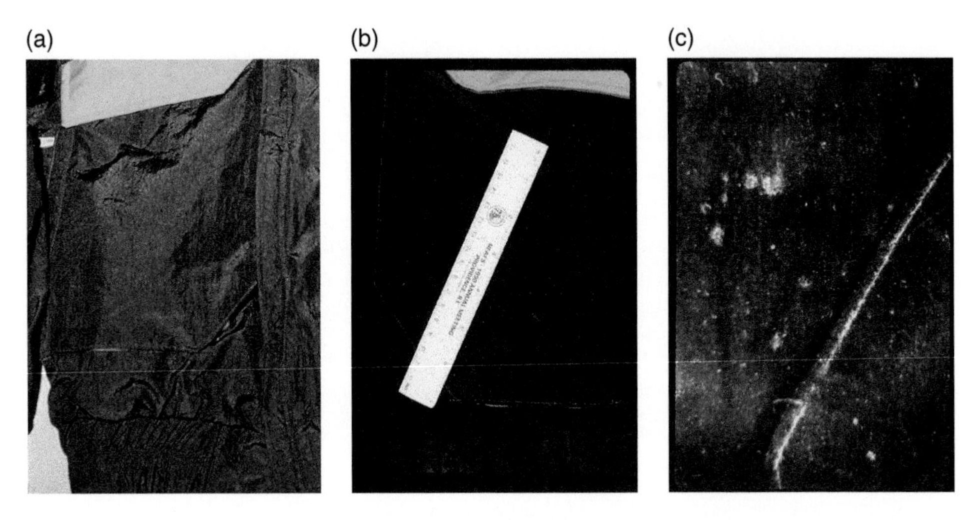

FIGURE 4.3 Photographs of an evidence jacket from a bludgeoning case, showing an image of a primarily black nylon (a). With ambient illumination, it was very difficult to discern blood traces because of the dark color and glare from the textile. Blood traces were visualized using photography with crossed polarizing filters, both using color film (b) and with black and white film (using a red color contrast filter) (c). *Source: Images courtesy of Kirby Martir & P. Pizzola.*

father became important evidence when he was identified as a suspect in the stabbing. Although there was some indication that the jacket had blood on it, the configurations could not be visualized on the dark leather surface. The defendant offered an explanation for the possible presence of blood on his jacket by stating that his daughter would get frequent nosebleeds and liked to take naps on his jacket while it was lying on his bed. Unfortunately, in this case, swabbing for DNA by laboratory scientists was performed prior to visualization of the deposits. As a consequence, the swabbing for DNA, shown in blue in Figure 4.4, was done "blind" because the blood trace configuration (in red) had not been visualized prior to sampling. Of note, the elliptical airborne blood droplet deposit located at the midline of the jacket crossed the zipper (Figure 4.5), indicating

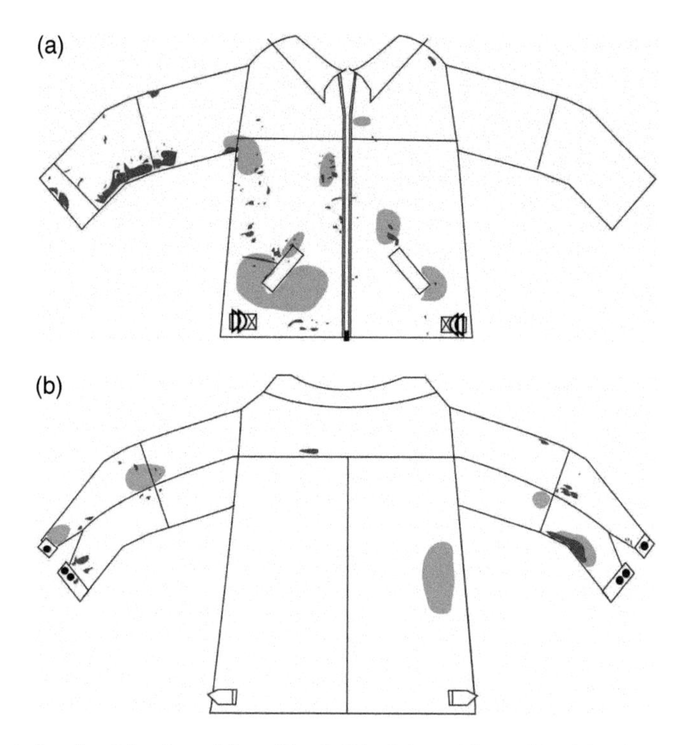

FIGURE 4.4 A sketch of the front (a) and back (b) of the evidence black leather jacket with the locations of swabbing in blue and the actual positions of the blood traces in reddish-brown. *Source: Figure prepared by Dr. Rebecca Bucht.*

FIGURE 4.5 A close-up photograph of the midline region of the jacket, taken using crossed polarizers which enabled the visualization of the elliptical airborne blood droplet deposit across the zipper.

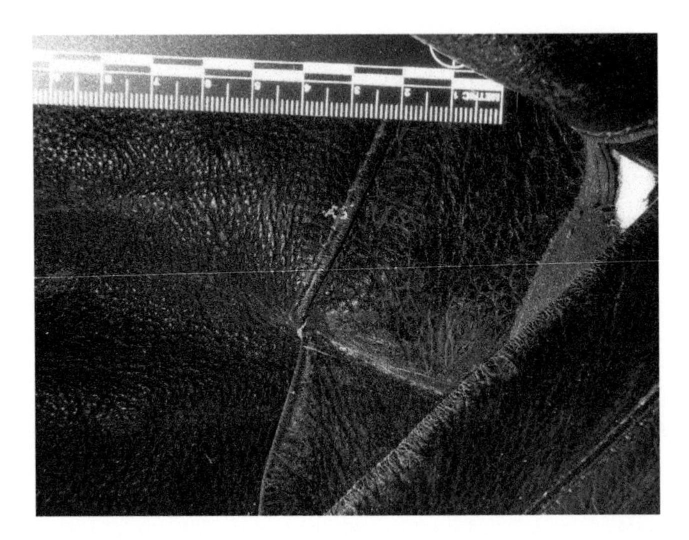

FIGURE 4.6 A close-up photograph of the right pocket area of the leather jacket, taken using crossed polarizers which enabled the visualization of the dynamic blood transfer pattern.

that the jacket was zipped up when the blood trace was deposited. Further, a dynamic transfer pattern, thought to have been deposited as a result of an incised wound on the suspect's right hand, was visualized by crossed polarized photography (Figure 4.6). This case demonstrates two points concerning the importance of visualization proceeding sampling. First, by visualizing a trace prior to sampling, one can target the desired specimen area and avoid inclusion of extraneous regions of the substrate which may result in unnecessary mixtures. This has the potential to eliminate the necessity of DNA mixture deconvolution in subsequent laboratory testing. The second is the value of understanding the nature of the deposition in the context of case reconstruction. In this leather jacket case, the mechanism of deposition was critical for making a clear distinction between deposits from a nosebleed and those resulting from a stabbing. The airborne droplet configurations on the sleeves and jacket torso were inconsistent with the girl napping on the jacket and instead supported the contention that the jacket was being worn with the zipper closed during the stabbing. An unequivocal reconstruction of events was only made possible by the visualization of the blood traces via polarized light photography. The ability to appreciate the extent and configuration of the blood deposits greatly facilitated the resolution of this case.

4.2 Catalytic Tests

A catalytic blood test is one where the heme and protein within the questioned blood specimen catalyzes a pseudo-enzymatic (peroxidase-like) activity which initiates a color change or chemiluminescence with selected reagents. The presumptive catalytic tests for blood include the following: Kastle-Meyer (i.e., phenolphthalin), leuco-crystal violet, leuco-malachite green, benzidine, ortho-toluidine, and luminol (chemiluminescent). Where the identification of a specific trace is important, it is often recommended for these tests are performed in two stages: first application of the reagent followed by the oxidizer (i.e., hydrogen peroxide). Using this two-stage application method reduces the possibility of false positives (e.g, hypochlorite bleach). However, when visualization of overall blood trace configurations is more important than confirming the identification of the trace, it is advisable to apply the two reagents in one solution to prevent dilution and diffusion of the deposit. These are mentioned here as there has been some confusion of the mechanism, for example, of leuco-crystal violet. Occasionally, this is referred to or thought of as a protein stain which it is not. An excellent reference on this topic is the "Sourcebook in Forensic Serology, Immunology, and Biochemistry" by Robert E. Gaensslen published by the National Institute of Justice in 1983.

Documentation of blood traces visualized via chemiluminescence (i.e., luminol or BlueStar®) is not trivial. Scientific investigators must be ready to photo document the blood traces as soon as the reagent is applied. Thus, it is advised to have the camera mounted on a tripod and set for a long exposure (30 seconds) before application of the reagent. A very useful method that enables quality photography of chemiluminescent-enhanced blood traces is to have one long exposure (~30 seconds) in complete darkness immediately followed by a short duration flash (e.g., rear curtain).

4.3 Protein Stains

Three commonly used protein stains for the enhancement of the visibility of blood traces in the field and in the laboratory are amido black, Hungarian red, and Coomassie blue. As protein stains, they will enhance the contrast of many materials containing protein. This may include vegetable and animal proteins, depending on the amino acid content. Thus, they cannot be considered presumptive tests for blood and instead are only used for the visual enhancement of blood traces. Important features of protein stains are that they provide enhanced contrast between the blood traces and the background substrate, and ideally they should incorporate a protein denaturant to prevent the stained protein from solubilizing in the reagent solution and as a consequence diffusing, spreading, or running.

4.4 Blood Typing and DNA Technology

Forensic serological examinations of blood traces were formerly standard practice in forensic science laboratories, but with the advent of DNA typing technology, unfortunately these have been eliminated without an awareness of their inherent value for providing complementary information in select nonroutine cases. Classical serological analyses include cellular antigen (e.g., ABO typing), serum protein, and iso-enzyme typing. For details on these historically valuable tests, the authors recommend references such as HC Lee's chapter in the Forensic Science Handbook (Saferstein, ed., Lee 1982, 267–337) and in Forensic Science: An Introduction to Criminalistics (De Forest et al. 1983, 230–263). These can produce valuable information for investigations, but presently few laboratories have maintained capabilities in this area. For example, serological testing may be of value in missing person cases where there may be no expectation of DNA information being available in national databases. Existing hospital and medical records for a suspected missing person may contain blood typing information that can be useful. Potentially, these serological tests may be able to provide inclusion or exclusion of an individual, thus providing valuable investigative information.

DNA technology has undoubtedly changed the forensic science landscape in profound ways. With respect to blood trace deposits, DNA typing provides potential source identifications. One of the most significant advantages of this technology is the use of DNA databases to provide investigative leads in cases where there is no known suspect. There are numerous texts that provide in-depth details relative to forensic DNA typing (i.e., Butler 2005), and where more information is desired, it is recommended that the reader review the science of DNA analysis by referring to one of them.

An essential consideration at crime scenes and in laboratory analysis of exhibits with blood trace deposits is for scientists to maintain an awareness of the importance of the integrity of the blood trace configuration so that they are not compromised in the sampling for the DNA typing analysis. It is always unwise to prioritize the source attribution perspective of a blood trace over the interpretation of the mechanism of deposit. Once a blood trace has been sampled for DNA analysis, it is commonly impossible to go back to evaluate the method of its transfer. This is especially important for blood traces at a crime scene, where there are rarely opportunities to get a second bite of the apple due to the constraints of a scene examination. The serious repercussions of the failure to recognize the potential of the blood trace configurations are illustrated in the leather jacket case described above, in Section 11.1, among many others.

Avoidable complexity and controversies regarding the interpretation of DNA mixtures may be addressed with informed sampling of blood traces. Mixtures can be eliminated if a careful scientific interpretation of deposits at the scene prior to sampling allows one to select deposits from a single source. The desirability of this selective capability cannot be overemphasized for its utility in circumventing the issues associated with problematic DNA mixture interpretation.

4.5 A Limitation of Laboratory SOPs

Although standardization of the analytical protocols for evidence examination in the form of standard operating procedures (SOP) is essential, at times a high degree of mandatory adherence to regimentation and uniformity can be too restrictive. This may thwart completion of a thorough scientific investigation which could lead to a misinterpretation or false conclusion. It is impossible to have an SOP for every eventuality that may be encountered in forensic science casework. Forensic science laboratories are distinctly different from clinical laboratories and must not be viewed as testing facilities with a finite "menu" of available examinations. De Forest (1998) stated "The fundamental difference goes deeper than differences in the nature of the samples and the analytical schemes applied. Rather, it goes to the question of scientific assessment of the problems to be addressed." Not all forensic problems can be anticipated, and thus conventional examinations that are conducted by a forensic science laboratory may not address critical case-specific questions that a complete interrogation of the physical evidence demand. Therefore, there must be a means of incorporating valid testing procedures into a case examination that are not part of the laboratory's SOPs. This flexibility has been successfully implemented in some laboratories by inclusion of a statement in the procedure manuals stating that alternative analyses may be conducted if the scientist demonstrates a rationale for the deviation from the SOP which is appropriately documented.

The Dew Theory case, discussed in detail in Chapter 11, is an example where non-traditional forensic testing was needed to answer a relevant question. In this case, there was a need to improvise, i.e., apply a technique in a new way to properly problem solve which is the essence of science. This is one of the great ironies that we are witnessing. Accreditation is designed to improve the quality. However, it appears that accreditation can actually cause the stagnation of science. Today, if a laboratory scientist decided that they wanted to do something novel, the attempt would likely be highly criticized by laboratory supervision and management. Nonetheless, it is possible to validate novel techniques so that their use does not violate accreditation programs. The Dew Theory case necessitated an innovative application of established protein quantitation methods to estimate the albumin hemoglobin ratio. This was used to determine whether the specific regions within a given blood trace, corresponding to different colors, consisted predominately of a cellular or serum fraction. In this case, techniques for quantification of proteins in blood, such as albumin and hemoglobin, are well established. The novelty in this case was using the ratio of the albumin to hemoglobin within a single trace to support the tentative conclusion that the distinctive appearance was the result of clotting. The resolution of this question required a novel application of established methods for which there were no SOPs in the particular laboratory.

A commonly encountered need in forensic laboratories is the differentiation of menstrual from venous blood. This has been a problem in medicine for over 150 years (Robin 1858; Gaensslen, Sourcebook), and to date, there is no universally accepted method for their differentiation. All proposed methods have their shortcomings, from microscopical analysis to lactate dehydrogenase isozyme identification and RNA analysis. Several high-profile cases, including the 1986 Woodchipper and the 2005 Catherine Woods murder investigations, would have benefited from methods that enable this distinction. In the Catherine Woods case, the defendant claimed a handprint in blood that was linked to him via fingerprints was composed of menstrual blood. Although identifying the deposit as being from menstrual blood would have been valuable in this case, it was not attempted. Although there is published literature on a variety of methods for menstruum identification (Robin 1858; Markert and Ursprung 1962; Asano et al. 1971; Whitehead and Divall 1973; Gaensslen 1983; Bauer and Patzelt 2002; Gray et al. 2012), the lack of an SOP ended the scientific investigation. While it is essential that valid methods be used by forensic scientists, oftentimes there are situations where it would not take long to validate a new technique that is not part of the laboratory's SOPs.

4.6 Ongoing and Future Research

New technologies continue to evolve that increase the capabilities for extracting relevant details from blood traces; however, it is critical to not lose sight of the potential risk of compromising the information derivable from the physical configuration of the deposit. In this regard, there are two currently emerging technologies that may have an impact on the integrity and interpretation of blood trace deposits: hyperspectral imaging and rapid DNA technologies.

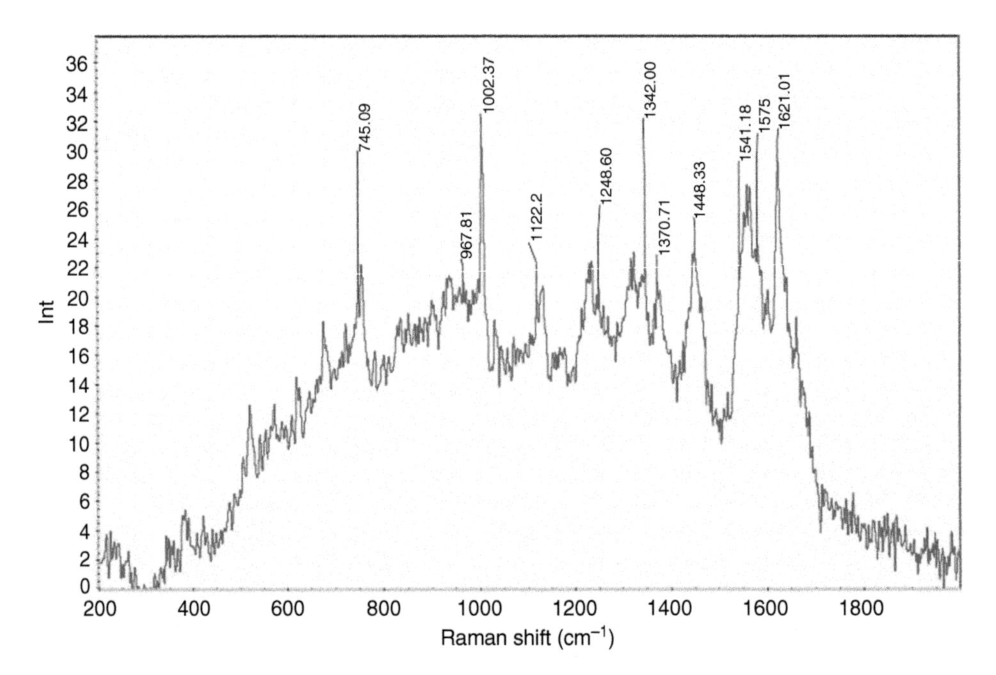

FIGURE 4.7 Raman spectrum of blood from a ~40-year-old female (Thermo Scientific DXR Raman microscope, equipped with a 780 nm excitation laser source, 20 mW laser energy, 10 s exposure time with 10 co-added exposures). Blood is a complex mixture, and the labeled peaks are those from components previously identified in the literature (Virkler and Lednev 2010; Boyd et al. 2011): hemoglobin (1122, 1368, 1542, 1620 cm^{-1}), tryptophan (744, 1342, 1448 cm^{-1}), fibrin (967, 1248, 1575 cm^{-1}), and phenylalanine (1000 cm^{-1}). *Sources: Virkler and Lednev (2010) and Boyd et al. (2011).*

Hyperspectral imaging can be thought of as having each pixel of the resulting image contain a range of wavelength-dependent information. Although there are a variety of hyperspectral instruments with different capabilities, infrared and Raman hyperspectral imaging have great potential for providing additional or complementary information about blood traces. Infrared and Raman spectroscopy are used to identify the chemical composition of a sample based on the molecular vibrations of its constituent components (Figure 4.7). As advancements in hyperspectral infrared and Raman imaging continue, a complete spectrum from each pixel of an image will be able to be obtained from a portable instrument which could be used for its compositional characterization at a scene. Both infrared and Raman spectroscopy have been used to differentiate blood from other body fluids and identify it on a variety of substrates. Further, preliminary results from research into the analysis of blood by infrared and Raman spectroscopy suggest that a determination of the time since deposition and donor profile (i.e., age, biological sex, etc....) may be possible. Hyperspectral imaging has the potential to augment and complement the geometric interpretation of blood traces by providing nondestructive presumptive blood identification and a means of visualizing additional details of the nature of blood deposits on difficult substrates.

Rapid DNA technologies are portable and automated instruments which enable a DNA profile to be generated in 90 minutes which can then be searched against a CODIS

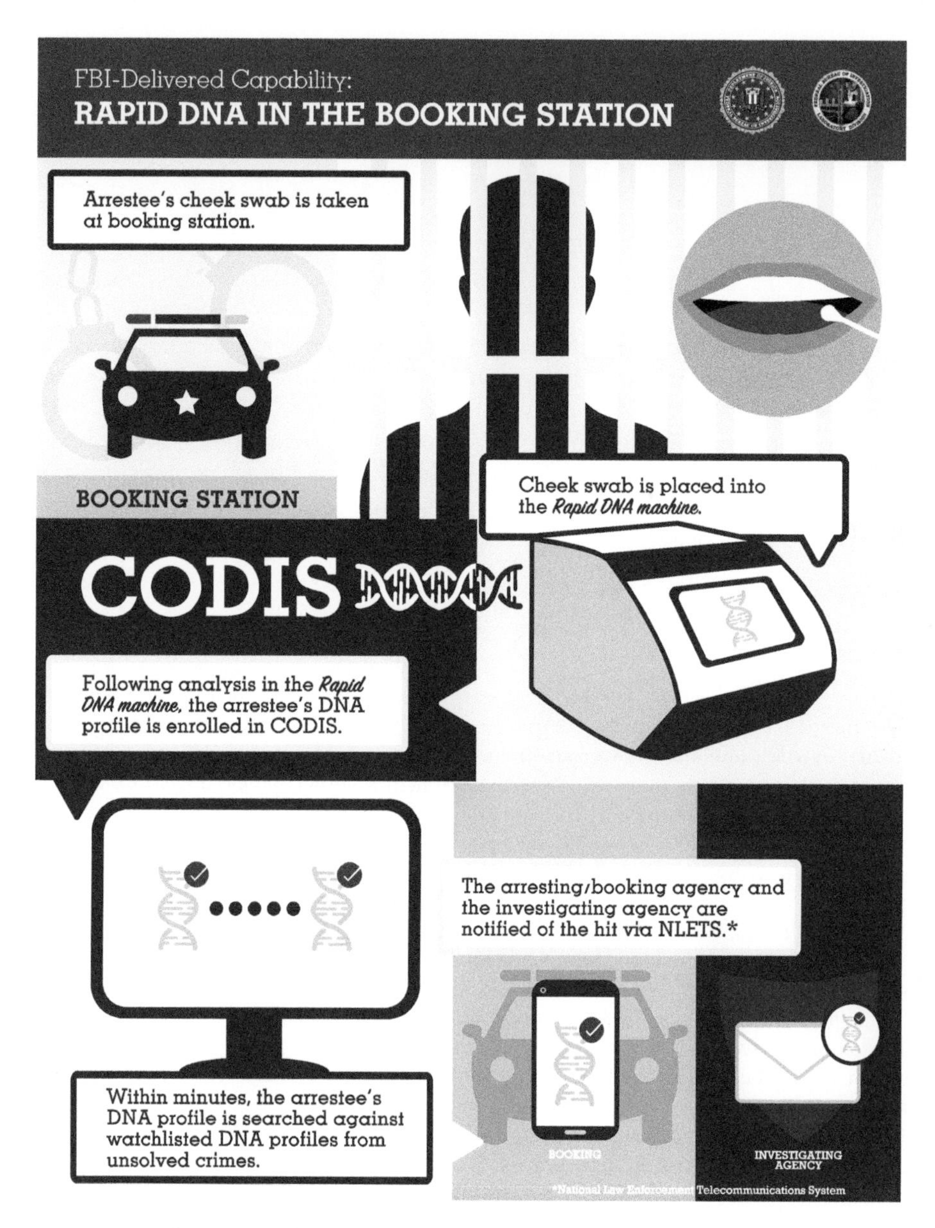

FIGURE 4.8 An illustration of the application of rapid DNA technology at a police booking station, in a simplified form. https://www.fbi.gov/services/laboratory/biometric-analysis/codis/rapid-dna. (Federal Bureau of Investigation (FBI), 2020). *Source: Rapid DNA General Information. https://www.fbi.gov/services/laboratory/biometric-analysis/codis/rapid-dna.Public Domain.*

(combined DNA index system) database (Glynn and Ambers 2021; https://www.fbi.gov/services/laboratory/biometric-analysis/codis/rapid-dna). Although it is currently in use only for the analysis of buccal swabs of arrestees (Figure 4.8), research is being undertaken to potentially apply this emerging instrumentation to the analysis of mixtures, crime scene, and other challenging samples (i.e., human remains). The ability to develop these investigative leads in a short time period will obviously be of great value;

however, the risk of prioritizing the source attribution information over the geometric configuration of the deposit must be explicitly recognized. It must be borne in mind that in the context of a given case, the biological source of the deposit may be far less important than the interpretation of its configuration, as discussed previously.

4.7 Conclusion

We cannot overemphasize that visualization using nondestructive techniques, perhaps something as simple as intense white light sources at various angles, should always precede any destructive enhancement or sampling for further laboratory testing. Of course, as noted above, more sophisticated visualization options are available, such as tunable light sources, infrared, or crossed polarized photography, and may be particularly useful in specific situations. In any case, the importance of visualization is paramount and should never be overlooked in favor of expediency of obtaining source information.

References

Asano, M., Oya, M., and Hayakawa, M. (1971). Identification of menstrual blood stains by the electrophoretic pattern of lactic dehydrogenase isozymes. Nihon hoigaku zasshi= The Japanese Journal of Legal Medicine 25 (2): 148–152.

Bauer, M. and Patzelt, D. (2002). Evaluation of mRNA markers for the identification of menstrual blood. Journal of Forensic Sciences 47 (6): 1278–1282. https://doi.org/10.1520/JFS15560J. ISSN 0022-1198.

Boyd, S., Bertino, M.F., and Seashols, S.J. (2011). Raman spectroscopy of blood samples for forensic applications. Forensic Science International 208 (1–3): 124–128.

Brown, K.C. and Watkins, M.D. (2016). Ultraviolet and Infrared Photographs. Evidence Technology Magazine http://www.evidencemagazine.com/index.php?option=com_content&task=view&id=2280 (acessed 1 October 2020).

Butler, J.M. (2005). Forensic DNA Typing: Biology, Technology, and Genetics of STR Markers, 2e. Academic Press.

De Forest, P.R. (1998). Proactive forensic science. Science & Justice 38 (1): 1–2.

De Forest, P.R., Gaensslen, R.E., and Lee, H.C. (1983). Forensic Science: An Introduction to Criminalistics, 230–263. McGraw-Hill Book Company.

De Forest, P.R., Bucht, R., Kammerman, F. et al. (2009). Blood on Black-Enhanced Visualization of Bloodstains on Dark Surfaces. US Department of Justice. NIJ, Award 2006-DN-BX-K026.

Eastman Kodak Company (1985). Kodak Filters for Scientific and Technical Uses (B-3), 3e. Eastman Kodak Company.

Federal Bureau of Investigation (FBI). *Rapid DNA. General Information: Laboratory Services.* https://www.fbi.gov/services/laboratory/biometric-analysis/codis/rapid-dna (accessed 4 October 2020).

Gaensslen, R.E. (1983). Sourcebook in Forensic Serology, Immunology, and Biochemistry. Washington, DC: US Department of Justice, National Institute of Justice.

Glynn, C. and Ambers, A. (2021). Rapid DNA analysis – need, technology, and applications. In: Portable Spectroscopy and Spectrometry, Volume 2: Applications (eds. R. Crocombe, P.E. Leary and B.W. Kammrath). Wiley.

Gray, D., Frascione, N., and Daniel, B. (2012). Development of an immunoassay for the differentiation of menstrual blood from peripheral blood. Forensic Science International 220 (1–3): 12–18. https://doi.org/10.1016/j.forsciint.2012.01.020.

Lee, H.C. (1982). Chapter 7: identification and grouping of bloodstains. In: Forensic Science Handbook (ed. R. Saferstein), 267–337. Englewood Cliffs, NJ: Prentice-Hall, Inc.

Markert, C.L. and Ursprung, H. (1962). The ontogeny of isozyme patterns of lactate dehydrogenase in the mouse. Developmental Biology 5 (3): 363–381. https://doi.org/10.1016/0012-1606(62)90019-2.

Miskelly, G.M. and Wagner, J.H. (2005). Using spectral information in forensic imaging. Forensic Science International 155: 112–118.

Perkins, M. (2005). The application of infrared photography in bloodstain pattern documentation of clothing. Journal of Forensic Identification 55 (1): 1–9.

Raymond, M.A. and Hall, R.L. (1986). An interesting application of infrared reflection photography to blood splash pattern interpretation. Forensic Science International 31: 189–194.

Robin, C. (1858). Mémoire sur la Comparaison Médico-légale des Taches de Sang Menstruel et des Autres Espèces de Taches de Sang. Annales d'Hygiène Publique et de Médecine Légale 10 (2nd series): 421–434.

Virkler, K. and Lednev, I.K. (2010). Raman spectroscopic signature of blood and its potential application to forensic body fluid identification. Analytical and Bioanalytical Chemistry 396 (1): 525–534.

Whitehead, P.H. and Divall, G.B. (1973). Assay of "soluble fibrinogen" in bloodstain extracts as an aid to identification of menstrual blood in forensic science: preliminary findings. Clinical Chemistry 19 (7): 762–765. https://doi.org/10.1093/clinchem/19.7.762.

CHAPTER 5

Terminology, Typology, and Taxonomy

5.1 History of Terminologies Applied to Blood Trace Configurations

The era of increased awareness of the value of bloodstain pattern interpretation, on the part of scene investigators, began with the publication of MacDonell's report, *Flight Characteristics and Stain Patterns of Human Blood*, in which he named and classified several types of blood deposit configurations produced by airborne blood drops (MacDonell and Bialousz 1971). Spattered, projected, splashed, and "cast-off" were terms used to describe different mechanisms of deposit configurations production; corresponding images and descriptions of deposit configurations characteristics were included. MacDonell further delineated two types of spattered blood: "medium-velocity blood spatters" and "high-velocity blood spatters" (1). Based upon the velocity of the object that impacted a source of liquid blood, these terms were used to characterize the size and distribution of small stains in a radial spatter configuration as evidenced by traces of dispersed airborne blood drops.

MacDonell's terms, definitions, and classifications have been criticized and have since been expanded and modified. In the mid-1990s, the International Association of Bloodstain Pattern Analysts (IABPA) produced a "suggested" terminology list (Robbins 1996). In early 2002, the Scientific Working Group on Bloodstain Pattern Analysis (SWGSTAIN) was formed, and in 2009, the group produced a "Recommended Terminology" list. As of this writing, the National Institute of Standards and Technology (NIST) Organization of Scientific Area Committees (OSAC) Bloodstain Pattern Analysis subcommittee has assumed the role served by SWGSTAIN; one task was the production of an updated and uniform terminology list. The American Standards Board (ASB) of the American Academy of Forensic Sciences (AAFS) has been tasked with the responsibility of converting the recommendations of OSAC to standards.

Blood Traces: Interpretation of Deposition and Distribution, First Edition. Peter R. De Forest, Peter A. Pizzola, and Brooke W. Kammrath.

For over a decade, there has been a movement to create a taxonomic classification scheme for blood deposit configurations. In 2002, SWGSTAIN created several subcommittees including the Terminology subcommittee; this was quickly rebranded as the Taxonomy and Terminology (T2) Subcommittee with a goal of creating a "hierarchical taxonomy." As of the disbandment of SWGSTAIN and the succession by the OSAC Bloodstain Pattern Analysis subcommittee, no new "taxonomic" scheme has been published by the ASB.

In 2015, Arthur et al., suggested a new way of classifying blood traces in what they considered to be a more objective manner. Their concern is that many bloodstain pattern examiners use nomenclature when describing patterns that include the mechanism of formation, upon initial examination, rather than describing the traces' observable morphology and distribution; the inference essentially that the latter approach would be more objective and less ambiguous. Subsequently, the formation (mechanism) of the traces – indivdually and as a group could possibly be determined. This overall concept is not entirely new in that many examiners have for decades objected to the use of such terms such as "medium-velocity" and "high-velocity impact" to describe groups of traces. All too often we have been involved in cases where examiners have prematurely locked themselves into a specific mechanism for the formation of traces without even an adequate preliminary examination. We agree with Arthur et al. that each trace should be carefully observed, documented, and measured early in the examination. In addition to proposing the "mechanism-free language," those authors are compiling a prototype "Atlas" that will act as a visual aid to examiners in forming decisions regarding the classification of traces. Attinger et al. (2018) have compiled a set of images (61 sets of traces) that can be utilized for instruction and research and are available online. These configurations were produced under simulated bludgeoning conditions as the authors primarily studied impact velocity as well as the distance between the target and the source.

In a recent NIJ draft, Neitzel and Smith (2017) asserted that there is considerable overlap of the ranges of both Weber (We) and Reynolds (Re) numbers corresponding to dripping blood and blunt force trauma; in contrast blood traces generated by gunshot are typically much greater. Their table of data is recapped here:

Gunshot	$1 < Re < 12000$	$1 < We < 85000$
Dripping	$240 < Re < 6000$	$20 < We < 2100$
Blunt force	$10 < Re < 4800$	$1 < We < 3400$

While there is also some overlap of the Re & the We, in comparison to gunshot-generated traces, given sufficient data (both experimentally and in a casework context) these estimates could be of some value in evaluating causation or classification. For example, if the vast preponderance of numerous blood traces in a specific case were approaching the upper range of Weber #s (e.g., 85000), this might help one assess the mechanism of production. However, there still is the troubling problem of how the original drop size is accurately determined from the trace diameter. Hypothetically, one might be able to make some safe assumptions (at certain levels) about drop size from trace diameter based on experimental droplet spread data on a similar surface.

Liu et al. (2020) have proposed an automatic classification system for bloodstain patterns produced from blunt force trauma and gunshot incorporating machine learning, since current methodology is not sufficiently quantitative. According to the authors, one of the factors affecting classification is what they refer to as the "BT" or the distance between the deposited blood trace and its source. At short distances, i.e., 30 cm, the accuracy of the proposed method is excellent (99%), whereas at a distance of 60 cm the accuracy is reduced to 93% and at 120 cm it is diminished to 86%. As the distance from the source increases, there is a greater probability that the larger droplets will have traveled further resulting in a predominance of the larger blood traces. This is due to the fact that the flight of smaller objects, including droplets, will be more influenced by drag compared to larger droplets. They pointed out that even experienced and well-trained examiners have achieved only an accuracy of approximately 70% based on the Taylor reliability study (Taylor et al. 2016a,b). However, the authors of this book have looked at the Taylor study and are very concerned that a lack of an academic background of the participants could have had a profound influence on the error rate. This is discussed here in Chapter 9 (Practitioners). While the machine learning may in the future be a useful tool in the scientist's arsenal, we feel that automatic classification methods should remain an adjunct, rather than replace human reasoning.

A "taxonomic classification system for bloodstains" was introduced by Bevel and Gardner (2008). Unfortunately, an underlying problem exists with this particular bloodstain "taxonomy" and perhaps the desire for having such a scheme is misguided. The description and classification of bloodstains that have been created from known, controlled mechanisms is a relatively simple process. In most cases, however, the actions that produced bloodstain patterns at crime scenes will never be known with certitude. Rather than use of *taxonomy*, a word used primarily in the biological sciences to denote the classification/subclassification of organisms, a more appropriate word for our purposes may be *typology*. The concept of taxonomy was developed in the biological sciences, and inherent to the concept is a hierarchy of classification; typology, which is more general, imputes no such hierarchical structure. A simple typology of blood deposits and configurations was described in 1983 that remains useful and relevant. These configurations included radial spatter, arc (cast-off), arterial spurt, trail, flow, pools, contact transfers, and configurations from shootings (De Forest et al. 1983); nonetheless, it is naïve to assume that there is a finite limit to the number of different geometric configurations that can be encountered – many of which defy being definitively placed in any such category. Typologies can be useful, but any "typology system" risks oversimplifying the highly variable and dynamic processes involved in trace production in casework and attempts to force an open-ended situation into a framework where there would be a finite number of patterns; this is not realistic.

It should be realized and appreciated that the mere classification of blood deposit configurations or a series of them is the first step in arriving at interpretation of a complex reconstruction (De Forest 2018).

5.2 A Typology for Blood Trace Deposits

A typological categorization of basic bloodstain patterns, in conjunction with appropriate terminology and other forms of documentation such as photographic illustrations, can enhance the conveyance of information to other criminalists, forensic scientists, forensic examiners (CSU types), for the instruction of students and lay persons. An expanded version of the 1983 simple typology is presented in the following section.

5.2.1 Contact Transfers

Contact transfers with blood are classic examples of Locard's theory of exchange; they are created when two surfaces experience physical contact with each other and at least one of the surfaces bears wet blood. (It is also certainly within the realm of possibility that some degree of dry blood may be transferred between various surfaces). However, for the purposes of this discussion, we will be considering the classical transfer with wet blood, since this obviously predominates the production of blood traces of this type. Upon contact with the blood-bearing surface, blood is transferred to the other surface. In the absence of relative lateral or rotational movement of either or both surfaces, a *static contact transfer* is produced. Clear and conspicuous impressions of fingerprints, footwear outsoles, and other familiar surfaces or objects are examples of static contact transfers; however, static contact transfers may not possess the requisite characteristics to permit the identification of the source of the transferred blood. Amorphous blood traces may result from any number of scenarios and surfaces; their lack of recognizable spatial trait data typically precludes identification of the source surface or object. When present, class and individual (random) characteristics may yield critical information about the source of the blood and its manner of deposition (Figures 5.1–5.3).

Contact transfers from fabric are especially useful in tracking the movement of suspects and/or victims where blood has been shed. Often the importance of these is overlooked since their value is not as overtly conspicuous as footwear or fingerprints. Frequently, the precise location on the fabric, responsible for the transfer, can be located on the clothing (Figure 5.4). The impression must also be studied in detail for the transfer of fibers and other trace evidence onto the substrate which bears the impression. The position of the impression can be crucial in crime scene reconstruction perhaps in refuting or confirming claims made by suspects and witnesses.

FIGURE 5.1 Bloody footwear trace onto fabric.

FIGURE 5.2 Contact transfer of a bloody fingerprint onto copy paper. *Source: Photograph courtesy: Jessica Hovingh, Pennsylvania State University, Forensic Science Program.*

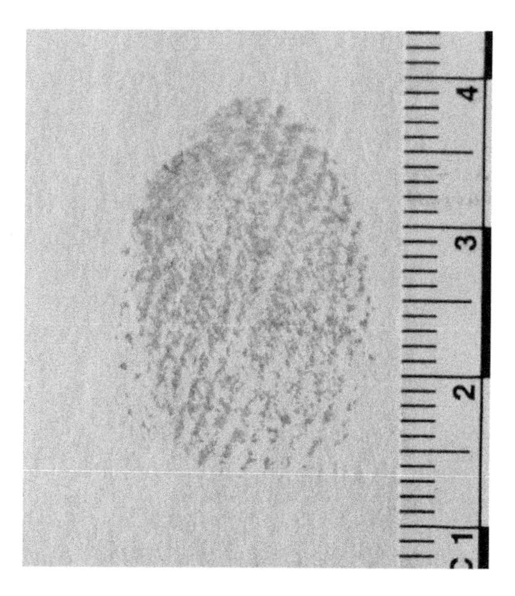

FIGURE 5.3 Contact transfer of a bloody print onto an index card. Little information in the form of fingerprint ridges is useful; however, some reconstruction information is present. Notice the vestige of a fabric impression within the fingeprint mark. When the blood was in a tacky condition on denim fabric, it was contacted by the finger; the blood that transferred to the finger retained the morphology of the fabric weave until the above mark was deposited.

FIGURE 5.4 Bloody fabric impression and overlying rivulets. *Source: Photograph courtesy of the Yonkers Police Department.*

5.2.1.1 Figure(s): Static Contact Transfers When one of the surfaces exhibits relative lateral or rotational movement while in contact, a *dynamic contact transfer* may result. Occasionally, the identity of the source may be apparent from the stain characteristics figure has been eliminated previously. In many cases, the source of dynamic contact transfers cannot be ascertained; however, the direction of motion may be discerned. In some cases, this information can prove valuable to a reconstruction.

Trail patterns may also consist of a series of interrelated contact transfers. These patterns result from repetitive contact transfers onto various surfaces as an individual moves through a scene. Footwear and footprint trails are examples of a trails comprised of a series of contact transfers. Contact transfer trails may consist of static, dynamic, or mixed (static and dynamic) transfers. Footwear patterns often demonstrate a dynamic aspect because the wearer may have a difficult time walking and running and may experience slipping in the wet blood. Commonly, the earliest traces in a trail made by bloody footwear may contain excessive amounts of blood obscuring fine detail and individual features, while continued walking will remove some blood and likely generate clearer and more valuable traces.

The scientist should make every effort to adequately photograph every such trace albeit they may only represent a small fraction of the entire outsole area. One should never be satisfied with documenting one seemingly representative element from a trail composed of multiple contacts. In Figure 5.5, every footwear pattern

FIGURE 5.5 White labels affixed to the floor adjacent to each bloody footwear trace as part of a continuous trail that was 40–50 ft in length.

in the continuous trail on the floor was photographed pre- and post-enhancement. Some individual characteristics were detected in some patterns but not others. Considered holistically, a positive association could be made between the continuous trail and exemplars made from the suspect's footwear. Thus, potentially, failing to document one or more footwear patterns could have precluded an association that was achieved.

In another example, similar to the situation described above, numerous bloody footwear traces can be seen on the linoleum floor (Figure 5.6). The photograph is presented below. Many of the outsole traces were of poor quality due to excessive blood and relative movement. This underscores the necessity of properly documenting each individual trace. A total of 20 bloody traces were observed, 16 of these lacked adequate detail for a positive association with exemplars made from a suspect's footwear, whereas all 20 had similar class characteristics. Ultimately, four of the traces were positively associated with the known footwear. Of significance, a single trail of traces was present leading from the victim's kitchen, out into the hall and into a nearby apartment. These observations were reinforced once the traces in the building hallway were enhanced chemically. The defense attorney asserted that the 16 patterns in the victim's apartment originated from the "real" killer rather than his client. However, since it was clear that there was only a single trail of traces and each had the same class characteristics, it was concluded that it was unlikely that a second person had left the footwear traces, i.e., all the traces originated from one set of footwear. Although this kind of reasoning is relatively straightforward, it is nevertheless exceedingly important. It was the most plausible hypothesis based on a logical process utilizing both abductive and inductive reasoning. Examination of impressions in Figure 5.7 demonstrate contact of a bloody hand along with a bloody sleeve cuff. If recognized, useful inferences can be drawn about the position and orientation of the hand and forearm at the time of deposition.

FIGURE 5.6 Trail of bloody footwear patterns. *Source: Photograph courtesy of the Yonkers Police Department.*

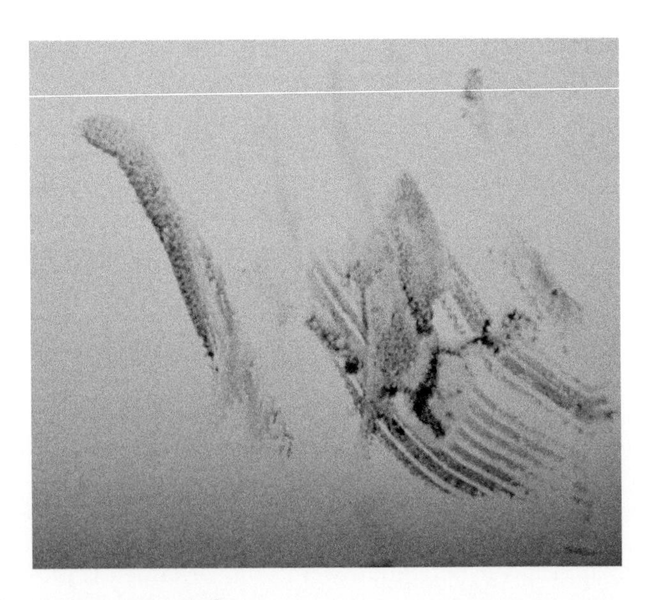

FIGURE 5.7 Dynamic contact transfer.

5.2.2 Noncontact Deposit Configurations

Deposits formed from airborne blood droplets interacting with a target surface create what should be classified as a noncontact pattern. This is because there is no physical contact between the donor and receptor surfaces. Blood may become airborne due to a variety of circumstances; it may be dripped from an object or it can be dispersed through the air. Caution must be exercised when analyzing patterns created by airborne blood drops. Patterns generated under controlled laboratory conditions are identified effortlessly; it may be difficult, if not impossible, to identify the mechanism of deposition or creation of patterns present at crime scenes (Figures 5.8 and 5.9).

5.2.3 Arc ("Cast-off") Deposit Configurations

Arc or cast-off patterns are produced when liquid blood is released from an object that is swung in an arc. Baseball bats, hammers, hands, and other objects that are swung through the air undergo a rotational motion and can produce arc patterns. According to Marin et al. (2010, 60), "The preponderance of the blood that is projected from a weapon is done so on the upswing or as it is moved backward. The droplets that are produced will be projected tangentially to the arc of the swing as a result of inertia. Inertia can be described as the tendency for an object to resist change or continue its motion in a straight line, or state of rest unless acted upon by some additional force. The projection of a droplet at a tangent (straight line) to the arc of the swing is an example of this principle." Williams et al. (2019) point out that the motion of the separated blood drops is tangential to the trajectory of the object's end where the droplets form from the disruption of a filament (p. 414, 418). An example of the formation of droplets from cast-off can be seen from the high-speed video collection by Laber et al. (Midwest Forensic Center 2007). The collection has many other blood droplet videos that may of interest to both students and scientists alike.

The resultant arc patterns may possess sufficient characteristics to identify the plane of the arc of the swinging object and the direction of the swing. This information

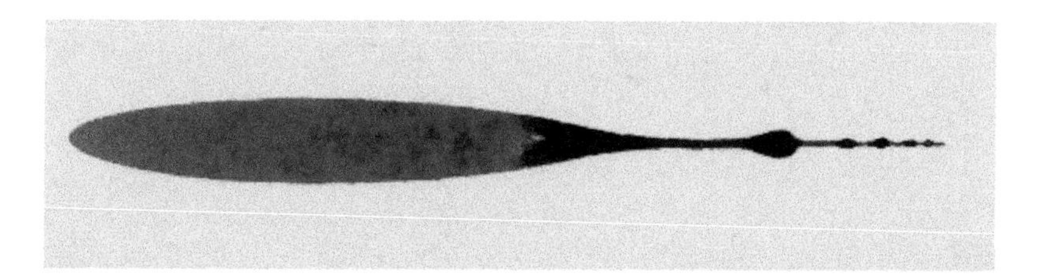

FIGURE 5.8 Blood trace produced from droplet falling vertically onto stationary inclined surface (Angle of Incidence = 80°). *Source: Photo courtesy of Norman Marin & Jeffrey Buszka.*

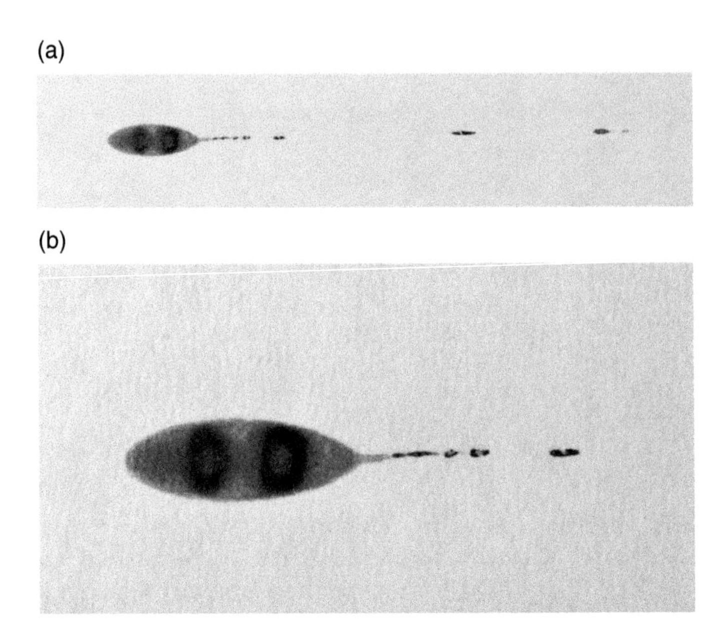

FIGURE 5.9 Blood trace produced from a blood drop projected horizontally and impacting the horizontal paper substrate (angle of incidence: 68°), depicting the entire deposit including several traces created by the impact of satellite droplets toward the right (a) and an enlarged view of the primary portion of the trace (b).

may assist in positioning the person who is swinging the object (Figure 5.10). A determination of handedness and number of swings may be possible in some cases; however, extreme caution must be exercised because a given arc can be generated independent of handedness. Arc patterns may be simple (single-linear pattern of droplet stains) or complex (several single-linear patterns, possibly overlapped, or wider linear patterns); patterns will vary dependent upon the object. A large number of overlapping arc patterns, created by several swings, will sometimes make analysis more difficult.

5.2.4 Arterial Deposit Configurations

Arterial patterns may result when an individual receives a wound deep enough to either partially or completely transect an artery. Arterial pressure, created by rhythmic contractions of the heart muscle, fluctuates throughout the cardiac cycle and may project blood forcefully from the wound in spurts or sprays. If the blood exits the wound unimpeded, *arterial spurt patterns* may be produced on nearby surfaces such as walls. Arterial spurt patterns may exhibit a curvilinear appearance and are often characterized by the projection of larger masses of blood with varying velocities (Figures 5.11 and 5.12). If the projected blood is impeded at the wound by an intermediate object, an *arterial spray pattern* may result. With time and in the absence of medical intervention, as the injured person exsanguinates, the quantity and force of the projected arterial blood will diminish.

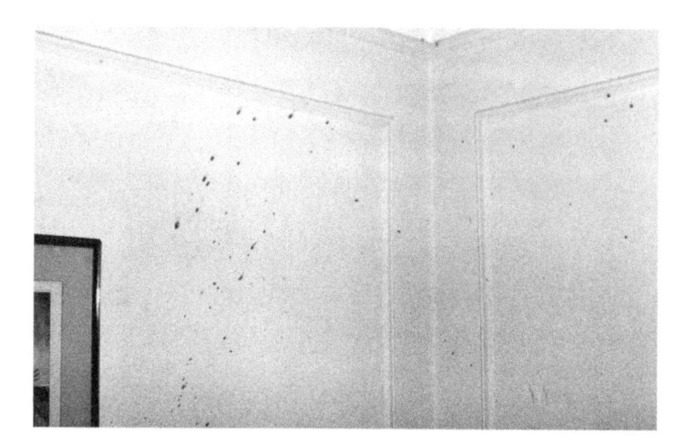

FIGURE 5.10 Arc or cast-off patterns from a bludgeoning. The uppermost stains in this group are approximately 8–9 ft from the floor.

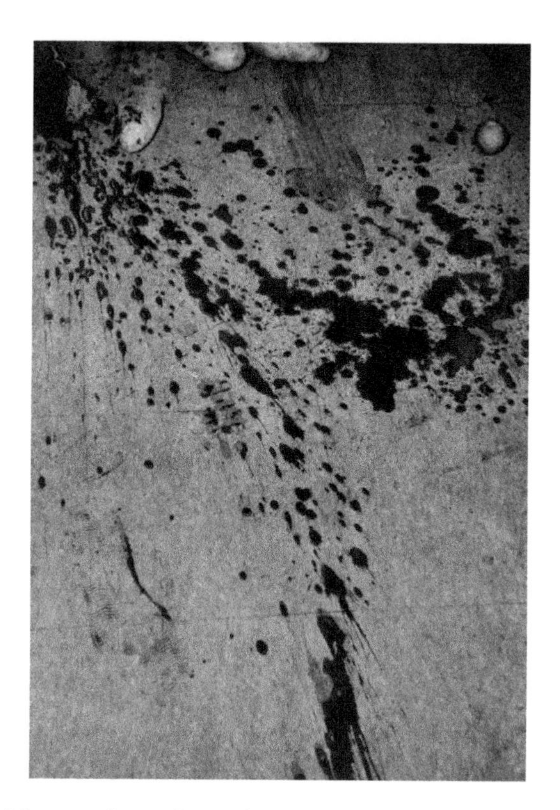

FIGURE 5.11 Arterial spurt where a femoral artery was breached.

5.2.5 Droplet Trail Deposit Configurations

Trail patterns produced by droplets of blood are a series of blood traces often circular in shape. There are several types of trail patterns including those created by blood dripping from a bleeding wound in a moving person or animal. These trails can become extensive if the injured individual continues to bleed while in motion.

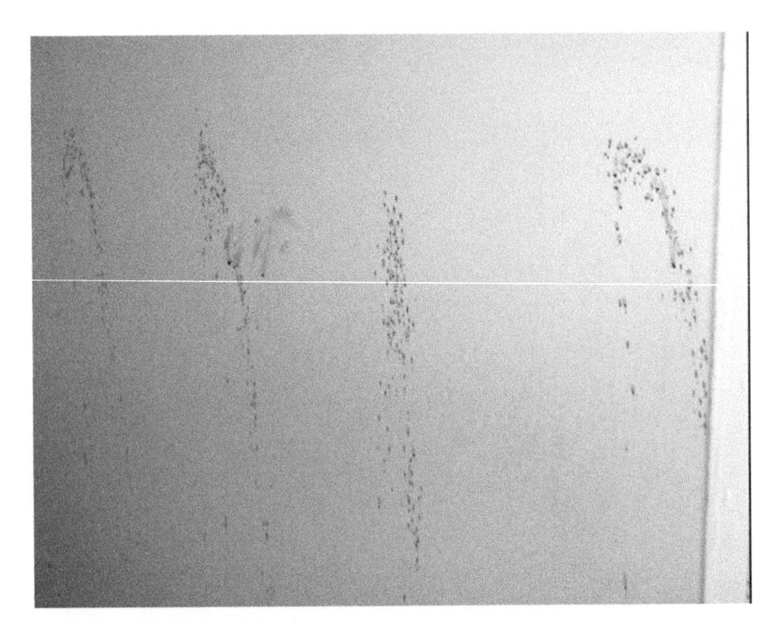

FIGURE 5.12 Configurations resulting from a failed ateriovenous (AV) graft. *Source: Photograph courtesy Denis Cavalli.*

An object with liquid blood on its surface may also drip blood while being carried. These trails are typically shorter and the distance between drops is greater as the amount of blood on a weapon is limited. The direction of movement of an individual, whether bleeding or carrying an object releasing liquid blood, may be ascertained if the velocity of horizontal motion is great enough to provide a sufficient horizontal vector to the falling blood drops to cause them to impact at a somewhat oblique angle. The resultant stains will appear slightly elliptical and may exhibit satellite droplets or fingering dependent upon the target surface.

5.2.6 Airborne Droplets in Respiratory Airstreams

Airborne blood droplets can forcibly exit the body in airstreams from the oral and nasal cavities as well as directly from airstreams at external wound sites to the respiratory system such as penetrating chest wounds. The airstreams from nasal or oral cavities could result from coughing, sneezing, or medical intervention. In the blood traces formed by some mechanisms, blood may be admixed with saliva from the oral cavity, which is often detectable via laboratory analysis. Deposits can consist of hundreds to thousands of airborne droplet traces, many of which are very small (<1 mm). The reader is referred to studies done by Donaldson et al. (2011) and by Denison et al. (2011) for additional details regarding spatter related to respiration (Figure 5.13).

5.2.7 Radial ("Impact") Spatter (Include Close-Up)

Radial spatter patterns are formed when an object or liquid impacts liquid blood, which results in the production of airborne blood drops that are projected radially away from the impact site (Figure 5.14). These patterns are often observed in blunt force trauma incidents, but they can also be associated with shooting incidents (projectile impacts to liquid blood). *Secondary spatter*, a type of radial spatter, results from the impact of blood into blood.

FIGURE 5.13 Blood traces resulting from droplets that possessed entrained air. *Source: Photograph courtesy of Brian Gestring.*

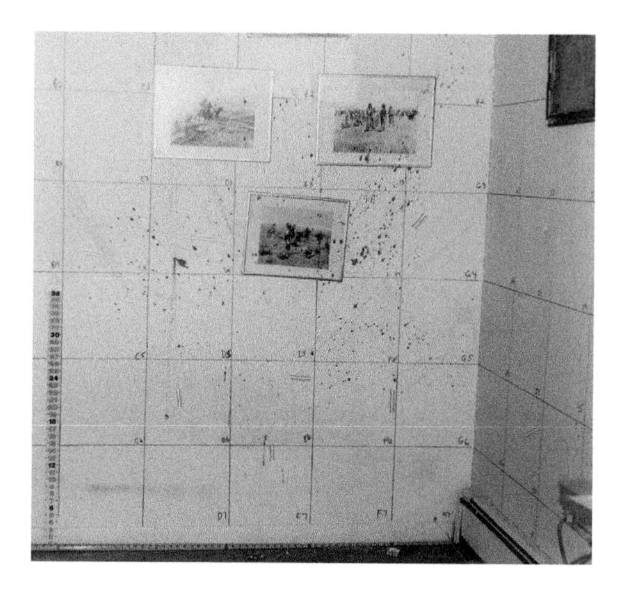

FIGURE 5.14 Radial spatter traces from bludgeoning that has been "roadmapped." *Source: Photograph courtesy of the Yonkers Police Department.*

The size and distribution of airborne droplet stains vary significantly. Hundreds to thousands of droplet stains may be present with dimensions that range from submillimeter to several millimeters in size. Previous attempts to classify radial spatter patterns as low, medium, and high velocity are flawed. They conflate velocity or energy of the impact with the velocity and size of the resulting droplets.

In bludgeoning cases, a single blow is unlikely to spatter blood. One or more impacts are generally required to lacerate skin and produce a source of pooled blood. Subsequent impacts to this blood may then produce radial spatter patterns. In general, the impacts that produce radial spatter patterns result in droplets being projected outward radially thus the name. The radial dispersion of the droplets is not normally expected to result in equal dispersion in every direction. This can greatly complicate some interpretations. For example, the absence of droplet traces on a suspect cannot necessarily be interpreted to exclude an individual as the assailant. In addition, similarly, voids in the radial pattern are not necessarily due to the presence of a person or object intercepting the droplets that would be expected to be deposited in the area where the void is observed. More symmetry would perhaps be expected where more blows are administered. The degree of symmetry or asymmetry can be highly dependent on the geometry of the implement as well as that of the area where the pooled blood has accumulated. Variations in the location of the site of impact and its relative movement can have a profound influence on the degree of symmetry.

5.2.8 Secondary Spatter

One of the most common occurrences at crime scenes where considerable blood has been shed is blood either being projected or falling into a pool of blood. When a volume of blood impacts a preexisting pool on some substrate, be it resilient or rigid, or in the absence of a pool involving the rapid collapse of a bolus of blood into itself, numerous individual droplet patterns are likely to be produced in the adjacent area surrounding the initial impact site. Some of the airborne droplets that are generated will not be visible as they will have been assimilated into the preexisting pool or the newly formed one. The spatter that is visible that was created from such an impact may be commonly referred to as secondary spatter (Figures 5.15 and 5.16). These patterns are very often a source of confusion in the reconstruction of crime scenes and may mimic patterns arising from bludgeonings, gunshot, and stabbings. Serious misinterpretations of these secondary spatter patterns have taken place resulting from a failure to recognize the prevalence of this phenomenon. The danger of misinterpretation is enhanced where there has been removal of objects from a scene or clean-up prior to scientific investigation.

Because of the common occurrence of secondary spatter at crime scenes coupled with the possibility of misinterpretation, Kodet-Sherwin et al. studied it in detail (1988). The following independent variables were studied: dropping volume, pool depth, dropping height, and type/nature of pool substrate; dependent variables included size distribution of droplet stain patterns, projection distances in addition to symmetry, and spatial distributions of droplet stain patterns.

FIGURE 5.15 Secondary spatter projected onto nearby sneaker. *Source: Photograph courtesy of Norman Marin & Jeffrey Buszka.*

FIGURE 5.16 Numerous blood traces on paneling and baseboard produced as secondary spatter from blood impacting blood on floor.

5.2.8.1 Dropping Height Experiments Kodet-Sherwin et al. (1988) concluded that when an 8-ml bolus of blood was dropped from 80 cm into a preexisting pool, blood onto asphalt tile blood traces were found at a maximum horizontal distance of 147 cm. However, no spatter was observed when the dropping height was decreased to 15 cm. In one experiment, spatter was observed as high as 56 cm when an 8-ml unit of blood fell 80 cm onto a pool formed on the ribbed rubber surface shown below (Figure 5.17). Similar experiments were conducted on a number of other substrates

FIGURE 5.17 Secondary spatter produced from blood impacting ribbed flooring. Note the heavier concentration of droplet traces aligned with the long axis of the ribbed flooring (the blood traces favor the "privileged direction"). *Source: Photograph courtesy of Lenore Kodet.*

such as carpet, glass, ceramic tile, velour, and sand with blood dropped from the following heights: 80 cm, 40 cm, 20 cm, and 10 cm. As expected, generally, more extensive secondary spatter occurred as the dropping height increased for all substrates.

5.2.8.2 Dropping Volume Experiments Kodet-Sherwin et al. (1988) also studied the effect of varying the volume of blood dropped on the formation of secondary spatter. The dropping height throughout this portion of the study was held at 80 cm and the substrate was asphalt flooring tile. The following volumes were dropped as single units: 8 ml, 4 ml, 2 ml, and 1 ml, and three single-drop volumes (0.08 ml, 0.05 ml, and 0.025 ml). Generally, as the volume was increased, the quantity of secondary spatter also increased. However, experiments conducted with 8 ml of produced droplets at lower numbers than 4 ml. No spatter was observed when the volume dropped was 0.025 ml (single drop). For the larger single-drop volumes (0.05 ml and 0.08 ml), a small quantity of spatter was observed. These stains were limited to regions in very close proximity to the pool.

At the larger dropping volumes, low "take-off" angles appeared to predominate. This observation is based on a comparison of the number of elliptically shaped stains to the number of circular stains counted. No generalizations should be made; however, since in several instances, the inverse was found to be true for the areas counted. Additionally, it should be recognized that since very small blood droplets are more elastic in nature than larger droplets, the smaller ones may resist drop collapse and not tend to form elliptical stains albeit they may have impacted the horizontal surface at a relatively grazing angle (resulting from low "take-off" angles). Thus, even though low "take-off" angles may have predominated, a negative bias (favoring the circular stains) may be present in the data that the examiner would not be aware of.

One way to obviate this lack of certainty would be to perform a series of high-speed photographic experiments.

As the dropping volume was increased, the average horizontal distance that bloodstains were observed increased. With 0.5 ml, the average distance was 40 cm, 71 cm with 1 ml, 83 cm with 4 ml, and 108 cm with 8 ml. These averages were obtained from a height of 80 cm and a 10-ml pool on asphalt tile.

5.2.8.3 Various Substrates Kodet-Sherwin et al. (1988) speculated that differences in secondary spatter (quantity and projection distance) might be observed due to differences in resiliency of the substrate. Specifically, the authors believed that less secondary spatter would be observed for the more resilient surfaces, i.e., carpet. These surfaces would tend to impart less energy to secondary droplets due to greater absorption of energy of the impact by the substrate. They compared three substrates: velour fabric, glass, and carpet (1-cm-deep fibers). An 8 ml of blood was dropped as a bolus (unit) from 80 cm onto a preexisting pool on each of these substrates. The pool volume for the glass and the velour fabric was 10 ml. Because the carpet absorbed a significant quantity of the initial 10 ml, additional blood was added in order to create an area of pooled blood (total volume of 30 ml was used). No significant difference in secondary spatter was found between the velour fabric and the glass substrates. This may be due to the fact that the fabric used was fairly thin and overlay a non-resilient surface. As some of the blood soaked through the fabric, it undoubtedly caused the fabric to adhere more readily to the underlying surface causing the substrate to be less resilient.

Marked differences were observed when comparing secondary spatter from the carpet to the spatter obtained with the velour and glass. Very little spatter was observed on either the horizontal or vertical surfaces for the carpet in comparison to the glass and velour. As indicated before, Kodet-Sherwin et al. (1988) concluded that this is due to absorption of impact energy resulting from resilience.

5.2.8.4 Secondary Spatter Discussion A suspect in a homicide by stabbing provided investigators with the clothing he was wearing when he discovered the body of a former girlfriend and claimed that he tried to render assistance. The laboratory found numerous small blood droplet stains on the rubber portion of the sides (foxing strips) of his athletic shoes ("sneakers"). No stains were found on the upper or canvas portion of the shoes. Two small stains were also found on the rear of one pant leg. An expert previously retained by the prosecutor claimed that the spatter on the suspect asserted that he was present during the incident. The prosecutor, to his credit, was somewhat skeptical about his expert's conclusion. Due to his skepticism, he invited the defense attorney to present his own expert to testify at the Grand Jury. The defense expert questioned the conclusion that the small spatters on the shoes, which were restricted to the lower rubber portion, proved that the defendant was present at the stabbing. The defense expert pointed to the possibility that the stains could have been the consequence of secondary spatter resulting from blood from the victim falling into pooled blood on the floor at the time he discovered the body and attempted to move her (Figure 5.18).

FIGURE 5.18 Illustration of some of the droplet traces that were present only on the lower portion of the footwear.

5.2.9 Spatter Associated with Gunshot Wounds

5.2.9.1 Patterns from Perforating (Through-and-through) Wounds

Projectiles (e.g., bullets) impacting or traversing materials can be expected to produce secondary projectiles. In perforating wounds, some of the secondary projectiles may be in the form of blood droplets that are ejected from the exit site in often a conical distribution. Other secondary projectiles will consist of tissue fragments. It needs to be noted that special conditions are necessary to result in the production of secondary projectiles in the form of blood droplets. Only if the path of the bullet intersects coalesced blood in producing a wound can we expect to see droplets in either the forward or retrograde direction. In traversing muscle tissue, for example, solely supplied by small blood vessels, such as capillaries and arterioles, no blood droplet spatter would be expected. For this reason, sometimes no forward or retrograde blood spatter is produced. Where present, such secondary projectiles (blood droplets) are commonly referred to as "forward spatter." Secondary projectiles emanating from an exit wound can also consist of fragments of tissue or fragments of other materials struck such as clothing, or fragments of the projectile itself. When a projectile perforates a target material, secondary projectiles that are produced can travel in a retrograde direction as well as in the direction of the primary projectile travel. In controlled experimental work, the secondary projectiles can be studied using high-speed photography and/or by placing "witness papers" near the point of impact or projectile exit. Witness papers are specially prepared surfaces, commonly white paper supported by a frame, used to reveal the path of a projectile or secondary projectiles. Of course, in "real world" case investigations, neither of these recording mediums is typically available. Thus, any record produced of such secondary projectiles will depend upon the fortuitous proximity of surfaces capable of retaining evidence of such secondary projectiles. Some of the earliest documented interactions with bullets producing secondary projectiles were recorded using high-speed

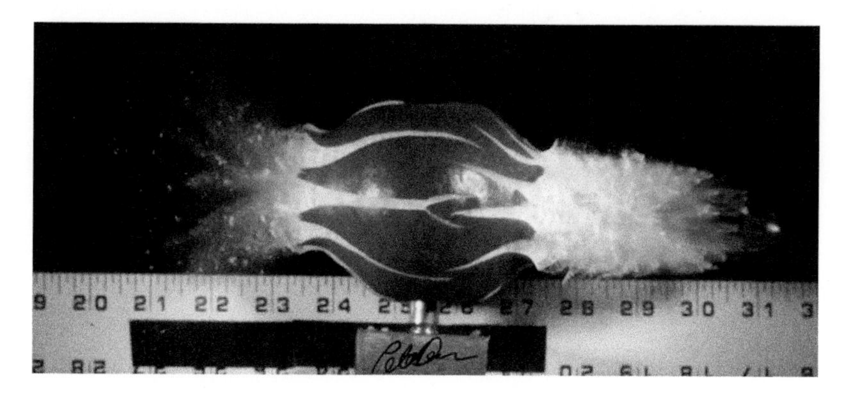

FIGURE 5.19 Forward and retrograde secondary projectiles as a result of bullet perforating an apple (cartridge: .45ACP, RN FMJ 230 grain). Muzzle-to-target distance ca. 15 ft. 10 μs shutter speed; 25 000 fps. Phantom VEO video camera. *Source: Frame grab image from video courtesy of Peter Diaczuk, John Jay College of Criminal Justice, CUNY.*

photography carried out by Dr. Harold Edgerton of the Massachusetts Institute of Technology (MIT).

Figure 5.19 illustrates such a high-speed photograph capturing the interaction of a handgun bullet perforating an apple. Secondary projectiles composed of small fragments of the apple pulp can be seen associated with the entrance hole and the exit by the bullet. The fragmented pulp can be seen to be projected in a retrograde direction from the entrance hole and in a forward direction from the exit hole. An apple contains a considerable quantity of juice, but this quantity of juice is not released when the apple is merely cut in half. The apple must be crushed in its entirety for most of the juice to be recovered. Prior to crushing, the juice is retained by vessels in the tissue of the apple. In the set-up for the experiment, if witness papers had been placed near the entrance hole and the exit hole in the apple prior to the shot being fired we would not necessarily expect to see droplet stains of juice on these papers. For properly placed witness papers we would expect to see fragments of apple tissue instead. The apple analogy is valuable but can only be taken so far. For obvious reasons experiments such as these with human beings cannot be conducted. Unlike the cut apple example, where little or no juice flows, a cut into capillary fed human tissue such as a muscle, blood will flow into the wound, but only after a delay of a few seconds. For people who take the time to make observations of an accidentally received cut, this delay, before blood flows into and accumulates in the wound, may have been noticed. This delay, although seemingly short, is many times longer than the time of bullet travel through tissue. The secondary projectiles that are produced by the primary projectile interaction with capillary fed tissue exclusively will not consist of blood droplets. These secondary projectiles will consist entirely of very small tissue fragments. There will be no forward spatter or back spatter as such, because in this example the bullet did not encounter any volume of coalesced blood. This can be illustrated experimentally using dialysis cartridges and other media as discussed subsequently in Chapter 13. Not unexpectedly, if a bullet perforates a blood-soaked sponge an unrealistic large amount of forward and retrograde spatter will be generated as illustrated in Figure 5.20.

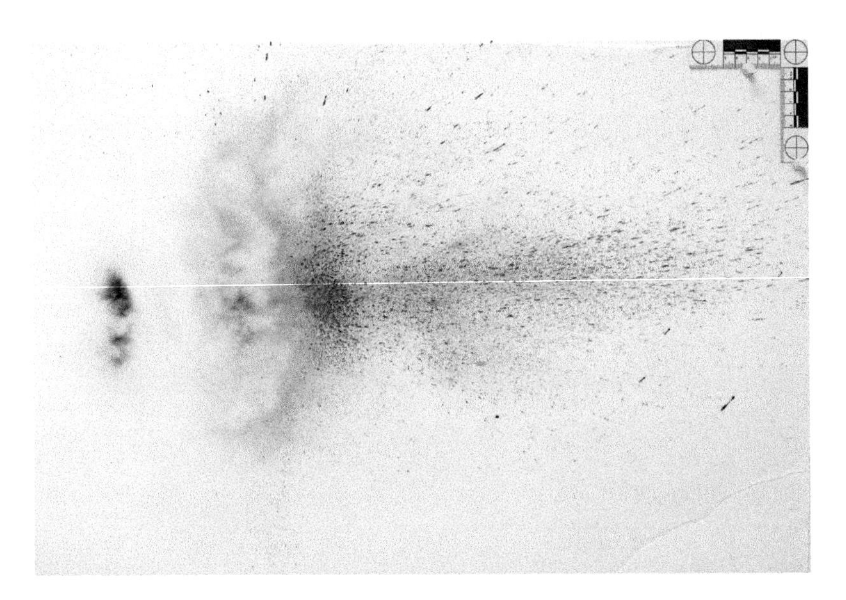

FIGURE 5.20 Traces deposited on witness panel situated parallel to firearm at time of discharge into bloody sponge. Predominately forward spatter accompanied by minor amount of backspatter and gunshot residue (GSR) deposits from revolver (cylinder gap and muzzle). *Source: Photograph courtesy Norman Marin & Peter Pizzola.*

5.2.9.2 Backspatter from Entrance Wounds with No Exit (Penetrating Wounds) Retrograde secondary projectiles as well as those traveling in the approximate direction of the primary projectile were described above for perforating interactions. Thus, it is no surprise that secondary projectiles can result from the impact of a projectile to surfaces where there is penetration only (Figure 5.21). In the case of gunshot wounds to merely capillary fed tissue, the secondary (tissue) projectiles will consist of tissue fragments but will likely not include small blood droplets (Figure 5.22). If the tissue deposits are onto a porous substrate, some blood may diffuse from the capillary fragments into the underlying surface creating blood traces however limited.

5.2.9.3 Blood Traces from Blowback Blowback is a very different phenomena than forward spatter or backspatter. It does not involve secondary projectiles as such. Blowback occurs only for extremely close-range shots (contact or close contact shots) to areas of the body where muzzle discharge gases can be injected into the body to form a temporary cavity. Following a delay, the cavity will collapse and in the process eject the gases through the bullet wound. Blood which has accumulated as a result of the disruption of the tissues in the wound can be ejected with the gases and form droplets. The delay allows the blood to coalesce and then form droplets when it is ejected. It should be noted that many practitioners consider this as simply another form of backspatter with a different etiology. However, we find it important to make the distinction between the two phenomena. Noteworthy is the experimental work done by Radford et al. (2015) to physically eliminate or separate blowback effects from backspatter. More recently, Comiskey et al. (2019) used cardstock as a diffuser in attempt to prevent or mitigate the muzzle gases from reaching the impact site.

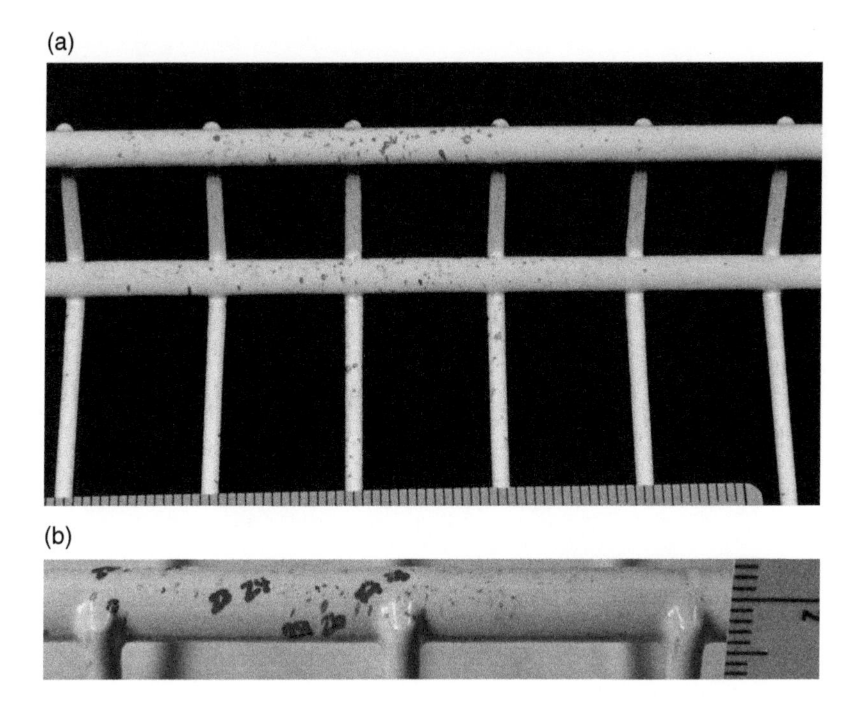

FIGURE 5.21 Photographs of a wire rack with backspatter from penetrating wound (a), and the flip side of the wire rack, also containing blood traces (b). *Sources: Photograph (a) courtesy of Keith Mancini, Westchester County Forensic Laboratory; Photograph (b) provided by Peter Pizzola.*

FIGURE 5.22 Tissue particles on headliner and door frame accompanied by no individual droplet traces. The lack of any deposits on the door indicated that the door was open at the time the firearm (shotgun) discharge occurred.

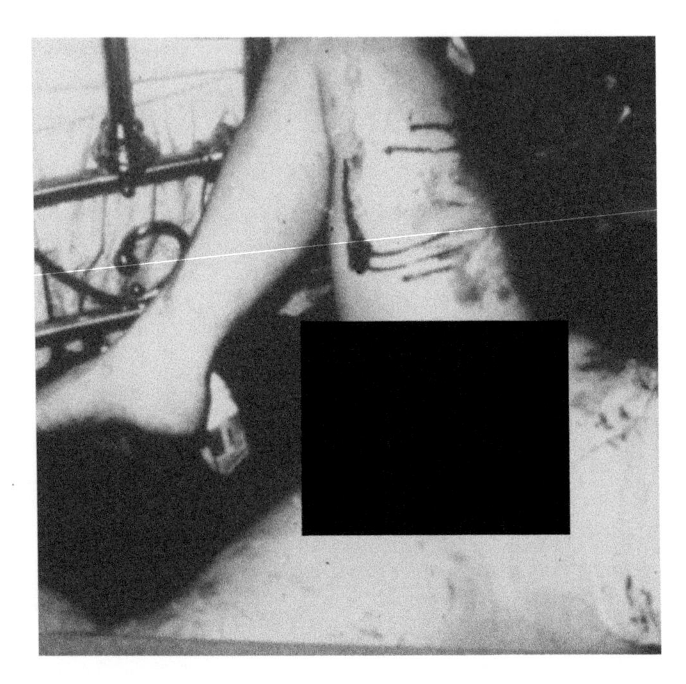

FIGURE 5.23 Change in flow direction of rivulets. *Source: Photograph courtesy Robert Shaler.*

5.2.10 Other Configurations

5.2.10.1 Flow Configurations The direction of flow of blood can be quite revealing at times to show, for example, a change in position of inanimate objects or that of human beings. As indicated in Figure 5.23, an examination of the right inner thigh of the victim demonstrates that the flow of two rivulets changed direction; clearly the leg is not in the original position when the blood flow initiated on the portion of the thigh not visible. The flow of rivulets can help establish the position of objects that may be able to move or rotate as indicated by the deposits on the door in Figure 5.24.

5.2.10.2 Pooling Configurations Pools of blood must be carefully examined for extraneous physical evidence. In particular, the periphery should be visually inspected for evidence of contact, especially where there is some indicia of disruption. Where blood has splashed outward, producing spatter may reveal that some object(s) have either fallen or come in contact with the pool – such as footwear. If there is some indication of the foregoing, the area surrounding the pool must be carefully examined for additional contact in the form of either static or dynamic transfers. Pools must also be carefully examined for partially or completely obscured forms of physical evidence such as bullets, bullet fragment, bullet holes, or impressions.

5.2.10.2.1 Clotting, Serum Separation and its Significance With time, clotting will take place in large pools of blood at a crime scene. With a small blood deposit, the blood may dry too fast to allow time for clotting. The clotting taking place in a large blood pool results in a formation with a central clot surrounded by serum. The timing of this depends on several factors such as the ambient temperature and the nature of

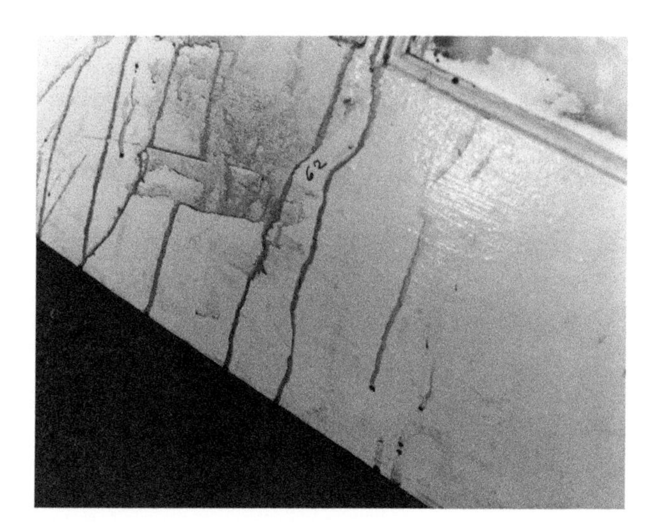

FIGURE 5.24 The flow of rivulets on the bottom of this door can help establish its position when the blood was shed assuming the blood continued to drip onto the floor in which case the traces on the carpet will be aligned with the rivulets. There has been a disturbance in the middle of the group which has altered the corresponding rivulet. *Source: Yonkers Police Department.*

the surface underlying the pool. The pooled serum can be transferred to footwear or clothing, and the presence of such traces would show that there had been a time lapse between bloodshed and the deposit of the serum and creation of the pattern. The serum stains may not only be on the footwear, but the outsole may also leave latent traces on the floor in the vicinity of the pooled serum that may require chemical or optical enhancement (see Ch. 4). This may show that the perpetrator remained at the scene for a significant period of time. Alternatively, depending on the case context, the serum stain could reveal that a person of interest was uninvolved and just happened upon the scene. Sometimes, the stains produced will be a mixture of serum and partially clotted blood, and these can also be useful in approximating a time frame for the event. The reader is referred to Chapter 10 regarding the "Dew Theory" case where such phenomena are discussed in more detail.

While clotted blood may be valuable in estimating the interval over which significant amounts of blood were shed, we must be exceedingly cautious in interpreting these traces. Clearly, the environmental factors which influence clotting time must be very carefully considered. Additionally, extreme caution must be exercised when attempting to interpret the presence of traces that appear to be the result of the distribution of airborne clots. Small clots in a nearby distribution are unlikely to have been formed in situ from blood droplets distributed earlier. On the other hand, in situ formation of clots from previously distributed blood for larger volumes may be possible (Figures 5.25 and 5.26).

5.2.10.3 Diluted Blood Deposits Blood may be mixed with a number of different liquids including saliva, urine, etc., which may make it more difficult to recognize or detect highly diluted blood traces as indicated in the far right tube in Figure 5.27. The presence of saliva in blood traces can be of value in reconstructions. Thus, where suspected, laboratory analysis for salivary amylase (AMY1) should be included. A trace, possibly of this type, is depicted in Figure 5.28.

FIGURE 5.25 Example of serum separation from clot in a pool of blood on a horizontal surface. *Source: Photo courtesy of Peter Valentin.*

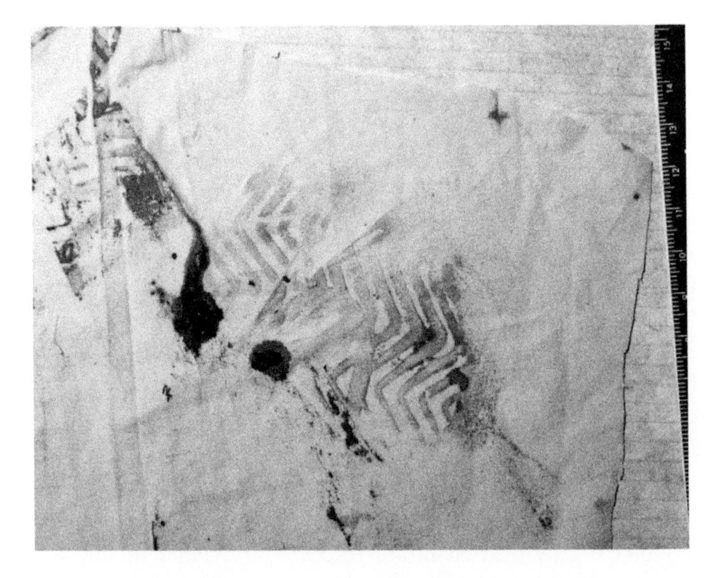

FIGURE 5.26 It may not be possible to determine whether or not the clotting took place on the fabric or whether partially clotted blood was projected onto the fabric. Interpretation of this kind can be exceedingly difficult as well as the sequence of clot versus footwear pattern.

FIGURE 5.27 Falcon™ conical tubes containing blood diluted with water (hemolyzed in tubes 2–5, reading from left to right). The range of 10-fold serial dilutions starting with undiluted on the left progressing to a dilution of 1×10^{-4} on the right. *Source: Photograph courtesy: Jeffrey Buszka and Norman Marin.*

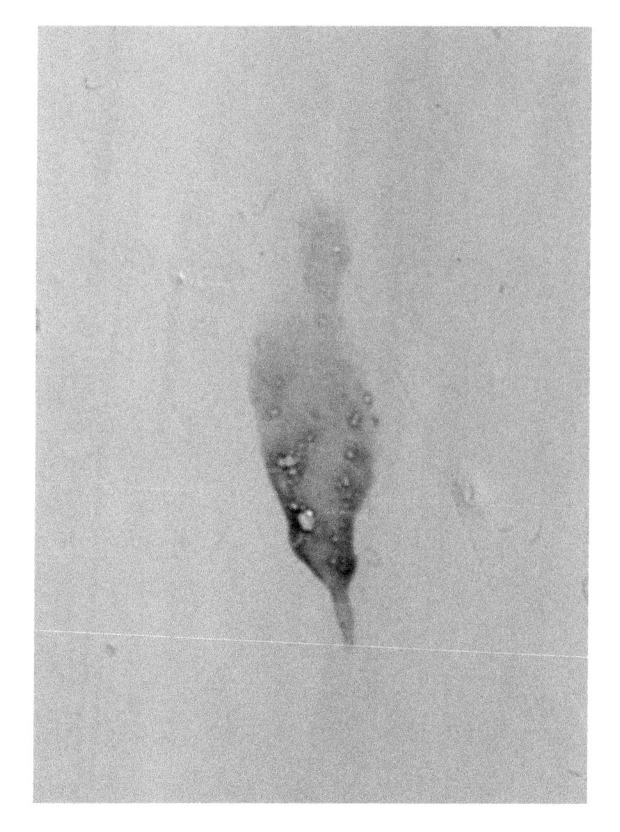

FIGURE 5.28 Blood trace possibly diluted with saliva. The presence of apparent entrained air bubbles – most of which appear to have burst on drying. *Source: Photograph courtesy of Jeffrey Buszka and Norman Marin.*

5.2.10.4 Significance of Voids The configuration of any voids may provide useful clues as to what intercepted the blood droplets as shown in Figure 5.29. On many occasions, the authors have taken advantage of voids to restore (after careful deliberation and documentation) furniture and other objects in their original locations when the blood was deposited. This process can be of considerable value when attempting to conduct a reconstruction. Some of the objects that may have been relocated could bear additional valuable traces. The position of the assailant can be sometimes be estimated as famously illustrated by Paul Kirk in the Sheppard case (Chapter 10).

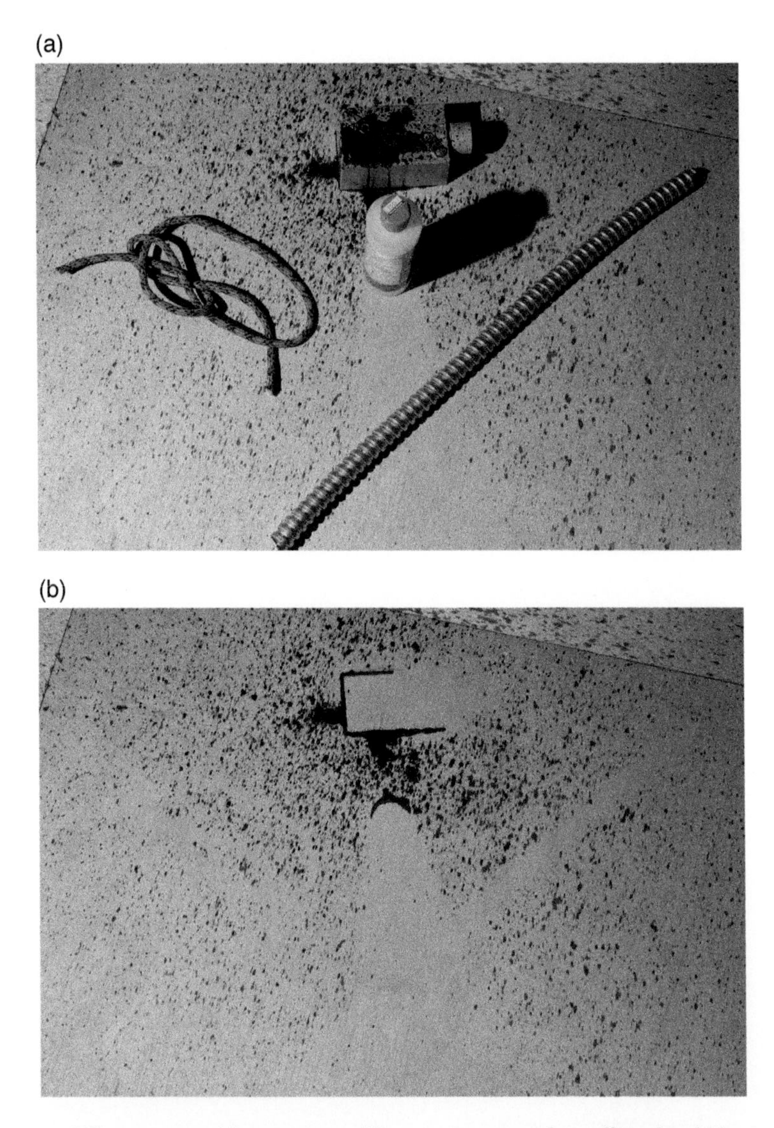

FIGURE 5.29 When traces with apparent voids are encountered an effort should be made to locate the possible intervening object(s) responsible for them. Photographs show the secondary spatter from blood impacting pooled blood on a gray electrical box. The objects responsible for creating the voids are shown in situ in (a) and removed in (b).

5.2.11 Post-Incident Events ("Artifacts")

5.2.11.1 Human Attempts at Clean-Up Configurations created during clean-up attempts must be recognized as not being indicative of activity during the incident (Figure 5.30). On occasion, the traces should be analyzed for the presence of detergent, bleach, etc.

5.2.11.1.1 Inhibiting and Obscuring Cleaning Agents The utilization of various cleaning agents may inhibit or obscure attempts to detect traces of blood. When there is evidence that such agents may have been used, as is sometimes the case, cautious and deliberate approaches to the potential detection of blood traces are necessary. We must avoid the temptation to employ chemical reagents to quickly discover the traces. This is the time when we should stand back and really think about what we are doing. Contextual information will need to be evaluated and an innovative approach developed. The onsite inspection could initially involve strong visual examination aided with adequate lighting and different light sources if deemed appropriate. With pressure, apply filter paper or Whatman Benchkote© (filter paper coated on one side with polyethylene) over the carpet to see if any blood traces, which may still be moist, will be absorbed. If this area is near a wall, we could pull up the carpet and

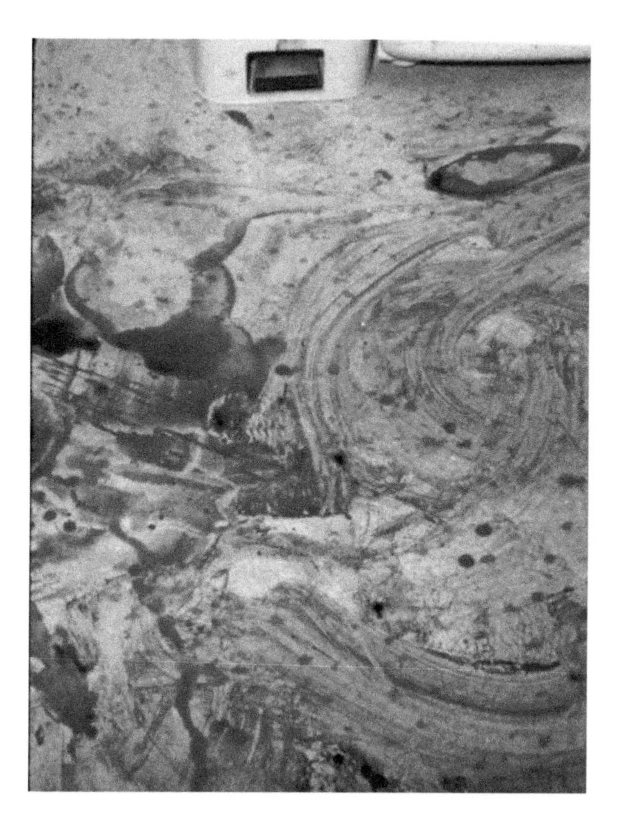

FIGURE 5.30 Attempted altering of the configuration of a large quantity of blood on floor evident in this figure. Notice also, the post-incident events, i.e., trail of single droplet blood traces and the continuous trail. *Source: Photograph courtesy of the Yonkers Police Department.*

FIGURE 5.31 Rear seat was treated with Luminol and the chemiluminescence documented photographically for 30 s in essentially complete darkness; rear curtain flash was then triggered to illuminate the substrate and surrounding objects. See Figure 7.13 in Chapter 7 for image of seat prior to application of Luminol. *Source: Photograph courtesy of Norman Marin & Jeffrey Buszka.*

look underneath. These authors have effectively utilized these approaches on several cases and saved much time, lessened exposure to chemicals, and also minimized the chances of diluting or degrading the blood traces. The presence of cleaning agents may be detected by their odor and possibly by fluorescence using both ultra-violet and visible excitation sources. If cleaning agents have been employed, it is essential that they be detected and it is understood that the resulting stain or trace configuration not unwittingly become misunderstood as being part of the initial assault. The laboratory could play a very important role and detect and identify the presence of such agents by detailed analysis subsequently. This laboratory analysis should commence with visual inspection in conjunction with stereomicroscopy and documentation.

5.2.11.1.2 Luminol and Investigative Leads When the nondestructive methods of detection are employed with limited success, we may have no choice but to employ more risky chemical detection methods such as chemiluminescence with Luminol or Bluestar® (Figure 5.31). However, we must be judicious in its use. We must always be aware of the possibility of a false-positive chemiluminescent response from materials other than blood. One such source may be common household bleach residues of which would not be unexpected in bathrooms, etc. See the discussion in Chapter 4 for more details. Chemiluminescence involving blood traces tends to be of a longer duration than some other materials such as bleach. Some of these other materials may produce a "flash" – very short duration chemiluminescence.

5.2.11.2 Animals and Insects Various kinds of insect traces such as from cockroaches, maggots (Figure 5.32), flies (Figure 5.33) may be present and should not be unexpected. The appearance of individual traces comprising fly patterns should

FIGURE 5.32 Maggot trails. *Source: Photograph courtesy of Jason R. Sanderson, Penn State Forensic Science MPS Graduate Program.*

be obvious to the examiner that their origin is not blood spatter. Context, too, can augment the reasoning of whether one is dealing with spatter verus fly deposits. For example, a cluster of droplet traces around a ceiling or wall lighting fixture is suggestive of fly deposits. As in the case depicted by the above images, which include a light fixture, the height of the overall pattern or traces is inconsistent with having originated at a point lower in the room. And the presence of a decomposing body in the room should also alert the examiner to the strong possibility of fly deposits from the regurgitation of the blood. The careful examiner would note the peculiar morphology of many of these individual traces some of which appear to have a filament drawn out from the main body (perhaps to some extent mimicking spatter) but then turns in odd directions. Many individual (if they are mistakenly treated as spatter) traces appear to have originated from different locations. The nearly circular traces lack any sign of "crowning" or "fingers" on this fairly rough textured surface which is inconsistent with droplet impacts. There are also signs of dilution of some of the individual traces. If the overall configuration only consisted of two or three stains, and lacking some of the anomalies just mentioned, an inexperienced examiner might mistakenly assume that the source was the result of some sort of violent impact. However, strong opinions must not be drawn from a limited number of observations. Again, the context of the traces

(a)

(b)

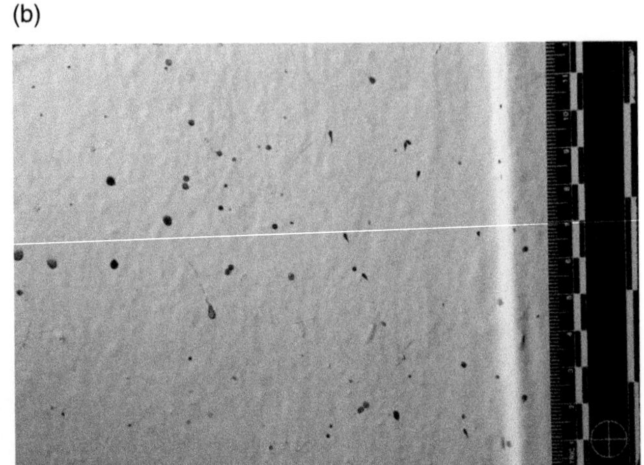

(c)

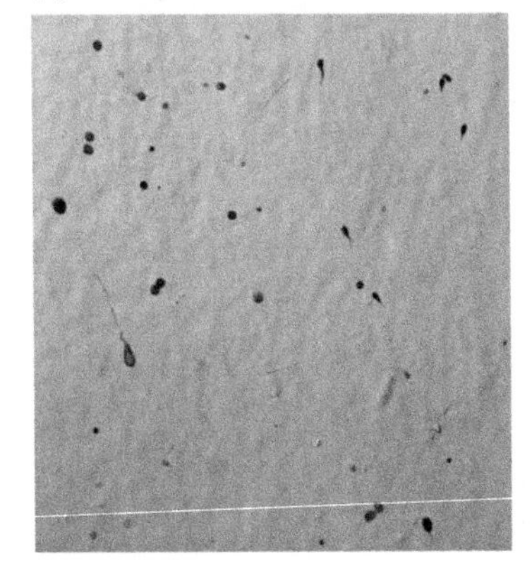

FIGURE 5.33 Photographs of blood traces from flies, with an overall image (a), an expanded view of the configurations above and to the right of lamp (approximately 7 ft above floor) (b) and a magnified image of those same blood traces (c). *Source: Photographs courtesy of Centre Criminalistics Consulting.*

should be a strong clue to even the inexperienced examiner that fly deposits are quite likely given the circumstances (Figures 5.34 and 5.35).

5.2.11.3 Unavoidable Environmental Events (i.e., Rain, Wind...)

There are times when considerations of the environment can shed some light on a matter under scientific inquiry. For example, in one homicide case, a mayonnaise jar was apparently thrown through a window of a third story apartment and was found on the ground about 30 ft below. Fallen, undisturbed snow was on top of the fractured jar (the last snowfall was three days before the incident occurrence). This case was an instance of where snow helped fix the time when the jar was used to break the window. One of the authors was requested to respond to the scene for the purpose of blood trace examination. The investigators and an assistant district attorney thought that the window might have been broken contemporaneously with the incident. The presence of dried mayonnaise on the fractured glass coupled with the snow on top of the mayonnaise bottle helped demonstrate that the glass had been fractured three days prior to the homicide.

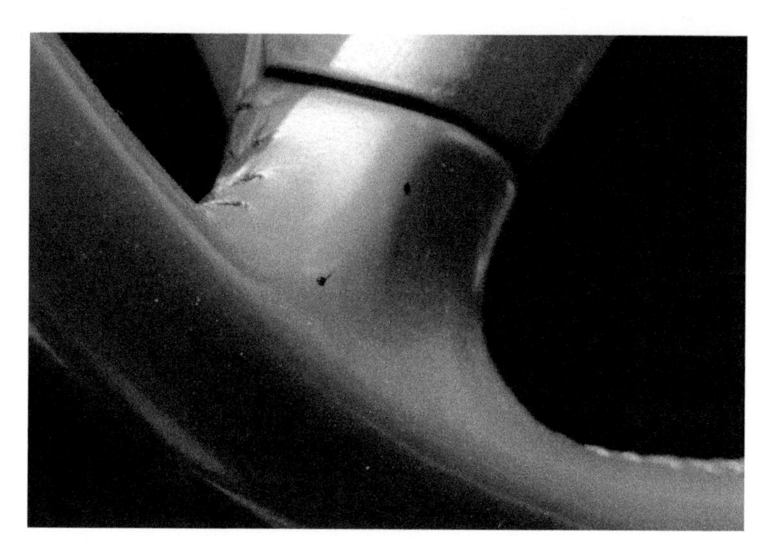

FIGURE 5.34 Two blood traces on a steering wheel apparently deposited by flies. The decomposing victim of this homicide was found in the trunk along with many maggots and flies. The cabin of the vehicle was locked with the windows closed without any sign of disturbance. A limited number of flies apparently managed to migrate from the trunk to the cabin. *Source: Photograph courtesy of Norman Marin & Jeffrey Buszka.*

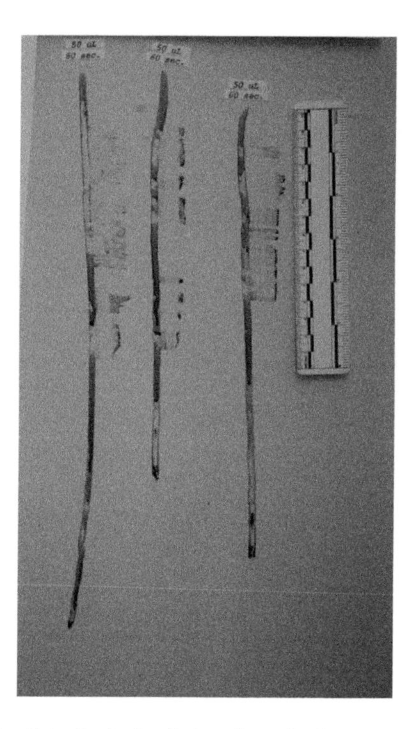

FIGURE 5.35 Three rivulets disturbed after being deposited. *Source: Photograph courtesy of JMR Forensics, Inc.*

References

Arthur, R.M., Cockerton, S.L., de Bruin, K.G., and Taylor, M.C. (2015). A novel, element-based approach for the objective classification of bloodstain patterns. *Forensic Science International* 257: 220–228.

Attinger, D., Liu, Y., Bybee, T., and De Brabanter, K. (2018). A data set of bloodstain patterns for teaching and research in bloodstain pattern analysis: impact beating spatters. *Forensic Science International* 18: 648–654.

Bevel, T. and Gardner, R. (2008). *Bloodstain Pattern Analysis*, 3e. CRC Press, Taylor & Francis Group.

Comiskey, P.M., Yarin, A.L., and Attinger, D. (2019). Implications of two backward blood spatter models based on fluid dynamics for bloodstain pattern analysis. *Forensic Science International* 301: 299–305.

De Forest, P.R. (2018). *Physical aspects of blood traces as a tool in crime scene investigation. CAC News*. 3rd Quarter, 30–33.

De Forest, P.R., Gaensslen, R.E., and Lee, H.C. (1983). *Forensic Science: An Introduction to Criminalistics*. McGraw-Hill.

Denison, D., Porter, A., Mills, M., and Schroter, R.C. (2011). Forensic implications of respiratory derived blood spatter distributions. *Forensic Science International* 204: 144–155.

Donaldson, A.E., Walker, N.K., Lamont, I.L. et al. (2011). Characterising the dynamics of expirated bloodstain pattern formation using high-speed digital video imaging. *International Journal of Legal Medicine* 125: 757–762.

Kodet-Sherwin, L., Pizzola, P.A., and De Forest, P.R. (1988). *A critical assessment of the phenomenon of secondary bloodspatter. In: Presentation at the 40th Annual Meeting of American Academy of Forensic Science, Abstract #B58, Philadelphia.*

Laber, T.L., Epstein, B.P., and Taylor, M.C. (December 2007). *High speed video*. Midwest Forensic Center (Ames; IA) www.mfrc.ameslab.govhttps://alvideo.ameslab.gov/archive/bpa-videos/ (accessed 23 December 2020).MFRC: No. 06-s-02.

Liu, Y., Attinger, D., and De Brabanter, K. (2020). Automatic classification of bloodstain patterns caused by gunshot and blunt impact at various distances. *Journal of Forensic Sciences* 65: 729–743.

MacDonell, H.L. and Bialousz, L.F. (1971). *Flight Characteristics and Stain Patterns of Human Blood*. U.S. Department of Justice, Law Enforcement Assistance Administration.

Marin, N., Buszka, J., and Pizzola, P.A. (2010). *Scientific Analysis of Bloodstain Patterns: Laboratory Workbook*. Special Investigations Unit (SIU), NYC Office of Chief Medical Examiner, Department of Forensic Pathology, Forensic Sciences Training Program. National Institute of Justice (NIJ) Grant 2009-DN_BX-K205.

Neitzel, G.P. and Smith, M. (2017). *The Fluid Dynamics of Droplet Impact on Inclined Surfaces with Application to Forensic Blood Spatter Analysis*. Department of Justice (NCJRS). Award # 2013-DN-BX-K003, Document # 251439, 4.

Radford, G.E., Taylor, M.C., Kieser, J.A. et al. (2015). Simulating backspatter of blood from cranial gunshot wounds using pig models. *International Journal of Legal Medicine* https://doi.org/10.1007/s00414-015-1219-x.

Robbins, K.S. (1996). Suggested IABPA terminology list. *IABPA Newsletter* 12 (4): 15–17.

Scientific Working Group on Bloodstain Pattern Analysis. (12–14 November 2002). *Minutes of Fall Meeting*. Quantico, VA: World Wide Web. http://www.swgstain.org/documents/Fall2002Minutes.pdf (accessed 28 January 2015).

Scientific Working Group on Bloodstain Pattern Analysis. (April 2009). *Recommended terminology*. World Wide Web. http://www.swgstain.org/documents/SWGSTAIN%20Terminology.pdf (accessed 19 May 2015).

Taylor, M.C., Laber, T.L., Kish, P.E. et al. (2016a). The reliability of pattern classification in bloodstain pattern analysis, part 1: bloodstain patterns on rigid non-absorbent surfaces. *Journal of Forensic Sciences* 61: 922–927.

Taylor, M.C., Laber, T.L., Kish, P.E. et al. (2016b). The reliability of pattern classification in bloodstain pattern analysis-part 2: bloodstain patterns on fabric surfaces. *Journal of Forensic Sciences* 61: 1461–1466.

Williams, E., Graham, E.S., Jermy, M.C. et al. (2019). The dynamics of blood drop release from swinging objects in the creation of cast-off bloodstain patterns. *Journal of Forensic Sciences* 64: 413–421.

CHAPTER 6

Blood Droplet Dynamics and Deposit Formation

An understanding of how droplets form and interact with surfaces and the basic physics underpinning it is of value to both the advanced student and practitioner of blood trace deposit interpretation. In this section, we will look at this from a vectorial perspective as well as a pictorial one.

6.1 Blood Droplet Motion and Velocity Vectors

Before drop dynamics is discussed in detail, a review of vector basics will be helpful with respect to trajectories, although a complete analysis is outside the scope of this book. We strongly encourage readers to review their college physics textbooks for a fuller explanation of vector calculations. Recall that vectors are composed of both direction and magnitude. Within a vertical plane, vectors can be resolved into two mutually perpendicular components. Velocity is a vector. When analyzing velocity, it is useful to resolve it into its two orthogonal vectors with respect to the Earth's gravitational field: the horizontal velocity (V_x) which is independent of gravity and the vertical one (V_y) on which the force of the acceleration of gravity acts. Using fundamental trigonometric functions, if two of the variables are known, the third can be calculated.

In a discussion of blood trace deposits, velocity vector analysis is initially concerned with the region very close to the impact point. The velocity vector of a blood droplet impacting a surface at some angle is represented by V_r. It should be understood that although the hypothetical droplet is moving in three-dimensional space, we are limiting this discussion to a trajectory in a single vertical plane, and thus our analysis of the two components of velocity is restricted to two dimensions.

Blood droplets falling in air are accelerated by gravity up to a point where the viscosity of the air has an appreciable effect and reduces the rate of acceleration. This resistant force from the air's viscosity, called drag, increases with the droplet's velocity, until the drag force balances the force due to gravity. At this point, terminal velocity

Blood Traces: Interpretation of Deposition and Distribution, First Edition. Peter R. De Forest, Peter A. Pizzola, and Brooke W. Kammrath.

is reached which is the maximum velocity attainable by an object as it falls through a fluid, or in this case air. Terminal velocity is dependent on the cross-sectional density of the falling object, thus larger droplets would potentially have a greater terminal velocity than smaller ones.

6.2 Angle of Impact

Where it is necessary to determine the region (or area) of origin for a configuration of droplet deposits, we need to select and analyze a carefully selected number of suitable blood droplet deposits. For each deposit, we must consider two angles with respect to the *impact plane* (wall, floor, ceiling, etc.) shown in Figure 6.1. This impact plane is the plane defined by the surface where an airborne droplet impacts and deposits an elliptical deposit. One of these two angles is an angle measured within the impact plane, known as the azimuth angle (a term commonly used in astronomy). The other is the elevation or out-of-plane angle, tangent to the flight path (or trajectory), the droplet makes at impact. The plane containing the trajectory of the airborne droplet is always a vertical plane (in the Earth's gravitational field) that has intersected the impact plane along a line passing through the long axis of the elliptical deposit. In other words, the line defined by the intersection of these two planes is parallel to an extension of the major diameter of the elliptical deposit. In fact, it is the extension of the major diameter.

Determination of the angle of the tangent to the trajectory with respect to the impact plane is one component of ascertaining the region of origin of the droplet responsible for the deposit. It is also necessary to determine the direction the droplet was traveling as defined by the projection of the trajectory onto the impact plane.

For determining the angle with respect to the plane of impact, two measurements are made on each deposit. These are the length and width of the elliptically shaped deposit. The length of the ellipse measured along its major axis is commonly referred to as the major diameter (D), whereas the width is referred to as the minor diameter (d). Using a ratio of these measurements (d/D), a useful approximation of the tangent of the trajectory of the droplet at the moment of impact to the surface bearing the blood trace can be calculated from a trigonometric function (either inverse sine or inverse cosine). These functions are appropriate as approximations because the projection of a sphere onto a planar surface will yield various elliptical shapes depending on the projection angle (Figure 6.2). The calculation yields an angle either relative to the impact plane (phi, ϕ) or relative to its *surface normal* (theta, θ) (Figure 6.1). Phi and theta are sometimes referred to as the angle of impact and the angle of incidence, respectively. The surface normal is an imaginary line erected perpendicular the plane of the surface. Measuring angles relative to the surface normal is the convention used in optics due to the fact that measuring angles from any surface other than a flat plane can be exceedingly difficult. With respect to bloodstain pattern studies, it has the additional advantage that there is a parallelism between the magnitude of the angles computed this way and the verbal descriptions such as "slight angle" or "highly angular," etc. However, the convention used by most investigators in the bloodstain pattern field is to measure the angle relative to the plane containing the deposit rather than the normal to it. For the above reasons, we advocate the use of

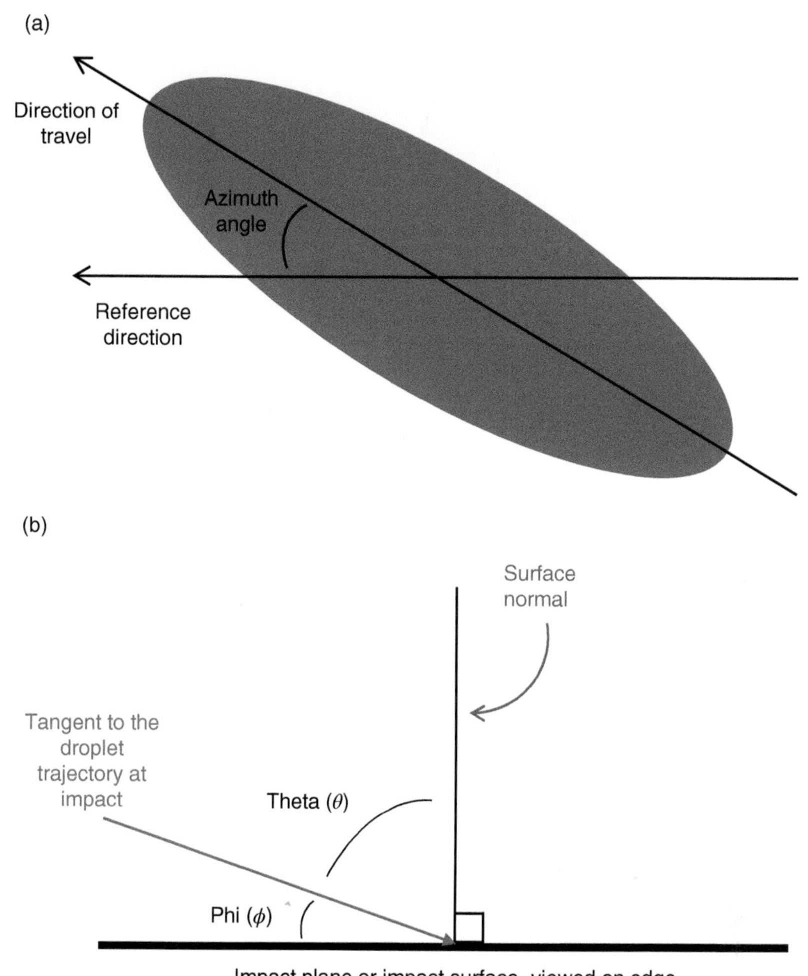

FIGURE 6.1 (a) Representation of an elliptical blood deposit in the impact plane. The reference direction refers to one of the cardinal directions (e.g., north, south, east, or west) or to a landmark at the scene (e.g., a door or wall). The line labeled "direction of travel" is the line depicting the intersection of two planes: the plane containing the droplet trajectory and the impact plane. (b) Representation of the plane containing the droplet trajectory, as viewed from the edge of the impact plane.

measuring angles relative to the surface normal (θ), which has been used in prior publications (De Forest et al. 1983).

In any case, the approximation using the trigonometric function gives useful results over a fairly broad range of angles impact, but it must be borne in mind that it is a useful but gross approximation. The angle for the tangent to the trajectory at impact measured relative to the impact plane is obtained by calculating the inverse sine (arcsine or $\sin^{-1}$) of the ratio of the minor diameter of the deposit to its major diameter. As noted above, this can be expressed as $\sin^{-1}(d/D)$. The angle relative to the surface normal is the inverse cosine of the same ratio, expressed as $\cos^{-1}(d/D)$.

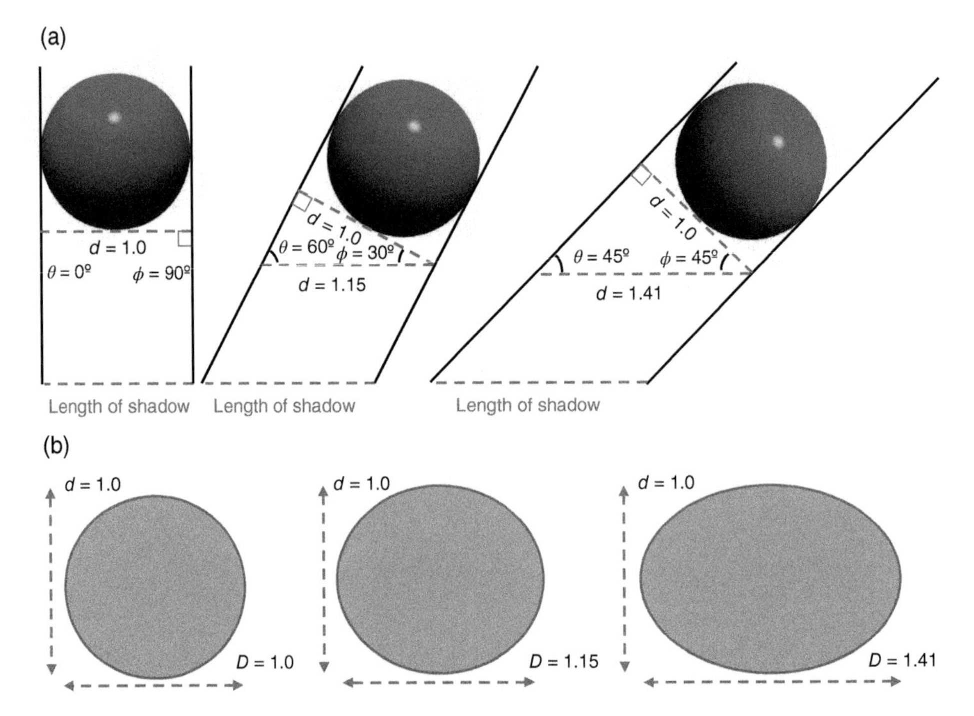

FIGURE 6.2 (a) Shadow of a sphere projected onto a planar surface at 0°, 30°, and 45° from a light source at infinity; (b) resulting elliptical shadows as viewed from a position normal to the surface.

6.3 Blood Droplet Trajectory and Resulting Impact Geometry

Where air resistance can be ignored, blood droplets in flight travel in a parabolic trajectory due to gravitational forces acting on its vertical velocity component (Figure 6.3). For short distances, this may approximate a straight line. Figure 6.4 is a high-speed multiple image stroboscopic photograph of a single drop of blood projected horizontally. It impacts the horizontal surface at some velocity (resultant vector) and at some angle. In Figure 6.5, we see a typical individual deposit arising from the impact of such a single droplet. The primary portion of the deposit (the elliptical area) is a function of the angle of incidence, while the distorted tip points forward in the direction of travel before and during the impact.

If we allow drops of blood to fall and impact stationary inclined surfaces, we can observe deposits become more and more elliptical as the angle increases (Figure 6.6). The degree of ellipticity of the primary portion of the deposit can be used to approximate the angle of incidence, if a series of controlled experiments are done on a given substrate; different substrates will often yield different results. In the work illustrated in Figures 6.6, 6.12, 6.13, 6.16–6.21, and 6.23–6.25, all experiments used a fixed volume of approximately 26 μl that was produced from a disposable glass Pasteur pipette. It should be clear that if a different type of pipette was utilized, such as volumetric pipette, a different size drop would be produced. This concept is discussed in detail in Chapter 3.

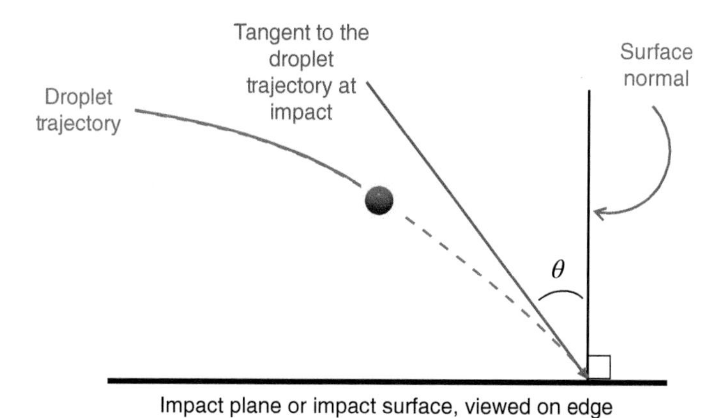

FIGURE 6.3 Representation of the parabolic trajectory of a blood droplet, ignoring air resistance. Also shown is the tangent to the trajectory at the point of impact.

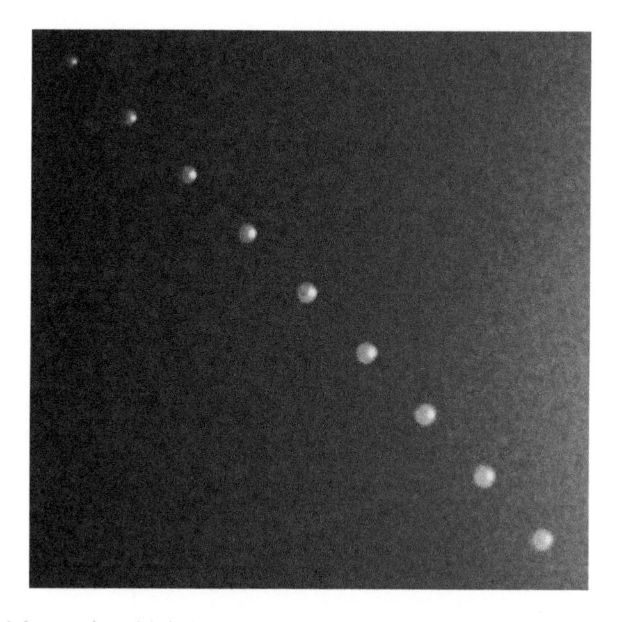

FIGURE 6.4 A high-speed multiple image stroboscopic photograph of a single drop of blood projected horizontally. The parabolic nature of the trajectory is barely discernable (pulse rate of strobe is approximately 15 000 fpm). *Source: Pizzola et al. (1993).*

The blood trace deposits in Figure 6.6, with corresponding measurements plotted in Figure 6.7, demonstrate the limitations of the angle of impact calculation and shadow projection model. One shortcoming of this model is most evident at the small angles of impact, as seen in the blood deposits at 0° and 10° which show little differences in their ellipticity. In addition to the inexactitude of the trigonometric model, there is also measurement error in determining the d/D ratio. Other shortcomings are the failure of the model to account for the properties of the substrate (e.g., smoothness and post-impact spreading due to absorptivity) and exclusion of the blood outside of the ellipse (e.g., forward projected or sometimes-called wave cast-off traces). The latter further complicates measurements due to ambiguity in determining the distal boundary of the ellipse.

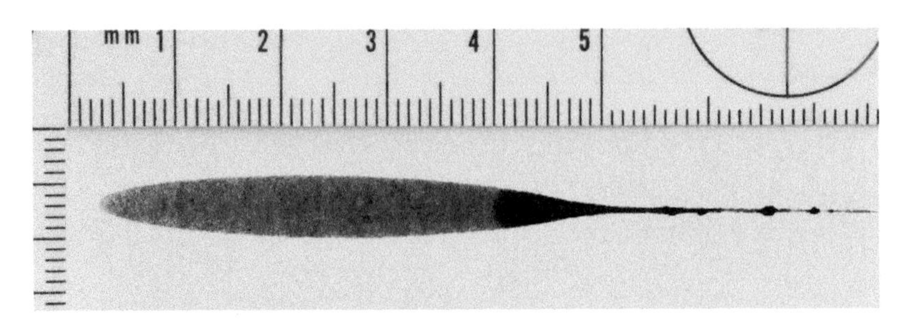

FIGURE 6.5 Deposit from blood droplet impact at 80° angle of incidence.

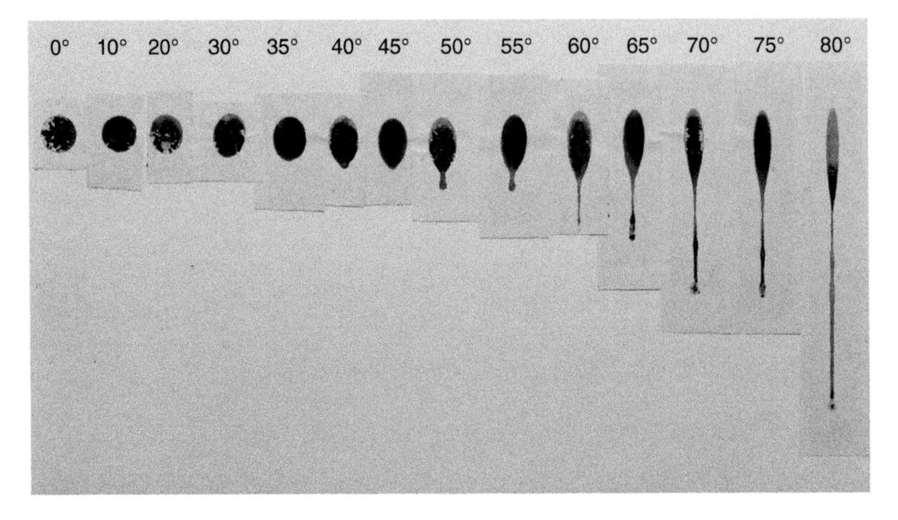

FIGURE 6.6 Blood trace deposits created by droplets falling onto an inclined plane at angles from 0° to 80° with respect to the normal.

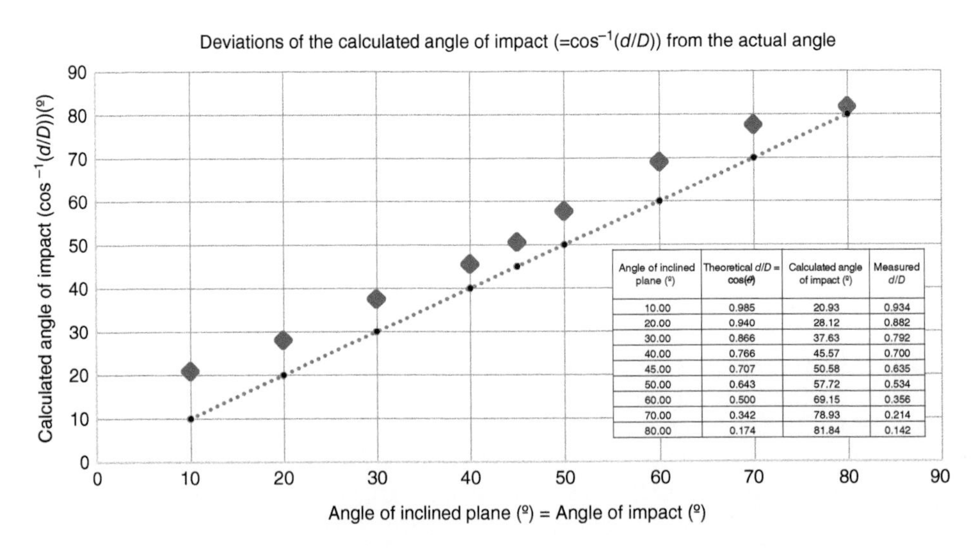

Angle of inclined plane (º)	Theoretical d/D = cos(θ)	Calculated angle of impact (º)	Measured d/D
10.00	0.985	20.93	0.934
20.00	0.940	28.12	0.882
30.00	0.866	37.63	0.792
40.00	0.766	45.57	0.700
45.00	0.707	50.58	0.635
50.00	0.643	57.72	0.534
60.00	0.500	69.15	0.356
70.00	0.342	78.93	0.214
80.00	0.174	81.84	0.142

FIGURE 6.7 A graph showing the angle of impact as the independent variable and the calculated angle of impact of blood droplets (from an average of five replicate measurements of d/D). The measurements were made on blood dropped vertically onto a stationary inclined plane, with the angle measured from the normal.

6.4 Region of Convergence and Region of Origin

The area in two dimensions, or region in three dimensions, where imaginary lines drawn through the long-axis intersect, is commonly referred to as the area or region of convergence as depicted in Figure 6.8. This figure shows the "point" of convergence in two dimensions; however, this oversimplifies the issue. In practice, due to numerous uncertainties, one cannot expect convergence at a single point from the back projections of the major axes of several elongated blood trace deposits. Instead, there is an area or region of convergence, as shown in Figure 6.9.

To back-project into three dimensions, a protractor or other similar device is necessary to establish the elevation angle. Manually placed strings (MacDonell and Bialousz 1971) have been used for this purpose, as shown in Figure 6.10. More recently, this process has been modernized using laser scanning (Hakim and Liscio 2015; Dubyk and Liscio 2016). In the manual method, the protractor's vertex is aligned with the blood trace locus where the drop initially impacted the surface. Strings are then attached to the position where the drop contacted the surface and the string pulled taut and moved until it was aligned with the angle of incidence (or angle of impact depending on practice) and superimposed over the lines representing the long axis. The stringing method can be augmented by use of a laser to assist with the placement of the string, as shown in Figure 6.10b. When using a laser positioned at the appropriate out-of-plane impact angle, a mark can be made on the adjacent plane (such as the floor in Figure 6.10b) and the string subsequently attached to it as well as the other end to the individual droplet trace. The region or area of origin is where the strings intersect in three-dimensional space (Figure 6.9).

The use of strings, no matter which way it is done, is a very time-consuming process with some issues. Early computerized systems ("Droplets" and "Trajectories") were developed by the Royal Canadian Mounted Police (RCMP) and the Departments of Physics and Mathematics of Carleton University (Carter and Podworny 1989, Carter 1991). "Droplets" was designed for teaching purposes accounting for both drag and

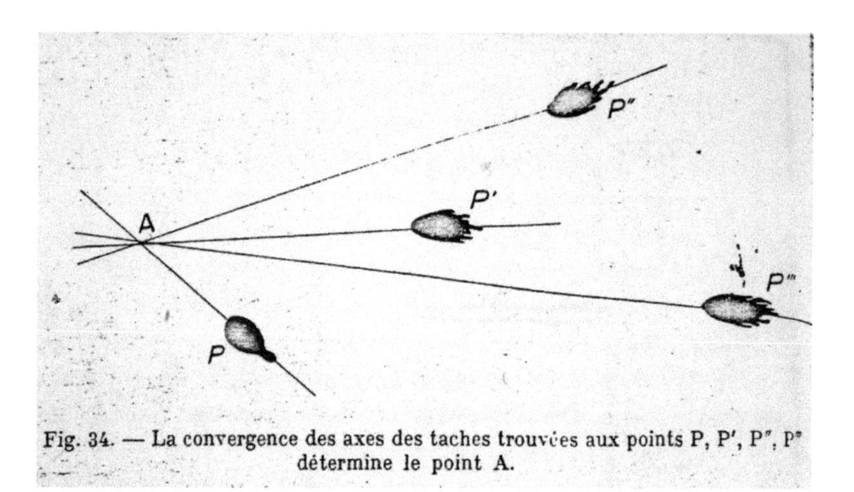

FIGURE 6.8 Depiction of the area of convergence (original caption in French has been retained in the figure). *Source: Balthazard et al. (1939). © Paris: J.-B. Baillière. Public Domain.*

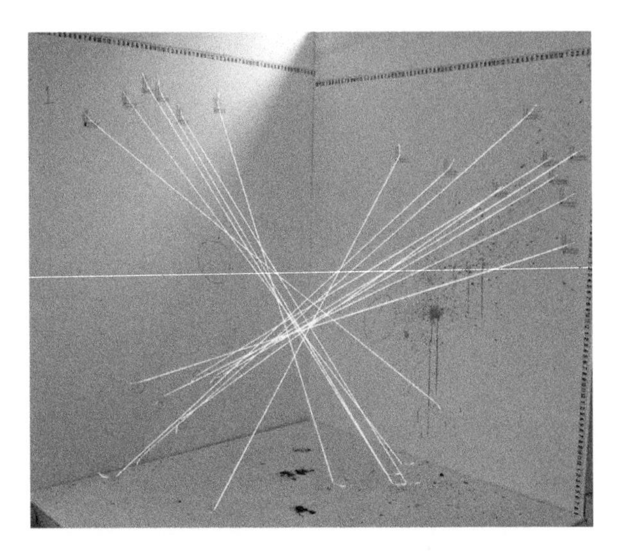

FIGURE 6.9 A demonstration of the visualization of the region of origin, made by using the stringing method. *Source: Photograph courtesy of Norman Marin and Jeffrey Buszka.*

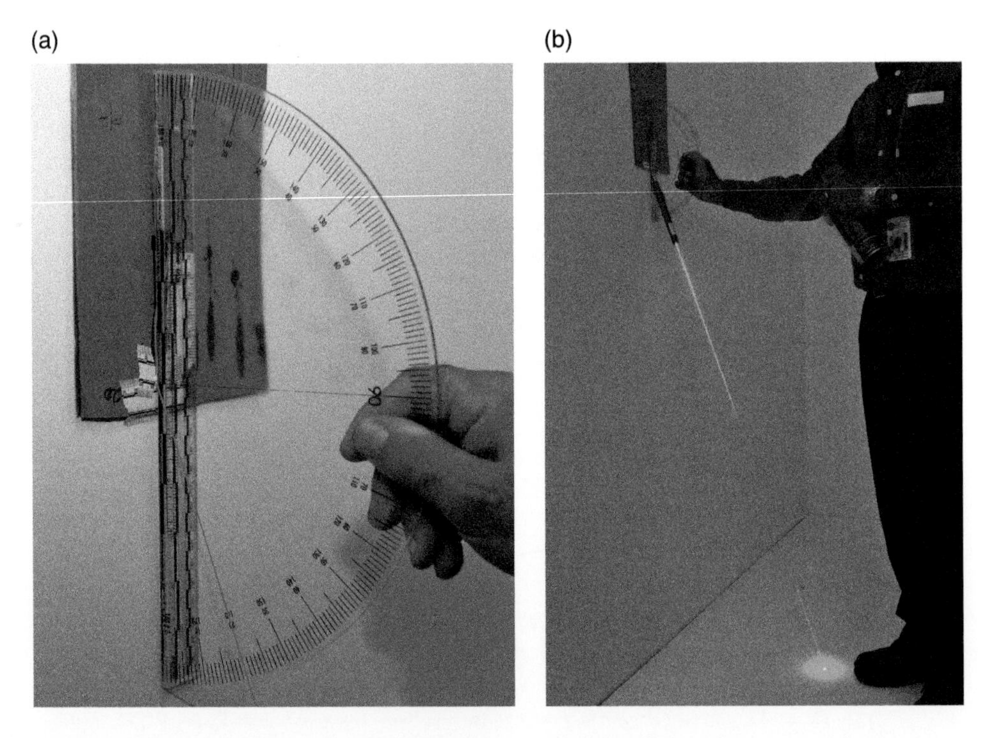

FIGURE 6.10 Manual determination of elevation angle (out-of-plane impact angle) using a protractor (a) and laser (b). *Source: Photographs courtesy Norman Marin.*

gravity. The program "Trajectories" was designed for the reconstruction of droplet trajectories. Subsequently, Carter introduced another program referred to as "Directional Analysis" to act as virtual strings (2001). A detailed explanation of "directional analysis" and the utilization of computer programs "BackTrack Suite" for estimating the area of origin were described by Carter (James et al. 2005).

It must be recognized that droplet trajectories are influenced by both gravity and drag (air resistance); the effects of both contribute to nonlinear paths. The use of strings or algorithms for determining the region of origin of a group of traces does not take nonlinear trajectories into account. Linear paths may be appropriate if the blood droplets possess relatively straight trajectories at the time of impact. This is only applicable where the droplets are generated near a substrate and travel at relatively high velocity. Attinger et al. (2013, p 383) demonstrated the difference between trajectories reconstructed with linear paths, those affected only by gravity and those where gravity and drag were considered. Under the conditions those authors utilized, the height of the area of origin was dramatically overestimated for the linear trajectories versus when gravity alone or when both drag and gravity are considered.

Multiple, very different, trajectories can result in blood droplet impacts at the same location on a substrate; some of the resulting traces may have very similar shapes (Balthazard et al. 1939; MacDonell and Bialousz 1971, pp. 13, 53–54; James et al. 2005, p 225; Bevel and Gardner 2008, pp 136–137). For example, a drop of blood may fall straight down onto the floor yielding a circular trace. In another instance, a droplet may have been projected upward in a nearly vertical path and turned in a parabolic arc and fallen far enough that its path is essentially straight down at impact. It is quite likely that an examiner could not determine which trace originated from which trajectory. This distinction should be clear from the context, but as indicated by Principle 8, extra care is necessary when attempting an interpretation from a limited number of blood trace deposits. For isolated blood deposits, the absence of additional contextual traces can lead to misinterpretations.

Comiskey et al. (2016, p, 2) amply, but somewhat illustrated misconceptions and potential errors involving droplet trajectories, if one mistakenly assumes that blood droplet trajectories follow a linear path. This concept is far from new in the criminalistics discipline, as it has been previously presented in 1939 by Balthazard et al., MacDonell in the 1970s, and several others. In fact, Comiskey et al. (2016) were provided with the material used in one of their figures by MacDonell to illustrate the differences between trajectories where both drag and gravity are taken into account, those where gravity is considered but not drag, and finally where both gravity and drag are ignored resulting in simplified linear trajectories. Despite these concepts being understood by many criminalists, there have been practitioners who have not recognized or applied these concepts and treated blood droplets as if they behaved in the same manner as bullets fired over short distances. The Comiskey et al. paper explains the issue in depth (2016, p.2). The more distant an impact site is located from the receiving surfaces, the more likely errors in region of origin interpretations will occur since the droplets will have more opportunity to be influenced by drag and gravity. Experienced scientists should appreciate that prior to even attempting to estimate the region of origin from a radial spatter configuration, additional observations regarding projection distance are necessary. For example, if there is physical evidence of significant blood traces, on the floor, 5 feet from the vertical area, it should be clear that the droplets that resulted in blood traces on the wall likely followed markedly parabolic trajectories. Thus, it would be naïve to use strings or other methods including computer programs that assume linear trajectories to back-project into three-dimensional space in estimating the region of origin unless the distances are short range (i.e., within a few feet). These authors have been involved in casework where supposed experts have used strings from circular traces back-projecting clear across a room assuming horizontal trajectories when it is not reasonable to make such assumptions – in other words, circular traces could also have been readily produced by droplets impacting the substrate orthogonally but from decidedly parabolic trajectories.

Some researchers have demonstrated the use of a laser-assisted computerized system that utilized ballistic trajectories, incorporating the influence of drag and gravity, for estimating the region of origin for blood trace deposits (Buck et al. 2011). However, a significant weakness of this approach was the built-in assumption that the trace's diameter (corresponding to its width or minor diameter) corresponded directly to the droplet's diameter. The basis for concern was stated at that time (Pizzola et al. 2012) and more recently (Attinger et al. 2019). If the aforementioned computerized system was modified to incorporate an accurate method for deducing the drop's original diameter, as suggested, for example, by Hulse-Smith, then it would have more value than those systems developed to date. Because of the inherent imprecision in the overall process irrespective of the constituent methods utilized, the region of origin of projected droplets cannot be known exactly. This is due not only to the limitations in reconstructing trajectories but also the potential repositioning of the source during the event and post-event activities. The victim is not always perfectly stationary nor is the wound site or sites a point source. Thus, a common problem of reconstruction procedures, including computerized models, is the expectations of a level of exactitude which exceed the capabilities of the science.

6.5 Equivalence of Relativistic Motion

In order to experimentally study the production of blood trace deposits from a vectoral point of view, it is necessary to maintain careful control of the horizontal component (V_x) and the vertical component (V_y) of velocity (Pizzola et al. 1986a). Further, it is critical that only single droplets are produced for impaction. This is not easily accomplished. A special apparatus is required to project single droplets of blood of known preselected volume horizontally. Alternatively, to accomplish this without utilizing a special apparatus, one can use a concept from basic physics involving Newtonian relativity. If a person runs with a pipet and allows a drop of blood to be released, it strikes the ground at some velocity and angle (measured from the tangent to the trajectory at impact). In a relativistic sense, if you suspend this person at his original position above the ground and allow a drop of blood to be released from the pipet while the "ground" is moving in a direction opposite to his or her original motion, an equivalent situation is obtained. This latter situation can be created and used for experimental purposes with a moving belt device (Figure 6.11). The correlation of these situations with a third situation (where blood falls vertically onto stationary inclined surfaces) can be studied with this device, as this device can be controlled to vary the horizontal component of velocity.

The results from the experimentation with the belt device in Figure 6.11 illustrated the trend of increasing ellipticity in the primary portion of each deposit as the belt

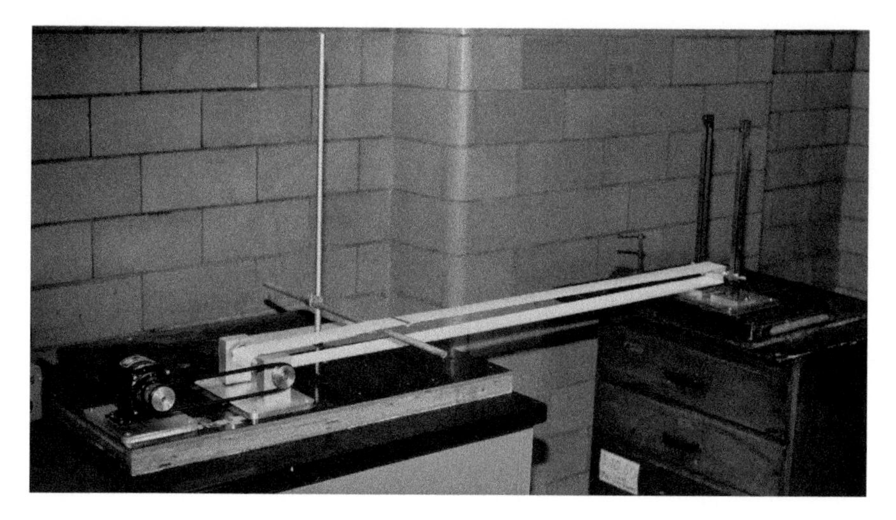

FIGURE 6.11 A belt device with the drive spool in the foreground on the left and the driven spool on the back right. The rod in the middle was added to dampen any vibrations in the belt, thus allowing for a stable impact site. The drive spool is connected to the drive motor with a rubber belt, which in turn is controlled via a variable transformer. The drive spool has a solid core, while the driven spool has a hollow core. Both spools are made of polished aluminum. The spools run ±5 μm concentric to the bearing holes. The device is arranged for operation utilizing a belt length of approximately 7 feet. The belt velocity is determined by counting the number of revolutions (at least 10) during a time span of 10 s or more. The length of the belt is measured and the velocity is calculated.

velocity increased. As seen in Figure 6.12, the distorted tip of each deposit is pointing in the direction opposite of the moving belt.

The vertical component of velocity was controlled via the dropping height. The height selected for this arrangement was 14 cm and was kept constant. This height was chosen because the velocity of the falling drop is not significantly affected by air resistance and can be calculated from the following rectilinear equation:

$$v^2 = v_o^2 + 2ay$$

where "a" is the gravitational acceleration, "v" is velocity, "v_o" is the initial velocity, and "y" is the height of fall.

If the drop is released in a gentle manner, as previously reported, it will generally have a velocity slightly less than that predicted by the equation for rectilinear motion (about 2% less than the velocity in a vacuum for this dropping height) because of air resistance (Pizzola et al. 1986, pp. 53–57). This was determined by high-speed stroboscopic photography (similar, but under different conditions, to that depicted in Figure 6.13). In this photograph, one can observe the multiple images of a single drop of blood and its acceleration due to gravity.

The "effective" angle of incidence (EAI) can be then calculated from the vertical component of the velocity (due to gravitational acceleration) and the horizontal component of velocity (the belt velocity) (Figure 6.14).

A graphical comparison of the two previously discussed situations (i.e., where a drop is falling vertically onto a stationary inclined surface and where it is falling vertically onto a horizontally moving belt), with d/D as the y-axis and the angle of incidence on the x-axis, the two situations are shown to be correlated (Figure 6.15). In fact,

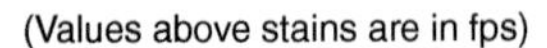

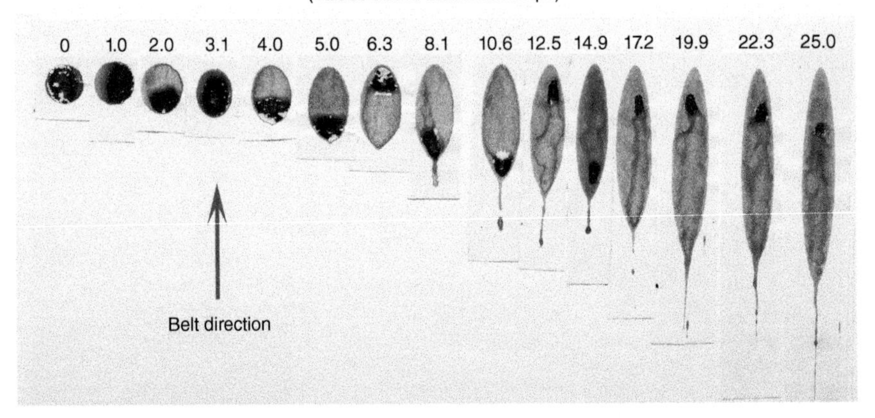

FIGURE 6.12 Resulting blood trace deposits produced using the moving impact site, showing the increased ellipticity of each deposit with increasing speed at a given dropping height for a vertically falling droplet onto the horizontally oriented belt.

FIGURE 6.13 High speed stroboscopic photograph (strobe pulse rate is 6000 pulses per minute) with multiple images of a single drop of blood illustrating its acceleration due to gravity. *Source: Photograph courtesy of Norman Marin.*

a third situation, where a drop of blood is projected horizontally and impacts a fixed horizontal object, is also equivalent to the other two, as shown in Figure 6.15. This third situation was confirmed to be correlated with the first two situations in 1993 by Pizzola et al. when a crude wind tunnel was fabricated (by Stephen Kwechin), positioned horizontally to project blood droplets horizontally. The droplet trajectories were photographed and in some experiments videotaped, with the angles of incidence upon impact determined with the corresponding d/D ratios.

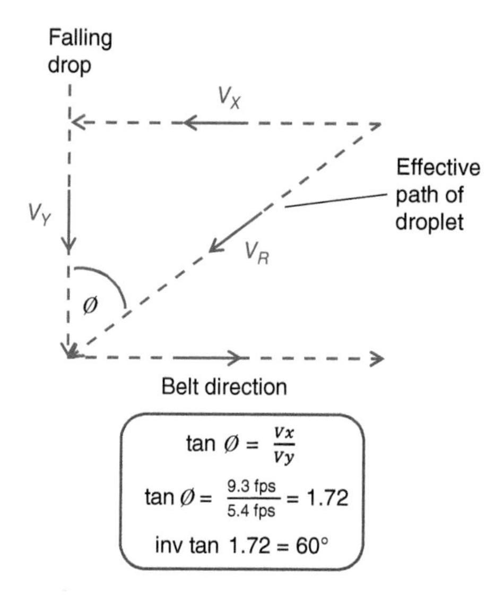

FIGURE 6.14 Diagram and resulting calculation of the EAI (Ø) using the trigonometric inverse tangent (*inv tan*) of the ratio of the vertical to the horizontal components of velocity (V_Y and V_X, respectively). Note that the resultant vector (V_R) of the blood droplet points away from the belt direction.

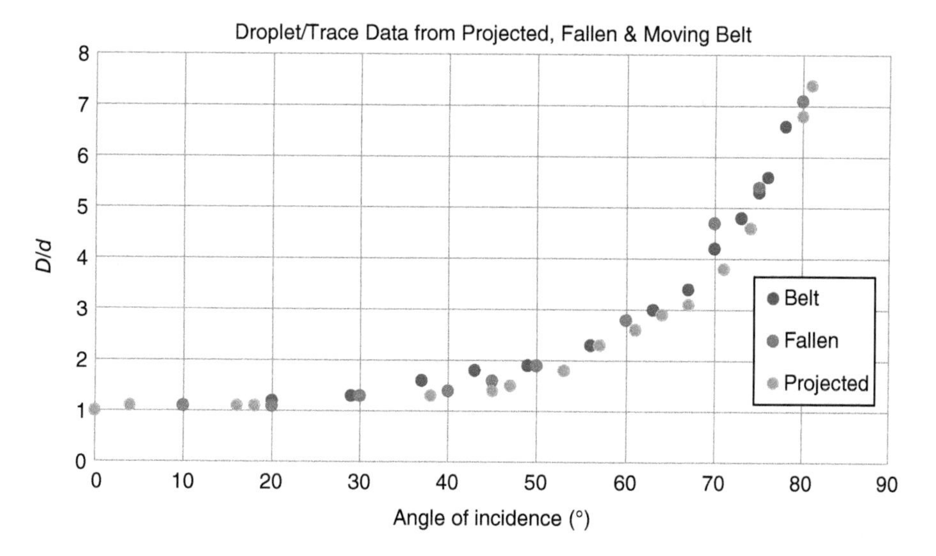

FIGURE 6.15 Plot of data from blood droplets that have been produced in three different ways: (1) falling vertically onto a stationary inclined surface, (2) falling vertically onto a horizontally moving belt, and (3) projected horizontally and impacting a fixed horizontal object. This allows a visualization of their equivalence. The small variations noted can be explained by experimental error.

It is not coincidental that the deposits produced with the fixed inclined surface (Figure 6.6) were considerably smaller than those generated from the moving belt device (Figure 6.12). Yet, graphically they appear correlated. This is because the ratios of the components of velocity were equivalent in each situation, although the magnitudes of the components were not. In order to obtain the exact same vector, not only must the ratio of the components be equal, but also their magnitude. The reader should recall that vectors are comprised of two attributes: both magnitude and direction. An example of this is depicted in Figure 6.16.

Figure 6.17 demonstrates the dramatic effect on deposit morphology by changing only one component of velocity.

Although it has not been discussed thus far, a necessary aspect of using the deposit ellipticity to ascertain the angle of incidence is that the ratio between the major diameter and minor diameter are relatively independent of the drop's impact velocity. Dr. Paul Kirk (1953) provided some evidence of this in the figures presented in his classic textbook *Crime Investigation.* When blood is dropped from different heights, but the angle of incidence is held constant, little change takes place between the *d*/*D* ratio. Only the

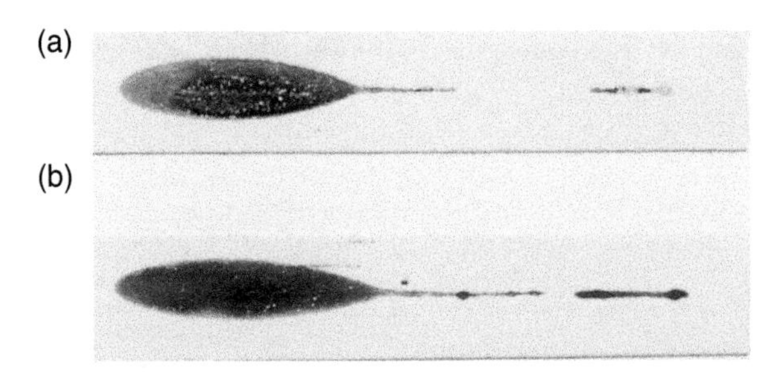

FIGURE 6.16 Traces produced by blood falling onto a fixed inclined surface (a) and generated from the moving belt device (b). They have similar shapes and sizes, but the vertical component in the two cases are dramatically different because of the difference in dropping heights. Although generated under different conditions, the droplets have equivalent resultant vectors of velocity (in both magnitude and direction).

FIGURE 6.17 Blood trace deposits made on a belt moving horizontally at a constant rate of 20 feet/second. The morphological differences of the traces are due to the vertical component of droplets velocity, having fallen from different heights. (a) Resulted from droplet falling 5–0.5 inches onto belt moving horizontally at 19.9 fps. (b) Resulted from droplet falling 72 inches onto belt moving horizontally at 20 fps.

size of the deposits should increase with increasing vertical velocity. And this is exactly what is depicted in the images in Figure 6.18. If this concept were not true, that that the d/D ratio did vary with dropping height, one could not utilize deposit ellipticity to approximate the angle of incidence. In practice, forensic scientists use tables of values, graphs, or a trigonometric function (cosine or sine, depending on whether the angle of impact, ϕ, or angle of incidence, θ, is measured). The independence of ellipticity from impact velocity can also be confirmed by a different method – utilizing the belt device.

The blood traces observed in Figure 6.19 were generated by varying the belt velocity as well as the dropping height being careful to maintain the ratios between the velocities in each instance. So, the EAI was kept constant, but the magnitude of the resultant vector was quite different in each individual experiment. Consequently, the expectations were met and each deposit is smaller than the previous one because the resultant vector is less, but the d/D is equivalent because the ratio of the components of velocity has remained the same.

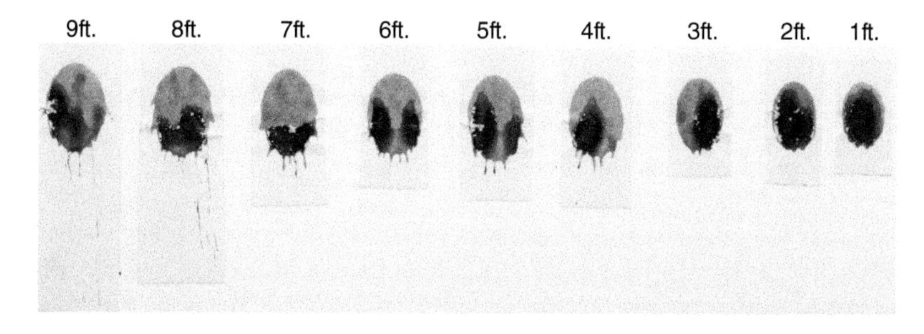

FIGURE 6.18 Photograph of blood traces resulting from droplets striking a stationary inclined surface ($\theta = 48°$) after falling from different heights (distance indicated above each trace). Observe how little the size differences of the ellipses are for droplets released from a distance of 5 feet or more, which is explained by the droplets approaching terminal velocity (see Section 6.1).

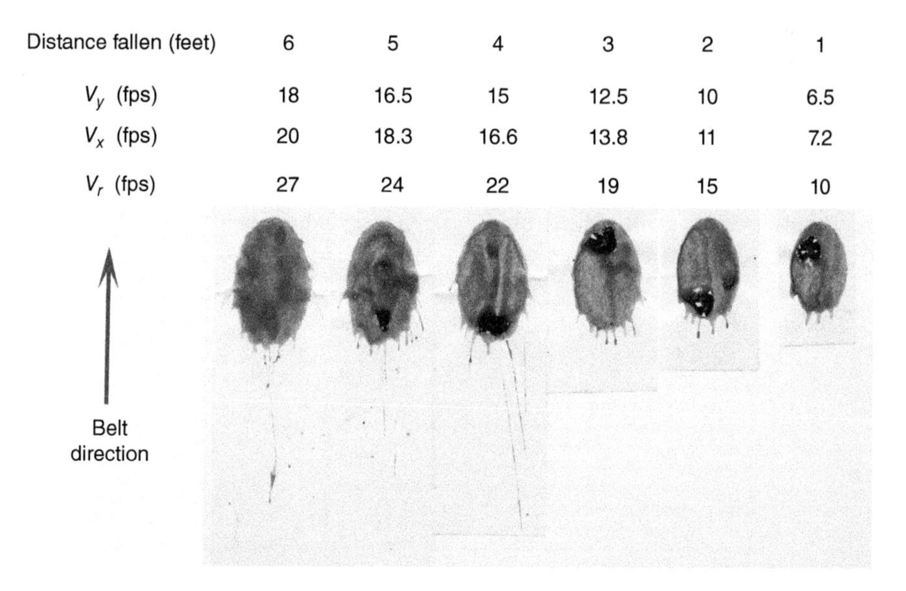

FIGURE 6.19 Blood droplet deposits made by varying both the belt velocity and the dropping height, in an effort to achieve a desired EAI (48°).

(a)

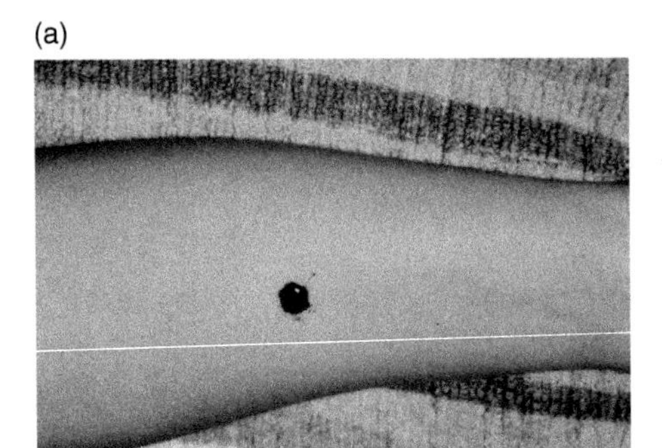

(b)

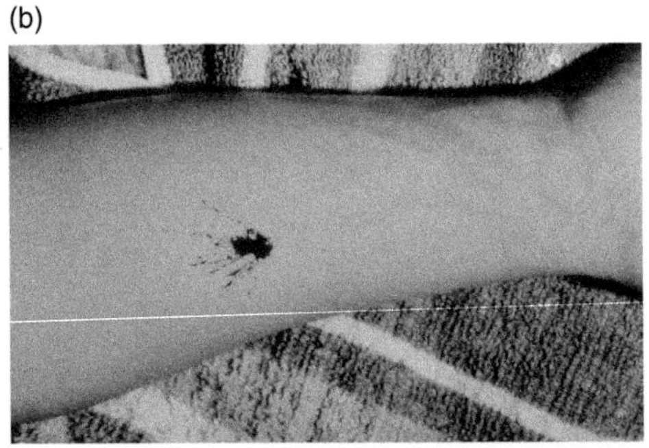

FIGURE 6.20 (a and b) The motion of the target may significantly affect the morphology of the blood trace deposit. The arm was stationary when impacted by the blood droplet that fell vertically resulting in the nearly circular trace (a). However, the arm was rapidly moving to the right when the blood droplet struck the surface resulting in the deposited trace (b).

Of considerable importance to actual interpretations at crime scenes is the possibility that the appearance of the blood deposits is a result of the motion of the substrate at the moment of impact, as illustrated by the above belt experiments. If the target surface is capable of its own motion, and one does not know with a reasonable degree of certainty that the target was motionless at the time of impact, it is risky to interpret the blood trace with regard to its directionality and angle of incidence. This factor has particular implications for traces found on human skin and vehicles (Figure 6.20). This fact is contrary to the claim by MacDonell and Panchou (1979) that blood traces found on human skin can be interpreted in the same manner as it is with walls and floors. A scientific investigator must be concerned also with inanimate objects at the scene that can be readily moved and acquire significant velocity, such as vehicles or chairs with wheels.

An extension to the previously discussed moving belt demonstrations, it is interesting to note the morphology of traces when the target surface is both moving and inclined. For one set of experiments, the belt device was arranged as shown in the photograph (Figure 6.21). The belt was oriented to move downward. In the Figure 6.21, one can see how successive droplet trace deposits change from an elongated ellipse to a circle and, finally, to a reversed elongated configuration as the belt velocity is increased in a downhill direction, while the dropping height is held constant.

6.6 Impact Mechanism and Blood Trace Deposit Formation

Diagrammatic and pictorial perspectives are useful to demonstrate how various blood trace deposits are formed from single airborne droplets (Figure 6.22). This has been described by Pizzola et al. (1986b). Specifically, the photographs in Figures 6.23 and 6.24 depict a drop falling vertically and impacting a stationary inclined surface (at 70° from the surface normal).

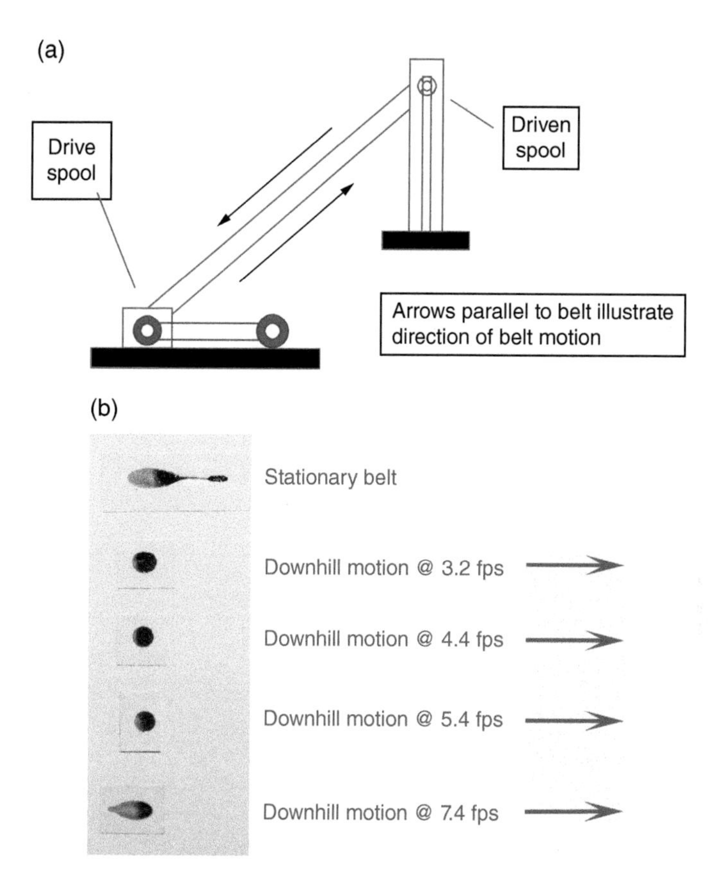

FIGURE 6.21 (a) Orientation of belt device for production of traces in (b); (b) droplet deposits produced when droplets released vertically onto inclined belt device under conditions specified alongside blood traces. Arrows indicated belt direction downward.

Figure 6.23 is comprised of a sequence of individual high-speed photographs of 26 µl drops approaching (Figure 6.23a) and then impacting an inclined surface at a velocity of approximately 10 fps. At this impact angle (70°), the distortion of the drop is limited to its lower area in contact with the impacted surface (Figure 6.23b). As the drop continues its travel (Figure 6.23c), it gradually collapses downward with respect to the target surface, accompanied by no observable change in the form of the upper hemisphere. The top of the upper hemisphere falls further, while the fluid displaced during the drop collapse is forced out radially, forming a rim at the circumference (Figure 6.23d). The surface tension opposes the lateral spreading of the drop. Shortly after the collapse, the center region is significantly depressed (Figure 6.23e). This depression coincides with that found in a drop that has impacted a surface zero degrees from normal. Balthazard et al., as discussed by Pizzola et al. in 1984 and 1986, were aware of this "involution" as were other researchers with other aqueous drops (Foote 1975). The latter author referred to it as a "depression" (1975, p. 400). Following the formation of this involution, the fluid forced to the rim retracts, coalesces, and progresses forward into a somewhat spheroidal form at the leading edge (Figure 6.23f). The lower portion of the spheroidal area adheres to the impact surface, while the upper

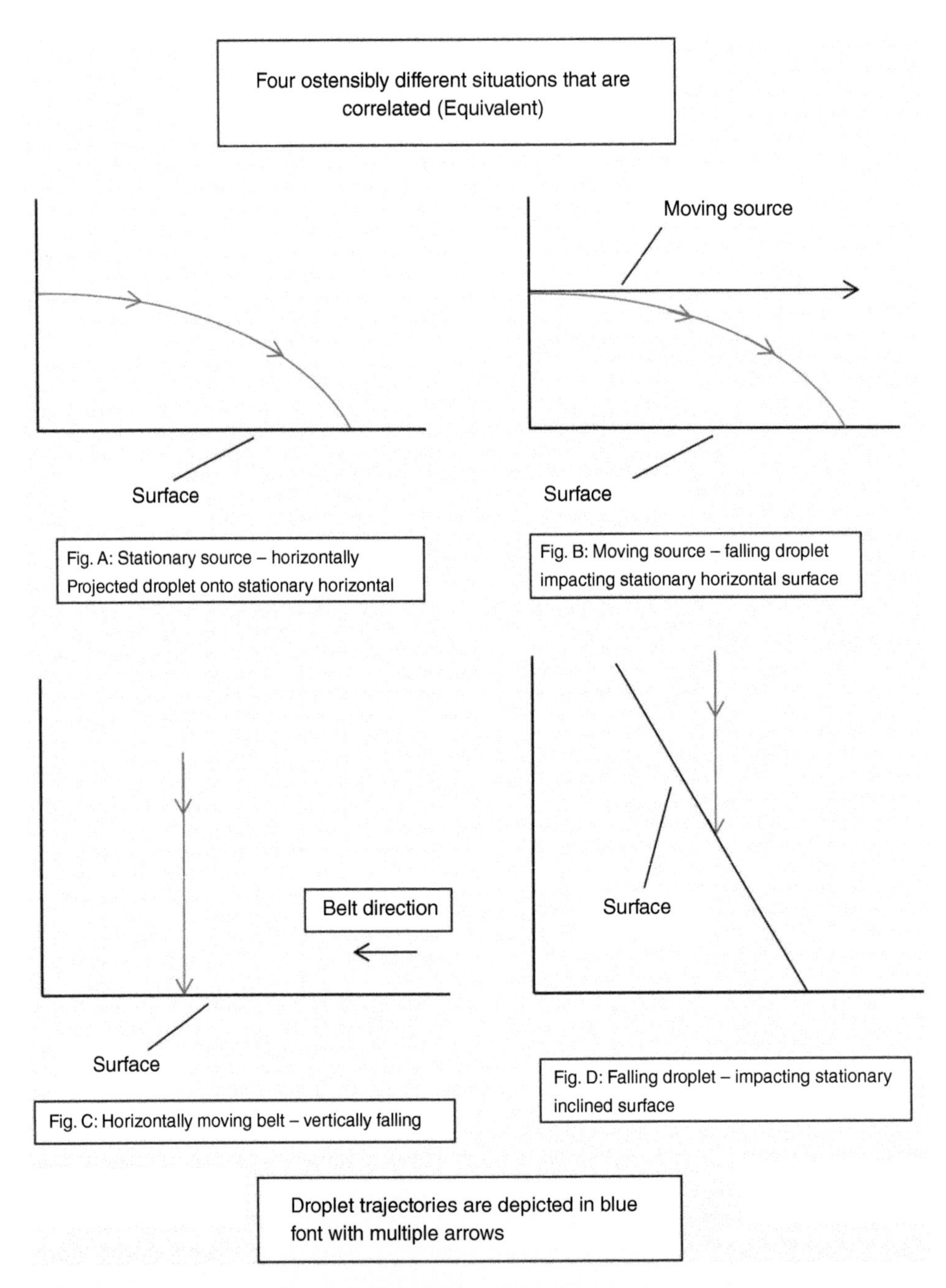

FIGURE 6.22 Diagrams of four primary situations that result in blood traces with equivalent components of velocity and angle of incidence or EAI.

portion grows or forms a droplet as it rises away from the impact surface. At the angle of incidence and dropping height employed in this experiment, the kinetic energy of the moving blood initially overcomes the surface tension and pulls away from the main body of blood to form a droplet which becomes spherical, drawing out a fine filament

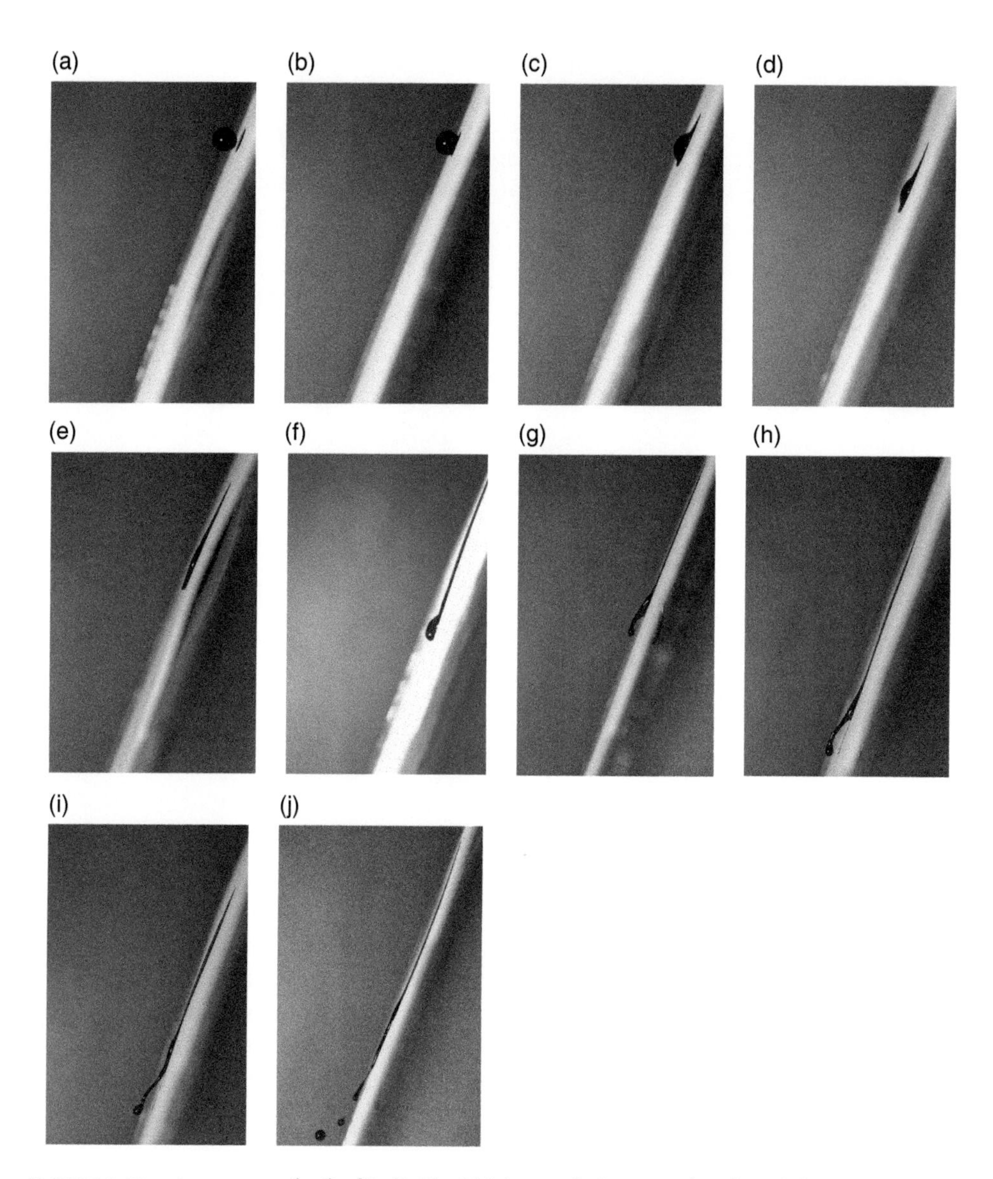

FIGURE 6.23 A sequence (a–j) of individual high-speed photographs of 26 μl drops approaching and then impacting an inclined surface at 70° from the surface normal at a velocity of approximately 10 fps. The actual target is a piece of paper taped securely to a glass base. *Source: Photographs courtesy John Perkins.*

in the process (Figure 6.23g–i). If the velocity is sufficiently high, the droplet formed can separate from the filament (Figure 6.23j). It is also interesting to observe the entire process as a video from a different perspective (Laber et al. 2007). The latter authors have a series of high-speed videos available online at the website cited here that provide valuable insights regarding various mechanisms of droplet production and trace formation (Figure 6.24).

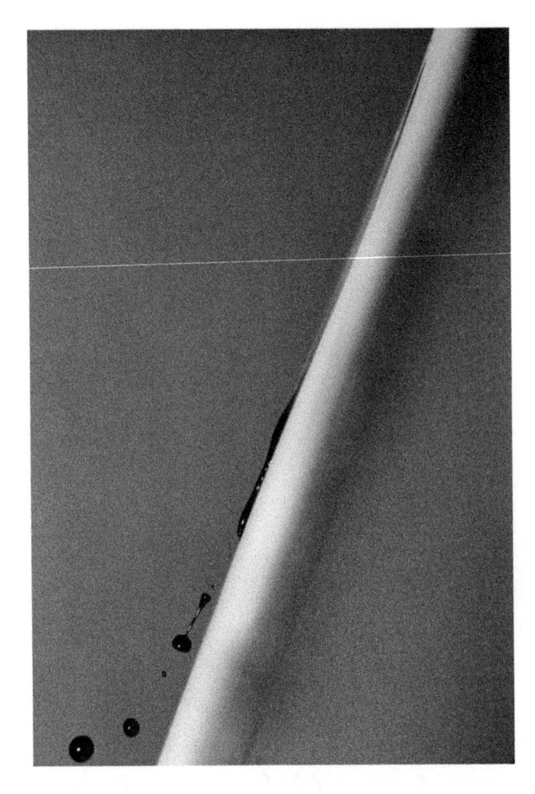

FIGURE 6.24 Some of the satellite droplets that are depicted in Figure 6.23 are clearly not spherical. One small droplet demonstrates a flattened side due to residual oscillations from impact. Flattened frontal areas of large spheroidal liquid droplets are commonly seen and described in the aqueous drop literature as a "cap cyclide". This is not what is seen in this image. Attached to the trailing edge of the flattened droplet is a short filament. Attached to the other side of the filament is another noticeably smaller droplet. This segment of the disrupted drop will make contact with the substrate within milliseconds. The resulting deposit configurations have shapes similar to these interesting droplet dynamics.

6.6.1 Impacts of Falling Droplets with Sessile Blood

Impacts of falling droplets with liquids as well as rigid surfaces involving splashes of fluids other than blood have been studied extensively, and a considerable body of knowledge has been developed. One of the most noteworthy researchers in this area was A.M. Worthington. Similar to C.V. Boys and Lord Rayleigh (John William Strutt), Worthington became well known before the turn of the twentieth century for his measurement of small-time intervals. His research included studies of solid spheres and liquid droplets impact liquids at rest (Worthington 1963). Little work has been done in this area utilizing blood. One of the most in-depth studies of blood was published in 1939 by Balthazard et al., but this publication did not address the formation of secondary spatter. The "crowning" effect noted by Worthington was also demonstrated by Balthazard et al. via high-speed photography. The dynamics of the impact of a falling drop of blood with a sessile drop were described by Kodet-Sherwin et al. (1988) and are reported here. The same photographic technique used by Pizzola et al. in 1983–1984 and in 1986 was used to produce

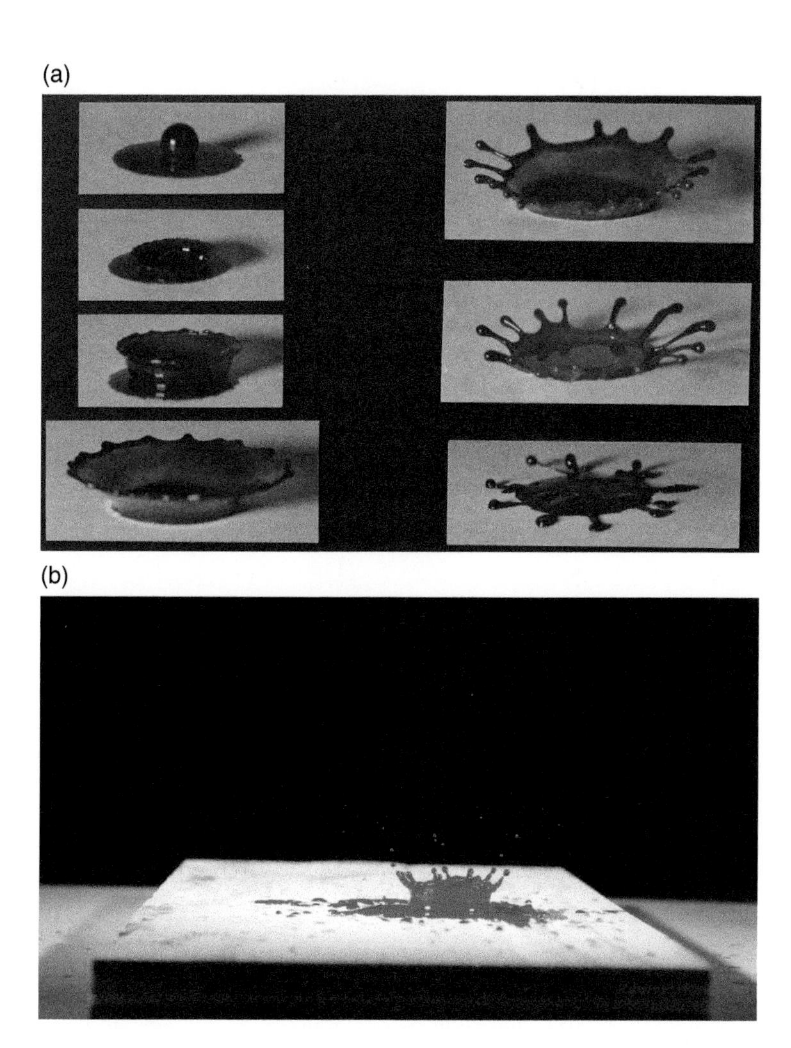

FIGURE 6.25 A montage of a synthesized sequence of a falling drop impacting a sessile drop. *Source: Courtesy John Perkins* (a) and a high-speed image of a drop falling into blood demonstrating secondary spatter with satellite droplet production (b). Extremely small or super fine satellite airborne droplets are visible against the black background and deposited traces can be observed at remote distances from the impact location. These may be obscured or not preserved.in the final trace. *Source: courtesy Norman Marin and Jeffrey Buszka.*

the synthesized montage in Figure 6.25. As the falling drop impacts the sessile drop, the upper hemisphere of the falling drop is relatively undisturbed in appearance. As it further descends into the drop at rest (sessile) on the substrate, a thin sheet of blood rises and forms a crater around the region where the falling drop had impacted. The impacting drop can no longer be resolved from the rest of the fluid. As the process continues, the wall of the crater spreads laterally becoming even thinner. As the crater retracts into the fluid, which is in contact with the substrate, the "crown" has significantly formed. Worthington referred to the points of the crown as "jets" (p.16). Fine filaments are drawn between the droplets composing the "crown" and the crater wall. The last high-speed photograph in the montage illustrates the crown after it has just come to rest. The foregoing process can be thought of as a microcosm of secondary spatter.

6.7 Conclusion

The experiments conducted and discussed in this chapter demonstrate the correlation between drops falling onto inclined surfaces, being projected horizontally, and falling onto moving and/or inclined surfaces, as summarized previously in Figure 6.22. It is important for scientific investigators to consider these possibilities before drawing conclusions from the blood trace configurations in a real-world scene investigation. For example, a moving source may produce a droplet trail where the movement may be difficult to detect. This is because the horizontal component of the velocity may be relatively small with respect to the vertical component being accelerated by gravity, which results in a nearly circular blood deposit. This can be seen in cases where a slowly walking person is dripping blood from waist height. For a much lower dropping height, the ellipticity of the deposit may be more easily detected. Additionally, the ability to discern small amounts of ellipticity in a nearly circular blood droplet deposit can be difficult to detect, as is evident in Figure 6.6 for angles 0°, 10°, and 20°.

References

Attinger, D., Moor, C., Donaldson, A. et al. (2013). Fluid dynamics topics in bloodstain pattern analysis: comparative review and research opportunities. *Forensic Science International* 231: 375–396.

Attinger, D., Comiskey, P.M., Yarin, A.L., and De Brabanter, K. (2019). Determining the region of origin of blood spatter patterns considering fluid dynamics and statistical uncertainties. *Forensic Science International* 298: 323–331.

Balthazard, V., Piédelièvre, R., Desoille, H., and Dérobert, L. (1939). Étude des Gouttes de Sang Projeté. *Annales de Médecine Légale de Criminologie, Police Scientifique, Médecine Sociale, et Toxicologie* 19: 265–323.

Bevel, T. and Gardner, R. (2008). *Bloodstain Pattern Analysis*, 3e. Boca Raton, FL: CRC Press, Taylor & Francis Group.

Buck, U., Kneubuehl, B., Näther, S. et al. (2011). 3D bloodstain pattern analysis: ballistic reconstruction of the trajectories of blood drops and determination of the centres of origin of the bloodstains. *Forensic Science International* 206: 22–28.

Carter, A. (1999). Bloodstain pattern analysis with a computer. In: *Scientific and Legal Applications of Bloodstain Pattern Interpretation* (ed. S.H. James), 20–21. Boca Raton, FL: CRC.

Carter, F. (2001). The directional analysis of bloodstain patterns: theory and experimental validation. *Canadian Society of Forensic Science Journal* 34: 173–189.

Carter, A. (2005). Bloodstain pattern analysis with a computer. In: *Principles of Bloodstain Pattern Analysis: Theory and Practice* (eds. S.H. James, P.E. Kish and T.P. Sutton), 241–261. Boca Raton, FL: CRC, Taylor & Francis.

Carter, A. and Podworny, E.J. (1989). Computer modeling of the trajectories of blood droplets and bloodstain pattern analysis with a PC computer. Presented at the Annual Training Conference, International Association of Bloodstain Pattern Analysis, Dallas, Texas.

Carter, A.L. and Podworny, E.J. (1991). Bloodstain pattern analysis with a scientific calculator. *Journal of the Canadian Society of Forensic Science* 24 (1): 37–42.

Comiskey, P.M., Yarin, A.L., Kim, S., and Attinger, D. (2016). Prediction of blood back spatter from a gunshot in bloodstain pattern analysis. *Physical Review Fluids* I: 1–20. https://doi.org/10.1103/PhysRevFluids.043201.

De Forest, P.R., Gaensslen, R.E., and Lee, H.C. (1983). *Forensic Science: An Introduction to Criminalistics*, 300. New York: McGraw-Hill.

Dubyk, M. and Liscio, E. (2016). Using a 3D laser scanner to determine the area of origin of an impact pattern. *Journal of Forensic Identification* 66 (3): 259–272.

Foote, B. (1975). The water drop rebound problem: dynamics of collision. *Journal of the Atmospheric Sciences* 32: 390–402.

Hakim, N. and Liscio, E. (2015). Calculating point of origin of blood spatter using laser scanning technology. *Journal of Forensic Sciences* 60 (2): 409–417.

Kirk, P.L. (1953). *Crime Investigation: Physical Evidence and the Police Laboratory*, 179. New York: Interscience Publishers, Inc.

Kodet-Sherwin, L., Pizzola, P.A., De Forest, P.R., and Perkins, J.C. (1988). A critical assessment of the phenomenon of secondary bloodspatter. Presentation at the 40th Annual Meeting of American Academy of Forensic Science, Philadelphia (20 February 1988).

Laber, T.L., Epstein, B.P., and Taylor, M.C. (2007). High speed video. MFRC: No. 06-s-02. http://www.ameslab.gov/mfrc/bloodstain_pattern_formation (accessed 24 December 2020).

MacDonell, H.L. and Bialousz, L.F. (1971). *Flight Characteristics and Stain Patterns of Human Blood*. Washington, DC: U.S. Department of Justice, Law Enforcement Assistance Administration.

MacDonell, H.L. and Panchou, C.G. (1979). Bloodstain patterns on human skin. *Journal of the Canadian Society of Forensic Science* 12: 134–141.

Pizzola, P.A., Roth, S., and De Forest, P.R. (1986a). Blood droplet dynamics – I. *Journal of Forensic Sciences* 31 (1): 36–49.

Pizzola, P.A., Roth, S., and De Forest, P.R. (1986b). Blood droplet dynamics – II. *Journal of Forensic Sciences* 31 (1): 50–64.

Pizzola, P., De Forest, P.R., Martir, K. et al. (1993). Blood droplet dynamics III. Presented at the annual meeting of the American Academy of Forensic Science, Boston, Mass (February 1993).

Pizzola, P.A., Buszka, J.M., Marin, N. et al. (2012). Commentary on "3D Bloodstain Pattern Analysis: Ballistic Reconstruction of the Trajectories of Blood Drops and Determination of the Centres of Origin of the Bloodstains" by Buck et al;. Forensic Sci. Int. 206 (2011) 22–28. *Forensic Science International* 220 (1–3): e39–e41.

Worthington, A.M. (1963). The splash of a drop – low fall. In: *A Study of Splashes*. With an introduction and notes by (ed. K.G. Irwin), ix. New York: The McMillan Co, 16.

CHAPTER 7

Blood Trace Interpretation and Crime Scene/ Incident Reconstruction

7.1 Principles of Blood Trace Reconstruction

The impetus for creating this list of principles is to present an overview of foundational information, including strengths and limitations, so that there is an organized format for scientists to consider in the practice of blood trace interpretation and its role in crime scene reconstruction. Table 7.1 contains the proposed eight principles of the analysis and interpretation of blood trace configurations.

> ***Principle 1:***
> ***Properly recognized and understood, blood traces MAY reveal a great deal of useful information during a crime scene investigation.***

Blood trace configurations can contain a rich amount of information. The relevance and significance of some of this information may be obvious, while other geometric arrangements may need to be decoded. Scientific knowledge combined with experience in science-based problem-solving is a critical component of recognizing and understanding blood trace configurations in the context of the overall physical evidence record. The various ways this information can be used are described in Section 7.2.

> ***Principle 2:***
> ***There will be cases where extensive blood traces are present but where a clear understanding of them does not address relevant issues in the case at hand.***

A physical analysis of the blood traces at a scene is a necessary and valuable part of a reconstruction; however, the significance of those deposits and their configurations

Blood Traces: Interpretation of Deposition and Distribution, First Edition. Peter R. De Forest, Peter A. Pizzola, and Brooke W. Kammrath.

TABLE 7.1 Principles of the Analysis and Interpretation of Blood Trace Configurations

1	Properly recognized and understood, blood traces MAY reveal a great deal of useful information during a crime scene investigation.
2	There will be cases where extensive blood traces are present but where a clear understanding of them does not address relevant issues in the case at hand.
3	The extent of the blood traces observed at the time of a crime scene investigation is commonly greater than that produced at the time of the initial wounding. This may seem like an obvious statement, but errors are made when this is not given proper consideration. Of course, the time of the initial wounding is often of the most interest in crime scene reconstruction.
4	The initial wounding may not produce any immediate or useful blood trace configurations.
5	Blood traces produced in the course of the initial wounding may be altered or totally obscured by the flow of additional blood from the wound.
6	Blood shed by the wound or wounds may be transferred by post event activities that may alter or obscure the blood trace geometry associated with the initial wounding event.
7	Configurations of blood traces consisting of a collection of airborne droplet deposits can be informative with respect to providing an understanding of events that have taken place at a crime scene. Although schemes for assigning such patterns to specific causes or production mechanisms can be helpful, they are often oversimplified. It is naïve to think that patterns encountered in a crime scene can always be assigned to one of a finite number of mechanisms as defined by a typological or taxonomical systems.
8	A collection of a few seemingly related dried blood droplets is not necessarily a pattern. It should never be treated as one. The number of blood traces may be inadequate to allow a meaningful interpretation.

must be evaluated as to whether they have value for addressing pertinent hypotheses and differentiating between competing scenarios.

A reconstruction of an event, even one that has copious blood traces, may yield details that are consistent with two or more competing scenarios. Analysis of the blood trace configurations, albeit done well, may not provide the answers necessary to make such a differentiation. An example of this might be a scenario where the victim is killed as a result of a single stab wound that produced no airborne droplet patterns (also see Principle 4). This could still yield contact blood traces on the clothing of either a perpetrator or a person discovering the body who tries to render aid. Analysis of contact blood traces on the clothing of a person of interest might be unable to differentiate between these two scenarios and thus not be able to aid in determining this individual's role in the crime. Other aspects of the physical evidence record should always be considered to test the various hypotheses and address relevant matters.

It is also important to distinguish between blood trace configurations arising during the event of interest from those stemming from subsequent blood flow and distribution (also see Principle 5) or activities of first responders and others that may be present at the scene (also see Principle 6). Some examples would be blood flowing from a wound obscuring spatter patterns, footwear outsole patterns in blood left by police or medical personnel, or a trail of blood droplets resulting from the victim's body being carried from the scene. If not recognized as such, these traces from post-event actions have the potential to contribute to confusion in understanding the relevant issues in the case.

Principle 3:
The extent of the blood traces observed at the time of a crime scene investigation is commonly greater than that produced at the time of the initial wounding. This may seem like an obvious statement, but errors are made when this is not given proper consideration. Of course, the time of the initial wounding is often of the most interest in crime scene reconstruction.

Blood will often continue to flow from a victim's wound(s) after the time of initial injury. This must be explicitly recognized because it can complicate interpretation of the blood trace configurations. This can lead to misunderstanding the initial injury or injuries and resulting blood traces, which is often the most critical aspect of the event of interest. This principle has several corollaries. Some of these are principles in their own right (Principles 4–6).

Principle 4:
The initial wounding may not produce any immediate or useful blood trace configurations.

In a bludgeoning, stabbing, or shooting, the initial injury may not necessarily produce any immediate blood flow or airborne droplets. An understanding of the anatomy of the human circulatory system can explain this. There are many areas of the human body where all of the blood contained in that region is in finely divided arterioles, small veins, venules, and capillaries. Approximately 50% of the blood volume is contained within these narrow tubular structures, with diameters as small as 5 μm and not in a large volume of coalesced blood. The dimension of a typical airborne blood droplet is substantially larger than the vessels described above, and these cannot be formed from blood contained in vessels with smaller diameters than the droplets themselves. Before droplets causing blood traces in the millimeter range of diameters can be formed, the blood must exit the vessels, accumulate, and coalesce into a larger volume before it can be spattered. Although there may not be blood traces formed from the initial wounding, there may be fragments of tissue spattered from the violent activity producing the injury.

It should be recognized that another explanation for the lack of an airborne droplet configuration from an initial wounding could be the absence of a suitably structured and situated surface capable of receiving blood droplets and recording a pattern. For example, a situation where a victim is standing during the initial through-and-through gunshot wound through the heart would be expected to result in blood droplets from the exit wound in the chest. This pattern would not be recorded and preserved without the presence of a suitably placed and textured surface. A nearby vertical surface, such as a wall, would be an ideal surface for preserving this physical evidence record. In the absence of a close-by vertical surface, the droplets would fall to the floor or ground and a blood trace configuration might be recorded depending on several factors, principally its texture. Examples of textured surfaces that would be poor for the recording and preservation of an airborne droplet pattern would be such things as grass, soil, gravel, carpeting, or a moving surface such as an escalator or stream of water.

Principle 5:
Blood traces produced in the course of the initial wounding may be altered or totally obscured by the flow of additional blood from the wound.

This seems self-explanatory, but is often overlooked in a blood pattern reconstruction. The physical evidence record can be almost immediately overwritten by events taking place after the wound is created (also see Principle 6). Much of this is unavoidable. If indeed there are airborne blood droplet deposits caused by the initial injury, some of this information contained in the blood trace configuration may be lost or altered due to subsequent blood flowing and spreading.

Principle 6:
Blood shed by the wound or wounds may be transferred by post event activities that may alter or obscure the blood trace geometry associated with the initial wounding event.

There is no escaping the fact that there is no such thing as a pristine crime scene. Before there is a recognition of an event warranting an investigation, it must be discovered, brought to the attention of the authorities, and initially dealt with by first responders including medical personnel and police officers. This is depicted in the physical evidence continuum (Figure 1.3). These post event activities unfortunately can have negative consequences with respect to the physical evidence record. Some aspects of these events are avoidable, such as the presence of extraneous people at the scene, while others are not, specifically life-saving efforts. To minimize the possible adverse consequences of these activities, it is important to have policies for the protection of the scene be in place and be consistently enforced (See Chapter 1). In a surprising and disappointing number of scene investigations, this is not the case.

There are activities of the perpetrator(s) that may contribute additional blood trace configurations and/or alter the initial ones. These actions can result in the production of secondary crime scenes, such as a vehicle or disposal site when a body is moved. Attempts at clean-up can often be readily recognized (Figure 7.1); if they are not easily visible with natural light, the use of an intense hand-held light at different illumination angles can be extremely useful.

As a last resort, just prior to the scene being released, chemical enhancement techniques such as luminol, amido black, and fluorescin (chemically reduced fluorescein) can be used. These should not be used before other approaches have been exhausted because to a degree, each of these is destructive. Several cases, such as those shown in Figures 7.1 and 7.2, demonstrate the advantage of a more critical and thorough examination of a scene prior to destructive testing. Even in situations where there is thorough cleanup, when this kind of reasoning and process is followed, laboratory testable quantities of blood that otherwise would not be recognized can be recovered.

Principle 7:
Configurations of blood traces consisting of a collection of airborne droplet deposits can be informative with respect to providing an understanding of events that have taken place at a crime scene. Although schemes for assigning such patterns to specific causes or production mechanisms can be helpful, they are often oversimplified. It is naïve to think that patterns encountered in a crime scene can always be assigned to one of a finite number of mechanisms as defined by a typological or taxonomical systems.

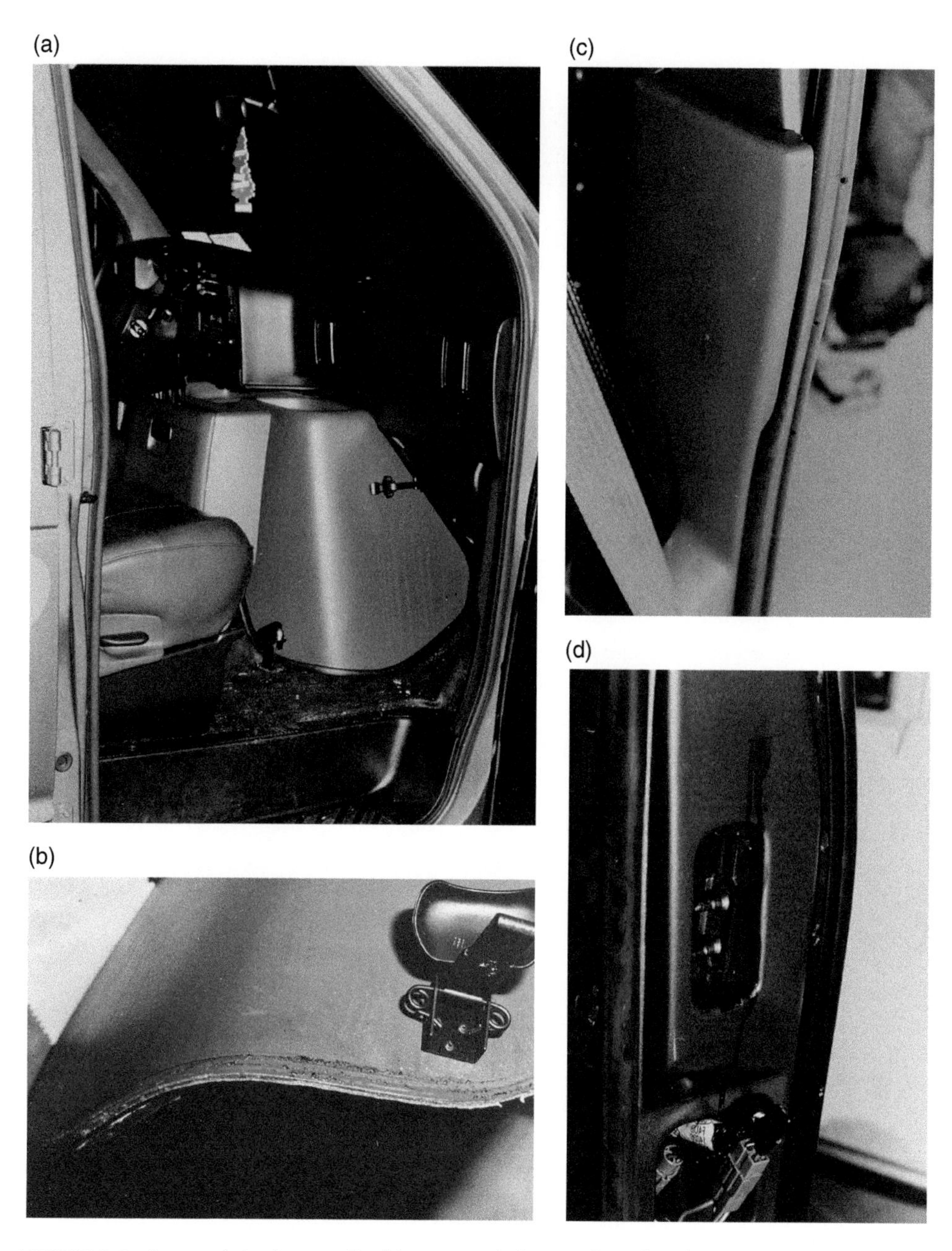

FIGURE 7.1 Incomplete cleanup. In this case, a victim was found in the street adjacent to a pool of blood, having been shot in the head. Crime scene investigation revealed secondary spatter near the blood pool and on the curb, suggesting that the victim bled some distance above the ground to produce the spatter. A suspect was identified, and his van examined. Upon opening the passenger side door, whitish drip-like residues, suggestive of the use of a cleaning agent, were seen on the engine cover when illuminated with a bright light source (a). After the van was seized, the engine cover was removed and a dried residue of diluted blood was found along the base seal (b). In addition, a faint contact transfer of possible blood was found on the upper surface of the plastic covering the pillar behind the front passenger seat (c). Instead of applying luminol or another enhancement reagent, the plastic cover of the pillar was removed and the thick crust of blood observed (d). *Source: Photographs courtesy of the Yonkers Police Department.*

(a)

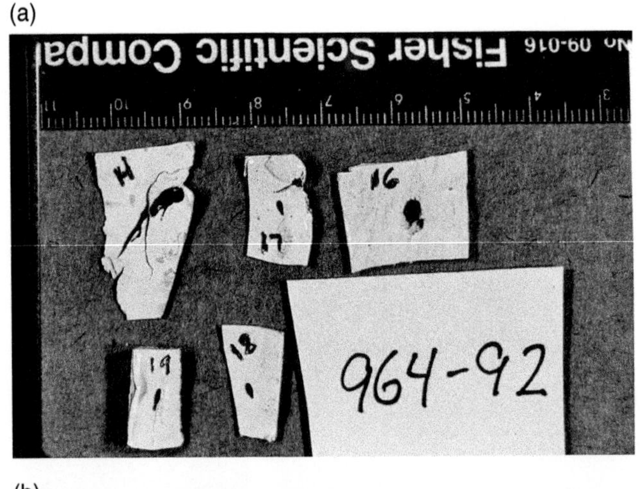

(b)

FIGURE 7.2 In this case, the victim was found in the backyard of an apartment building, in a state of decomposition in two contractor bags. When the first floor apartment was initially searched, the kitchen floor appeared to be clean to the police investigators. It was later hypothesized by forensic scientists that if the victim was beaten and stabbed in the kitchen, then some blood droplets may have followed a parabolic-like trajectory and impacted the wall behind the appliances (e.g., the stove or washer) unbeknownst to the assailant. Blood droplet-like deposits were found behind appliances after additional examination (a). It was further hypothesized, that if the kitchen floor had been cleaned, that blood may have seeped under the floor trim on the other side of the kitchen. Partially pulling away the floor trim partially from the base molding revealed thick crusts of blood (b). *Source: Photographs courtesy of the Yonkers Police Department.*

Much can be learned from blood traces in the form of airborne droplet deposits (see Principle 1). Typologies and taxonomies of blood trace configurations, as described in Chapter 5, are useful for classifying and organizing similar patterns, but should not be used as a flow chart for their identification. The assignment of blood traces at crime scenes into a specific typology or taxonomy is often oversimplified. Although this categorization for patterns from controlled laboratory experiments is easily accomplished,

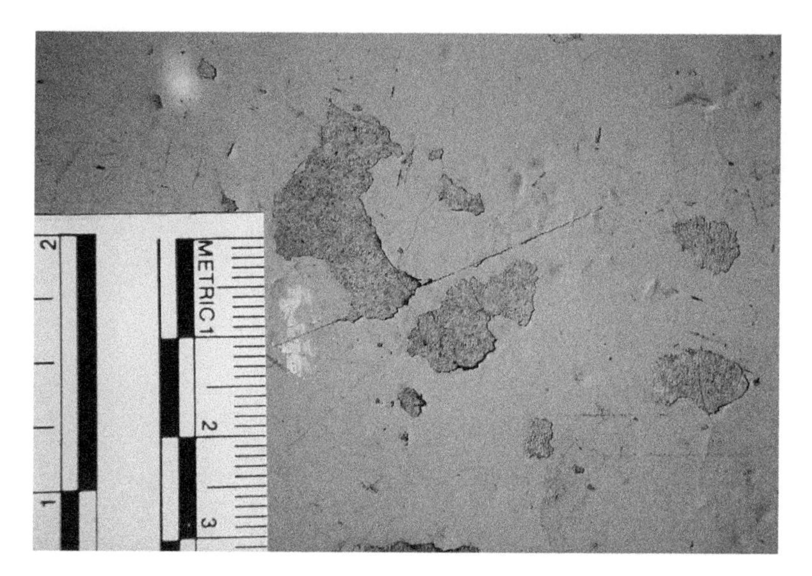

FIGURE 7.3 A blood trace deposit with an unknown and uncharacterizable mechanism associated with defects on painted concrete surface in a residential garage. One could hypothesize a mechanism of formation where the water in a dilute blood deposit slowly evaporated and dried with the blood gradually concentrating into a residue at the edge of the paint, but this is by no means certain. The earlier presence of diluted blood could be consistent with cleanup or an alternative explanation where blood traces were diluted by water, such as blood drops on snow that subsequently melted. Without further evidence, it cannot be known whether the water preexisted the blood deposit or followed it.

extreme caution is necessary in making such assignments in crime scenes. In the real world, there are many more possible mechanisms for producing blood trace configurations than are enumerated in standard typologies. A configuration resulting from a given mechanism may mimic those produced by another distinctly different mechanism. Configurations may be overlapped or even be superimposed. They can be far more complex than are generally appreciated. In addition, there is a risk of a forced assignment of a blood trace deposit to a specific category. Complexities and limitations of the interpretation of blood trace configurations are discussed later in this chapter.

> ***Principle 8:***
> ***A collection of a few seemingly related dried blood droplets is not necessarily a pattern. It should never be treated as one. The number of blood traces may be inadequate to allow a meaningful interpretation.***

There may be a temptation to make something out of nothing and assign a "pattern" classification to a limited number of blood trace deposits. Too few deposits would make it impossible to assign a mechanism and would eliminate the possibility of more than one explanation for producing the "pattern." For example, there is no way of knowing if the few droplet deposits even came from the same production mechanism or at the same time. This can lead to serious mistakes in reconstruction that can lead to either an erroneous exculpation or a wrongful conviction. This problem is discussed in a subsequent section on over-interpretation.

7.2 Utility

As stated in Principle 1, a great deal can be learned when blood trace deposits at the crime scene are scientifically examined and interpreted. A wealth of information can be encoded in configurations of such traces, and an understanding of this requires insight into the physical properties of blood and the mechanism with which it is transferred. It is the role of the crime scene scientist to extract the information that could remain hidden from those without this expertise and experience. From this information, it can be helpful to group the contributions to a reconstruction into five categories of value. These are: associative, action, positional, directional, and temporal.

In addition, understanding the three-dimensional geometries of blood trace deposits can be indispensible for directing sampling for laboratory testing, such as DNA analysis.

7.2.1 Associative

Associative information refers to evidence that may link (or associate) a person to a crime scene or two or more players to each other (i.e., a victim with a perpetrator).

An example of the associative evidence category with respect to blood trace configurations would be an imprint in blood left by a footwear outsole at a scene. Details of the outsole pattern could be reproduced in the evidence imprint and contain class and/or individual features for classification and/or an approach to individualization. In favorable circumstances, this can then be used to uniquely link, or associate, the imprint to the footwear. If the imprint is a part of a trail, it is important to realize that the value of each imprint component of the trail may differ. For example, the imprints closest to the blood source (e.g., a pool of blood which has been stepped in) are the most obvious indicators of a blood trace configuration, but may contain too heavy a deposit of blood that could obscure useful fine details in the outsole pattern. In these situations, it is valuable to follow the trail as the imprints become lighter and lighter. At some point, they will no longer be visible without employing a mode of enhancement, such as oblique lighting, protein staining, or presumptive catalytic testing that accentuates and preserves the configuration. The formerly invisible configurations may provide the best associative evidence to aid in the goal of approaching an individualization.

7.2.2 Action

Certain movements and activities at an event where blood has been shed are reflected in the physical configurations of blood traces left at the scene. These are what we are referring to as action information. Very often two distinct proposed actions can easily be differentiated from an examination of the three-dimensional geometry of the blood trace deposits. A suspect in a bludgeoning case may claim that blood traces on his or her person or clothing were the result of discovering the body and attempting to render aid rather than the presence during the crime. An examination of the blood traces can be used to support or exclude the suspect's account. The presence of airborne droplet deposits on the clothing of the person who discovers the body may raise questions about the veracity of the explanation offered. On the other hand, if all of the blood traces on the clothing of the individual who discovers the body are contact transfers that would

FIGURE 7.4 The investigative information that can be obtained from a bloody footwear outsole print in isolation is limited. This case, a double homicide, demonstrates an example of passive documentation of a crime scene and its limitations (a–d). In this case, what appears to be a partial footwear outsole pattern in a blood-like substance (c and d) was found in the external access path between the two floors of the building, where one victim was found in each of the upstairs and downstairs living quarters. To preserve the imprint, a white box was placed over it, which can be seen in the center of figure (a) and (b). Despite the laudable recognition and precautions to preserve and document the outsole imprint, the initial investigators did not go further than these perfunctory steps. In this case, there was no attempt at understanding the imprint or the significance of its isolation in the context of the reconstruction. Nor was there any attempt at enhancement to complete the full shoeprint and elucidate the fine details of the outsole pattern. There are possible explanations of a footwear outsole print in blood found in isolation. Examples could include stepping on transient materials, such as pieces of paper that are no longer present. However, none of these conceivable explanations was considered by the initial examiners. Had appropriate attention been given to this blood trace deposit, and had its features been enhanced to allow a comparison with the suspect's shoes, it is possible that this could have been used to demonstrate the suspect's presence at the scene, while there was wet blood present, thus refuting his account.

lend support to his or her version of events. It would be important in the situation where airborne droplet deposits are present to eliminate possible explanations that did not involve bludgeoning, such as blood aspirated from the airway or secondary spatter from blood dripping into blood. This question can often be explored and evaluated by careful examination of the blood trace configuration evidence and a review of the context at the scene. It is *always* critically important to consider and thoroughly evaluate alternative explanations.

Another useful illustration of an action being recorded in the blood trace configurations would be a trail of blood droplets from a bleeding wound. This was valuable physical evidence in the O.J. Simpson murder investigation (see Chapter 10). There were two blood droplet trails noted. The first was at the primary crime scene and lead from the area of the victims' bodies to the presumed location where the perpetrator had parked his vehicle. This was accompanied by footwear outsole patterns in blood that provided complimentary information about the movement of the perpetrator. The second blood droplet trail originated at the defendant's white Ford Bronco that was parked in the front of his home. This blood trail led up the driveway to his front door and continued into his house. It was clear that both of these blood droplet trails were from a bleeding wound rather than a dripping weapon due to the consistency in the droplets with no indication of a lessening of the flow as at the trails progressed. DNA analyses of the blood droplets in both trails were matched to that of O.J. Simpson. The ability to distinguish between blood dripping from an open wound versus that from a bloody weapon, in addition to the movement as indicated by the trail of the perpetrator at the scene and at O.J. Simpson's home, was of great importance for this reconstruction.

7.2.3 Positional

A useful component of a reconstruction derived from blood trace configurations is positional information. Knowledge about the approximate location of various aspects of an event, such as the origin(s) of the blood traces and positions of weapons or people at certain points in the unfolding event, can be essential to a comprehensive reconstruction.

Positional information can be derived from a variety of blood trace configurations, for example, those created during the result of a bludgeoning (see Section 10.3). In this situation, two general patterns can result from the action of the perpetrator striking the head of the victim with a club or other instrument. Following the accumulation of blood in the wound in the area struck previously, succeeding blows can produce distinctive radial impact spatter. Second, an arc pattern is commonly generated from blood adhering to the weapon being separated from it during the swing (as discussed in Chapter 5). Although this may not inform us about the action producing the first wounding (see Principle 4), it can be very informative about the general locations in three-dimensional space of the assailant and the victim's head during the administration of subsequent blows. Knowledge of the positioning of the perpetrator and/or the victim, such as whether either was standing or on the ground, could provide critical information for reconstruction of the event. Two important caveats must be kept in mind. First, reconstruction scientists should constantly be aware that these types of blood trace geometries do not necessarily indicate where the victim was during the initial blow, but instead provide information as to where he or she was positioned during these subsequent hits. Second, the location of the origin of the blood traces cannot be precisely determined from the pattern, although useful approximations are possible. A discussion for the reasons of this uncertainty is presented in later in this chapter.

An illustration of the importance and potential of positional information that can be derived from blood traces was recognized by Dr. Paul Kirk in his analysis of the blood trace configurations in the well-publicized Sheppard murder case from 1954 (detailed in Chapter 10). In this case, Dr. Samuel Sheppard was accused of killing his pregnant wife Marilyn by bludgeoning which left a plethora of blood trace evidence (Kirk 1955; Thorwald 1966). When analyzing the blood trace patterns, Dr. Paul Kirk was able to show the positional relationships between the assailant and the victim based on where blood traces were present and where they were not. This was accomplished by recognizing the void in a radial impact pattern caused by the body of the perpetrator blocking a sector of the spattered blood. This was the first documented case where a scientist recognized the significance of the void in a radial blood droplet pattern and used it to reconstruct the geometry during the commission of a bludgeoning.

7.2.4 Directional

Directional information refers to knowledge about the orientation and/or direction of motion of the source of a blood trace.

There are several examples where directional information can be derived from blood traces. An individual elongated blood droplet deposit can be analyzed to determine its angle of incidence and the direction of travel of the droplet that created it (discussed in Chapter 6). This angular information can be used to back-project to reveal the direction from which the blood droplet originated. A separate example is the asymmetry seen in a blood trace produced by a droplet falling from a moving source. This asymmetry is most apparent in blood traces formed by blood droplets that have fallen from a relatively low height and/or from a rapid and horizontally moving source, as discussed in Chapter 6 (Figures 6.17b and 6.18a). This has been shown by Pizzola et al. (1986) to be equivalent to blood falling vertically from a stationary source onto a moving surface, such as a moving limb. This asymmetry can be useful when interpreting a blood droplet trail or pattern to reconstruct the motion of the source of the blood droplets or the surface on which they landed.

7.2.5 Temporal

Temporal information is that which informs us about the timing and sequencing of component activities taking place during the course of an event. Some of these component activities leave a record in the physical evidence including blood trace configurations.

One can learn about the sequence of events from the temporal information latent in certain blood trace configurations. An example of this is with overlapping and overlying deposits. It may be important to attempt to discern the sequence in which blood trace deposits were created in certain cases. This may not be possible if the time interval between the two deposits is relatively short because the damp blood traces may have partially merged. A case example where this was important was the investigation of a murder which took place in the office of a retail establishment. A blood droplet deposit resulting from the stabbing of the victim was co-located with a footwear imprint in the victim's blood with details of the pattern matching that of the defendant's outsole. If it could be shown that the footwear outsole pattern was on top of the droplet pattern, it may have indicated that a significant amount of time elapsed between the stabbing event and the production of the bloody footwear imprint. The defense proposed that the defendant came upon the scene long after the stabbing event transpired and stepped in a pool of blood when discovering the body. Although this sequence of events could not be established from these blood trace configurations, it demonstrates the potential for providing temporal information from blood deposits. An illustration of this concept is depicted in Chapter 5 (see Figure 5.4).

7.2.6 Pattern Directed Sampling

A very important contribution of the recognition of blood trace configurations is its role in providing knowledge about which deposits or portions of deposits in the case of possible mixtures should be sampled for subsequent laboratory testing, such as DNA analysis. There are two fundamental benefits to deposit configuration or pattern directed sampling. The first is that it can provide economy of sampling, meaning that oversampling would be avoided and both efficiency and economy maximized. For example, in a configuration of airborne droplet deposits that has been recognized as originating from a single source, it is not necessary, inefficient, and foolish to sample every droplet deposit in the pattern. Similarly, collecting multiple samples from a large blood trace deposit is often excessive and could be wasteful of resources. The second is that pattern directed sampling of larger blood deposits can potentially avoid issues that can arise from the collection of mixtures. A single sampling of a large deposit would suffice if it were recognized that the blood trace itself was produced from a single event or contact. Sampling blood deposits for subsequent laboratory testing without consideration of the three-dimensional geometry of the scene could be destructive of evidence and detrimental to the case solution.

It must be noted that it can be important to document the blood trace configurations both before and following sampling, with photographs and notes about the exact portion of the deposit being sampled. This is too often not done in crime scene investigations but is laudable forensic practice.

7.3 Limitations, Problems, and Common Acceptance of the Status Quo

The analysis and interpretation of blood trace configurations is one of the most misunderstood and abused aspects of crime scene reconstruction. It is fraught with unrecognized limitations and problems that are commonly ignored or dismissed. An experienced and qualified scientist will be aware of these issues and take them into consideration when conducting or evaluating a blood trace pattern-based reconstruction. However, most crime scene investigations are done without the essential timely input from scientists. This unfortunate situation is commonly accepted by criminal justice policy makers and the public. The acceptance of this status quo thwarts investigations and prevents recognition of "the ground truth."

7.3.1 Lack of Teamwork and Potential Synergism Between Criminal and Scientist Investigator

Important contributions to a crime scene investigation involving blood trace configurations can be made by both criminal (non-scientist, law enforcement crime scene specialists) and scientist investigators. This is especially true in the case where a teamwork relationship has been cultivated. These investigations work best when there is cooperation and mutual respect between the criminal investigators and the scientist investigators at the crime scene that can lead to an optimized or even synergistic result.

The criminal investigator, or law enforcement crime scene specialist, may be a police detective or civilian employed by a police department or other government agency charged with the criminal investigation. A law-enforcement crime scene specialist may be responsible for obtaining search warrants, interviewing witnesses in addition to helping to recognize and document evidence at a crime scene. At times, the law enforcement crime scene specialist will see him or herself as an advocate for the victim and aggressively pursue obtaining information for conviction. This attitude should be avoided in favor of being an advocate for the truth. The scientist investigator should bring a scientific approach to the investigation. This approach is developed through a combination of formal science education and extensive experience at crime scenes. The scientist need not concern him or herself with some of the activities of the law enforcement crime scene specialist, which do not pertain to the interpretation of the physical evidence record.

Working as a team at the outset of an investigation is advantageous because both have a huge amount to offer in understanding an event that has taken place at a crime scene. These are complementary domains. The non-scientist criminal investigator brings knowledge in criminal procedure, skills in interviewing potential witnesses, as well as expertise in routine crime scene techniques such as fingerprint dusting and lifting, and documentation instrumentation such as total stations or laser scanning. In addition, the law enforcement crime scene specialist would be responsible for securing the scene and freeing up the scientist investigators to focus on the nuances of the physical evidence problem. Another advantage of this teamwork model is that the law enforcement team member would have an acute appreciation for the physical evidence and the importance of securing the scene, thus could aid in excluding non-essential personnel from the scene from, such as police supervisors, attorneys and politicians (refer to discussion in Chapter 1). A law enforcement crime scene specialist with a science education is an ideal that is rarely realized. Although there may be some individuals who possess this rare combination of attributes, they are not the norm. If this was to be generally accepted as the archetype of a crime scene scientist, this would obviate the need for non-scientists at the scene.

7.3.1.1 Lack of Appreciation for the Contributions of the Scientist (or Undervaluing of the Scientist) It is a common misperception of law enforcement crime scene specialists that scientists are not valuable at crime scenes. This undervaluing of the contributions of the scientist is likely based on the fact that currently, the vast majority of forensic laboratory scientists do not have crime scene experience. Laboratory scientists do not automatically have knowledge of crime scenes based on their proficiency with submitted physical evidence. As a scientist gains experience with crime scene analysis, their contributions to a physical evidence investigation at the scene would increase.

If the unrealized potential of the scientist to contribute to the early stages of an investigation was exploited by the law enforcement establishment, there would be a more thorough understanding of the physical evidence record at the outset of a case. However, the utilization of most current forensic scientists would not be an easy transition. Gradually, the experiential education of local laboratory forensic scientists would bring added value to crime scene analysis and be well worth the effort. This value would supplement the skills of the law enforcement crime scene specialist and stems from the wealth of their background science knowledge, their hypothesis-based reasoning process, and their knowledge of the evidence analysis capabilities of the laboratory. The crime scene experience gained would complement the scientific expertise provided by traditional science education and knowledge of laboratory approaches for the analysis of physical evidence. Furthermore, experience in the field will make the laboratory scientists more acutely aware of situations existing at crime scenes and thus will enrich their benchwork.

Another consideration is the inexperience many law enforcement crime scene specialists have with working with novice scientist investigators. The cultivation of this potentially valuable synergism would require educating all involved parties but would ultimately pay huge dividends.

7.3.2 Potential Failures of the Scientist Investigator

Having a scientist investigator at a crime scene does not guarantee an optimal investigation and that there will not be any problems. Scientists are humans, and with that comes fallibility. Steps should be taken to minimize this, and if they occur, appropriate corrective actions should be implemented.

There are five significant potential failures of the scientist investigator model that are detailed below.

7.3.2.1 Investigator Inexperience As previously discussed, currently most forensic laboratory scientists lack the necessary crime scene experience to enable them to be valuable as scientist investigators. Many of the existing laboratory staff would require hands-on experience and mutual mentoring with law enforcement crime scene specialists to become valuable in the field. Potential failures due to the scientist investigator inexperience can be mitigated by early mentoring and recognizing that this is a life-long endeavor that requires continuing experience at crime scenes and is not a short-term or part-time job.

7.3.2.2 Neglect of Scientific Principles An unconscious disregard for scientific principles can be a significant failure on the part of forensic scientists. Constant vigilance is necessary to avoid this despite external and internal biases and pressures. One must be aware that there are several ways that the scientific method can be compromised; these are enumerated below.

7.3.2.2.1 Misunderstanding and/or Misuse of the Scientific Method The scientific method (see Chapter 1) is more than the series of steps outlined by many writers of introductory science texts. This is often overly simplified by the author of such a text or trivialized by non-scientists. The scientific method is a process for approaching an understanding of an event or phenomenon and is one of the principal hallmarks of science. These actions are not intuitive or natural and may run counter to human instincts. The scientific method must be learned, internalized, and conscientiously practiced. There are some scientists who may employ the scientific method without recognizing that they are doing so. However, failure to raise the process to the conscious level can result in a neglect or misuse of the scientific method.

Observation and data gathering are the initial step in the employment of the scientific method and also lead into the iterative process wherein more refined observations are be made as a result of insights developed while testing hypotheses. If observations are not made carefully, the entire process will be flawed. In the case of bloodstain pattern analysis, if inaccurate observations are made, the reconstruction is likely to produce a misleading result. In the experience of two of the authors with a case where a body had been discovered down the side of a cliff at a highway turnout, suspicion fell on the business partner of the victim. It was suspected that the victim had been stabbed or transported in the suspect's car, which was black with black leather upholstery, and a search warrant of the vehicle was obtained. Initial examination of the vehicle by investigators failed to reveal any signs of blood or other evidence of violence. When investigating the car interior with a bright light, the authors found airborne blood droplet deposits on the louvers inside the air conditioner vents and a blood trace in the form of a thin red line at the juncture of the upper surface of the leather passenger seatback with a plastic seatbelt guide attached to the seatback. This observation suggested that blood may have seeped under the plastic seatbelt guide and would have survived the extensive and thorough cleanup. With this observation, the seatbelt guide was removed, where upon crusts of blood were discovered between the plastic and leather. There were also slight blood traces in between the seatbelt anchor attached to the rear floor and the factory-made opening in the carpeting to allow the metal anchoring to be bolted to the floor. There was an indication that blood had flown from the passenger seat and entered this opening and accumulated under the carpeting. Cutting the carpet away led to the discovery of a blood pool in the space between the top of the sound deadening material and the backing of the overlying carpeting. This evidence was instrumental in conclusively connecting the victim with the suspect's vehicle and led to a reconstruction of the homicide taking place within the vehicle. Because of the meticulous cleanup, initial observations had led to doubt that the car was involved in the homicide. Had this blood trace evidence remained unobserved, the case might have gone unsolved. Certainly, thoughtful and detailed observations are very important and essential components of the scientific method.

In the process of employing the scientific method, tentative working assumptions may be indispensable initially but must be critically evaluated continuously. When assumptions are not robustly scrutinized, mistakes may follow. A case example which demonstrates the need to evaluate working assumptions was one involving the stabbing death of a man during an argument with a neighbor (Figure 7.5a–d). On initial inspection, there was a previously described "linear bloodstain pattern" on the ceiling of the suspect's apartment. This led to an initial assumption by non-scientist investigators that the pattern was an arc pattern being caused by a weapon being swung. Upon closer examination of the deposits by the authors, it was noted that the directionality of the individual blood droplet traces was not consistent with the plane of the postulated weapon swing. In addition, the apparent straight-line observation was incorrect, as the initial interpretation was not borne out upon closer inspection. Assignment to a particular mechanism was not possible because it was felt that the configuration was too incomplete. Had this initial assumption not been critically examined, the investigation might have continued to proceed down the wrong path.

One of the most serious failures to adhere to the scientific method is not recognizing, considering, or evaluating alternative hypotheses. Commonly, scientists, as human beings, may become invested in their initial hypotheses and then run the risk

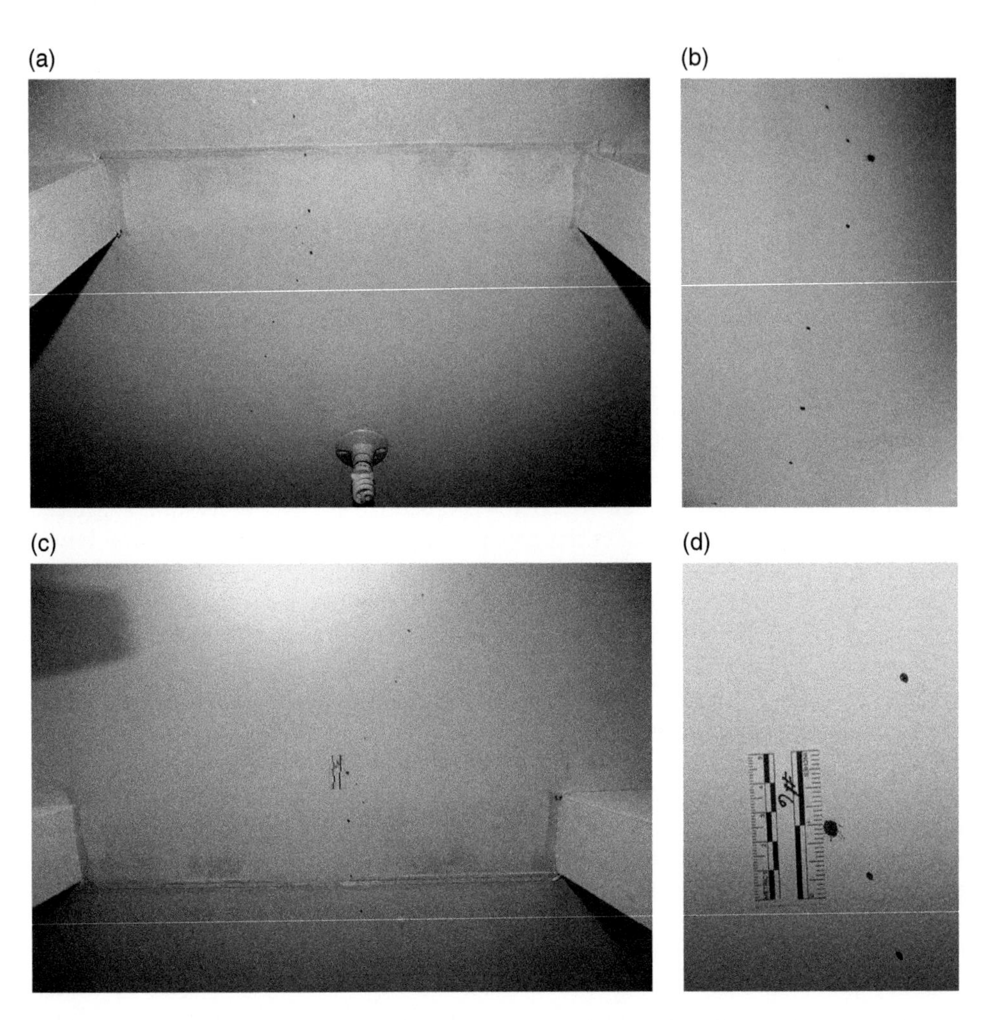

FIGURE 7.5 Initial responder crime scene photographs showing blood trace deposits on the ceiling of an apartment where a man had been fatally stabbed. The blood trace configuration was initially described as being "linear" and thought to be a cast-off pattern, but upon closer examination (Figure b–d), this working assumption was rejected.

of being unwilling to consider the possibility that there may be alternative explanations for the evidence. A case example that demonstrates this is one in which an expert for the prosecution utilized the presence of fine droplet blood traces on a suspect's athletic shoes to support the contention that he fatally stabbed the victim. The defendant in the case said that he found the victim lying in a pool of blood and tried to lift her up before realizing that she was dead. The expert retained by the defense found it difficult to explain how, in a stabbing, fine droplet deposits would only be found on the assailant's shoes and not on other areas of his clothing, thus alternative hypotheses were considered for the source of the fine droplet deposits. The victim's head was in a pool of blood when she was found, and her hair was saturated with blood. The consultant for the defense felt that a better explanation for the fine spatter on the shoes was blood from the saturated hair dripping into blood when the victim's head was lifted from the pool of blood by the defendant, but was not able to eliminate the possibility that the droplet traces were acquired during the stabbing. It must be borne in mind that even cases in which there is abundant physical evidence, the relevant evidence may not be adequate to support one hypothesis over another. What is of critical importance is to make a deliberate effort to consider alternate hypotheses.

Simply recognizing alternative hypotheses alone is not adequate; a scientist investigator must also thoroughly test hypotheses. Failure to rigorously attempt to falsify hypotheses is another possible failing even of a scientist investigator. A case example that shows the importance of scrupulously testing hypotheses was a shooting case with a claim of self-defense. At autopsy, two projectiles were found, one that had entered the scalp behind the left ear and traveled within the skull into the brainstem, and one that had entered the left chest. Consequently, it was reported by the medical examiner that there were two bullet wounds. In addition, the firearms examiner reported that the two projectiles were both fired from the same firearm. When examined in more detail by a defense expert, it was clear that there was a single projectile and the bullet had split into two fragments on a nearly tangential impact with the skull. Thus, the two wounds were made by the different fragments of the same bullet, where only one of the two fragments entered the skull. Consistent with the defendant's statement that the victim was charging him with his head down, the bullet entered downward into the head near the ear, in the mastoid area, split, and part continued on into the skull and then skidded along the bottom of the brain pan and severed the brain stem. The other half, which was recovered in the left pectoral muscle, remained outside the body and re-entered the body in the left chest. The autopsy pathologist and the firearms examiner assumed that there were two bullets causing two bullet wounds and did not test this hypothesis. Additionally, neither considered the alternative hypothesis of one bullet splitting upon contact with the skull. There was physical evidence not considered by either the pathologist or the firearms examiner that should have been recognized and would have suggested to them their mistake. The pathologist should have recognized that both of the bullet wounds were irregular, thus indicating that neither were simple entry wounds. The firearms examiner should have observed that the combined mass of the two bullet fragments equaled the approximate expected weight of a single 9 mm full metal-jacketed bullet. Furthermore, when the two fragments were examined by the consulting criminalist with a stereomicroscope, a physical fit was made with the copper jacketing (Figure 7.6). In addition, microstriae on the land impressions on each of the fragments showed that when compared to a test fired bullet, two of the land impressions had matching striae to one of the fragments, while the other four matched the other fragment. This was incontrovertible proof that the two fragments were one bullet at one time. The failure of the autopsy pathologist and the firearms examiner to test their initial apparent naïve assumption that there were two separate bullets led to a false reconstruction of events, and only after recognizing that the bullet fragmented and there was really only one shot fired was a correct reconstruction possible.

The formulation of untestable hypotheses is a potential trap for the scientist investigator. This notion of falsifiability as a demarcation between science and pseudoscience originated with the scientific philosopher Karl Popper and is an essential test of any scientific hypotheses put forward. A hypothesis is deemed to be scientific if it is possible to conceive of an observation that has the potential to demonstrate that the statement in question is false. The concept of falsifiability or testability is recognized by the courts and included as one of the Daubert standard's prongs for admissibility of

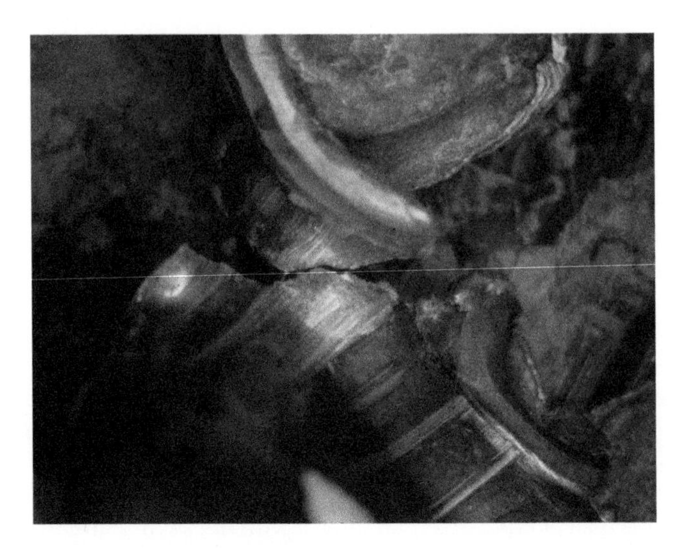

FIGURE 7.6 Physical fit of two bullet fragments.

scientific evidence (Daubert v. Merrell Dow Pharmaceuticals, Inc. 1993). For example, consider a case where it is believed that a bludgeoning occurred in a kitchen, however, there are no blood traces after a thorough scrutiny of the scene, including obscure and inaccessible recesses. It might be "hypothesized" by the scientist investigator that the suspect cleaned up the blood, thus leaving no detectable traces. This is not a scientific hypothesis because it cannot be falsified. One may consider that the scientist could test for cleaning products; however, considering the location, the presence of these chemicals in a kitchen would not have any probative value. Unfortunately, this type of misapplication of the scientific method is pervasive throughout criminal investigations. An example from one of the author's suspected arson cases included a theory by an investigator that the failure for the laboratory to identify any ignitable liquids in the collected residues was the notion that the accelerant used was water soluble and thus the suppression efforts of the fire department using water washed it all away. Untestable hypotheses have no place in conclusions drawn in scientific investigations and inquiries.

These misapplications of the scientific method do not occur in a vacuum, and these errors are more often than not interrelated. Consequently, an investigator can misuse the scientific method in several ways in a single case. A case example that shows this was the murder investigation discussed in Section 11.2.

7.3.2.2.2 Over-Interpretation There are two categories of over-interpretation that result from a number of human failings that can afflict even scientists. The first type of over-interpretation results from a neglect of proper hypothesis development and testing in the scientific method, specifically resulting from inadequate consideration of alternate hypotheses or not rigorously testing such hypotheses in an attempt to falsify them. The second type of over-interpretation results from extending the significance of observations beyond what is supported by the science. The latter has been exacerbated by the naïve desire to assign statistical values similar to those used in DNA analysis to other evidence types with more difficult populations and features that cannot be numerically expressed. There is a plethora of examples of over-interpretation in all forensic science disciplines, from DNA mixture interpretation to bloodstain pattern analysis. Several examples are discussed below, including the non-bloodstain pattern examples of microscopic hair comparisons and comparative bullet lead analysis that present published illustrations of over-interpretation.

Microscopic hair comparisons are a valuable forensic science tool when properly analyzed and interpreted by experienced microscopists; however, they do not generate objective data that can be expressed numerically and given statistical significance. Toward the end of the twentieth century, there were a number of examples of misuse of this science by assigning pseudo-statistical value of a "match" during court testimony in order to make it appear more objective. This over-interpretation was not made explicit in the reports, which concluded with an appropriate caveat to the effect that hair comparisons are not a means of personal identification. Recently, the FBI has reviewed more than 2000 of their cases involving microscopic hair comparison testimonies to determine if there was over-interpretation of the evidence. They found that there were false or exaggerated testimonies in >90% of those that had an inclusion. "In the most egregious example of this testimony, the examiner claimed to have an error rate of 1 in 5,000, and because there were four hairs, he took 5,000 to the fourth power and stated that the odds of those hairs coming from anyone besides the defendant approached one in a quadrillion" (Lentini 2015). The FBI's review of hair comparison testimony has led to a call for every state to conduct a similar analysis. In another non-FBI case, head hairs were microscopically matched with a frequency of 1 in 100 (which is a made up number, albeit deemed to be perhaps a conservative estimate) and pubic hairs matched with a frequency 1 in 100 (also unsubstantiated), and thus the examiner testified that there was a combined frequency of a 1 in 10000 chance of a random match. It is important to understand that this inexcusable over-interpretation problem must be dealt with, but should not obscure the validity of a properly conducted and interpreted microscopic hair comparison.

Another well-studied example of over-interpretation in the ultimate expression of the findings in a court of law is with comparative bullet lead analysis. Comparative bullet lead analysis was based on the ability of the analytical chemical technique that resulted in an extensive database of roughly 2000 analytically distinguishable compositions of bullet lead. The analytical chemical technique used (inductively coupled plasma – optical emission spectroscopy) to obtain the elemental composition of bullet lead was scientifically sound with high discrimination potential. Despite a good scientific foundation, over-interpretation occurred in testimonies when examiners attempted to assign a casework bullet to a specific box or one made at about the same time. For example, an FBI expert testified in *United States v. Davis* (103 F.3d 660, 8th Cir. 1996) that "the bullets must have come from the same box or from another box that would have been made by the same company on the same day." The FBI asked the National Research Council to conduct an assessment of bullet lead analysis, and in 2004, they published the findings of the Committee on Scientific Assessment of Bullet Lead Elemental Composition Comparison in the report "Forensic Analyses: Weighing Bullet Lead Evidence." The report was supportive of the scientific analysis but was critical of the testimony given in some cases. Ultimately, the following legal interpretation was made: "the committee found that CABL [*compositional analysis of bullet lead*] is sufficiently reliable to support testimony that bullets from the same CIVL [*compositionally indistinguishable volume of lead*] are more likely to be analytically indistinguishable than bullets from different CIVLs. An examiner may also testify that having CABL evidence that two bullets are analytically indistinguishable increases the probability that two bullets came from the same CIVL, versus no evidence of match status." Furthermore, "it is the conclusion of the committee that, in many cases, CABL is a reasonably accurate way of determining whether two bullets could have come from the same compositionally indistinguishable volume of lead. It may thus in appropriate cases provide additional evidence that ties a suspect to a crime, or in some cases evidence that tends to exonerate a suspect." Despite the positive aspects of the report highlighting the value of comparative bullet lead analysis, the FBI stopped offering this service and the technique remains unavailable to both law enforcement prosecutors and defense attorneys.

Blood trace configurations involving a limited number of airborne droplet deposits (discussed later in this chapter) can be difficult to interpret and lend themselves to over-interpretation. As previously detailed in Principle 8, "A collection of a few seemingly related dried blood droplets is not necessarily a pattern. It should never be treated as one. The number of blood traces may be inadequate to allow a meaningful interpretation." Unfortunately, assigning a mechanism for apparent airborne blood droplet deposits when there are too few of them to make a meaningful pattern appears to be too common an occurrence and has been a contributing factor to wrongful convictions (see the discussion in Section 11.1).

Another situation where a criminalist may over-interpret a blood trace configuration is where a non-descript deposit is wrongly attributed to transfer from a specific object or caused by a specific mechanism. A case example that demonstrates this type of error was a murder investigation reviewed by a consulting criminalist where there was no body or weapon recovered. In this case, relatively featureless blood traces in the fabric of an entryway throw rug were determined to be "hammer" imprint by a prosecution expert witness. As can be seen from the photograph in Figure 7.7, the blood deposit has no definitive shape and thus cannot be attributed to transfer from a specific

FIGURE 7.7 Throw rug from the suspected victim's residence, showing the distribution of droplet traces (circled by loose leaf punched hole reinforcements) and the alleged "hammer" transfer trace in the center.

object such as the face of a hammer. Additionally, due to the rug's texture, there are gaps in the deposit causing the apparent pattern to be incomplete. The assertion that this blood trace was made by the face of a hammer is a gross over-interpretation of the physical evidence and contributed to a spurious and unsupportable reconstruction of events.

7.3.2.2.3 Opinion of a Scientist vs. Scientific Opinion *There is a profound and critical difference between the opinion of a scientist and a scientific opinion.* Even scientists can lose sight of the fact that his or her personal opinions may not be grounded in science. It is important for scientists to keep this in mind throughout an investigation. For an opinion to be regarded as scientific, it must have undergone rigorous testing in accord with the scientific method and be relevant to observations of the physical evidence.

There are several examples where highly regarded scientists neglected scientific principles and inappropriately presented their personal opinions as scientific ones. One such case where this was guarded against was the case in Section 10.9 that took place in the stairwell of a 10-story apartment building.

Experts should know their limitations and understand the bounds of their expertise. Just because one is an expert in one thing, does not make him or her an expert in all areas of forensic science. One can have general knowledge of many areas of forensic science and have in-depth knowledge of some specialty areas. This forensic science generalist can be extraordinarily useful in providing a scientific oversight in a case; however, he or she cannot do or know everything. In the above case, the generalist criminalist was invaluable for the reconstruction because he possessed broad general knowledge and expertise in the interpretation of blood trace configurations, trace evidence, trajectory reconstruction, and firearms evidence. Generalists need to draw on the expertise of others, such as medical examiners and other specialists. Experts need to avoid the temptation of extending themselves beyond their areas of expertise. Incredibly some experts have claimed expertise in some evidence category based on watching reality television shows or being the victim of a shooting. For example, a criminalist should not make statements about cause of death, degree of disablement from wounds or other physiological determinations that require the expertise of a medical examiner. Similarly, medical examiners should not testify on subjects that may lie outside their own area of expertise, such as trace and bloodstain pattern evidence.

7.3.2.3 Deficiency in Scientific Integrity A scientific education does not guarantee that an individual scientist would always behave ethically. Although relatively rare, there have been instances of unethical actions by criminalists. There are a number of well-publicized examples of integrity transgressions by a scientist, with the most common ethical violations being drug use, stealing drugs, dry-labbing, lying about and/or forging credentials, falsifying results, and providing false testimony. However, it is important to remember that poor forensic practice does not necessarily mean unethical practice (Weinstock et al. 2000). Lack of knowledge, inadequate analysis, and other unintentional errors require appropriate corrective actions but do not necessarily constitute an ethical violation requiring punitive sanctioning. However, in one scenario when punitive action may be appropriate is when the individual is aware that they are incompetent and continues to examine physical evidence. The David Camm case is an example of this situation (see Chapter 11), as Robert Stites was "a plainly unqualified forensic assistant" who "has since admitted that he is not a crime-scene reconstructionist, has never taken a basic bloodstain-analysis course, and has almost no scientific background of any kind." (Camm v. Faith, 937 F.3d 1096 2019).

7.3.2.4 Cognitive Biases It is an inherent human failing to have biases, which are proclivities for presenting or maintaining a particular perspective, often paired with the refusal to consider alternative viewpoints. Forensic scientists are human and thus can suffer from biases. Some biases can be subconscious, while others can be more overt. It is important to recognize the existence of these biases and work to prevent them from affecting forensic analyses (Kassin et al. 2013). The scientific method when appropriately employed can help guard against overt and cognitive biases through rigorous hypothesis testing as a means of checks and balances.

There are a number of types of cognitive biases that can affect forensic scientists, including but not limited to confirmation bias, contextual bias, congruence bias, expectation bias, observer-expectancy effect, and the framing effect. There are numerous discussions in publications which detail these various concepts, and it is recommended that readers familiarize themselves with these resources (Dror 2020).

7.3.3 Pre- and Post-Event Artifacts

The possibility of pre- and post-event artifacts must be considered when doing an analysis of blood traces. Pre-event artifacts refer to items, bloodstains, or other evidence that exist prior to the occurrence of the event in question. Post-event artifacts are those referred to in Principle 6 at the beginning of this chapter, "Blood shed by the wound or wounds may be transferred by post event activities that may alter or obscure the blood trace geometry associated with the initial wounding event."

Not much is known or mentioned in the scientific literature about the prevalence of pre-event artifacts in blood trace configuration discussions. Frequency information about the presence of blood traces that pre-exist an investigation is unknown. For example, people may cut themselves while cooking, thus finding blood traces in a kitchen would not be an unusual occurrence. Observing cleanup efforts of blood traces in kitchens and overlooked smaller blood traces would also not be uncommon, and these would certainly complicate an investigation. One could think that persons responsible for cleanup would be able to find and remove all blood traces, but this is not necessarily true. An example demonstrating that subtle blood traces can go unrecognized was a case involving a homicide of a police officer in a hospital room. The hospital room was thought to have been thoroughly cleaned following the homicide and continued to be occupied by a succession of patients for several months. However, a detailed forensic examination of the room months later revealed significant airborne droplet deposits remaining on some areas of the ceiling and walls. The fact that such evidence can go unrecognized in a sterile hospital environment shows how blood traces can unknowingly exist in homes and other locations. Ascribing these pre-event blood traces to the event under investigation would be a serious error and lead to a gross misinterpretation and flawed reconstruction.

Post-event artifacts can also lead to serious misinterpretations. Awareness is necessary to avoid drawing the wrong conclusions. Some post-event artifacts are unavoidable, such as those caused by life-saving efforts, environmental factors or

FIGURE 7.8 Photograph of a homicide scene before (a) and after (b) removal of the body. The upper torso in the image has been blocked out. Traces on the floor were obscured or destroyed by the post-event blood, demonstrating the importance of properly documenting the traces before the body is removed. *Source: Photographs courtesy of Barbara Sampson, Chief Medical Examiner, NYC OCME.*

natural changes at a crime scene. For example, blood may be unwittingly transferred by people who come upon the body or 1st responders and emergency medical technicians who access the body for potential life-saving efforts. Life-saving efforts in themselves may cause blood traces to be created and blood to be transferred, such as by CPR creating small airborne droplets. In an outdoor scene, rain may wash away or spread blood traces or falling snow would alter or obscure a blood trace configuration pattern. Additionally, blood flowing from a body may pool and cover other blood trace configurations such as airborne droplets or imprints. Other post-event artifacts could be guarded against, such as blood trails being left by a body being carried from a scene or an investigator's footprints in the victim's blood. It is important to be aware that by the time the scene investigators arrive, many activities may have occurred to alter the evidence at the scene. Scene security is critical in minimizing post-event artifacts, as discussed in Chapter 5. To the extent possible, these post-event artifacts must be acknowledged and considered.

7.3.4 Risks Engendered by Limited or Erroneous Information

Contextual information supplied by eyewitnesses and investigators can be very useful in developing hypotheses to be tested during the course of undertaking a reconstruction with blood trace evidence. Helpful information provided by eyewitnesses and investigators includes, but is not limited to the number or people involved, the type or weapon, the duration of the event, and the locations of interest. Any known information provided concerning pre- or post-event artifacts would also be valuable. However, caution is necessary for evaluating the veracity of contextual information from eyewitnesses, as people can provide faulty information either purposefully or mistakenly due to the intensity of a situation. Additionally, information provided from investigators may be flawed or incomplete. A criminalist must critically evaluate the provided contextual information; otherwise, it could lead the investigation in the wrong direction. Fortunately, some erroneous information can easily be disproven by the physical evidence.

7.3.5 Problems with "Patterns"

7.3.5.1 General Problems The human brain is exceptionally proficient at pattern recognition, most notably with human facial recognition. However, the human mind is also capable of creating patterns where none exist, such as "seeing" animal shapes in clouds. Robert Park, in Voodoo Science: The Road from Foolishness to Fraud, stated while the most powerful computers are inferior to humans in detecting patterns we are so focused on finding patterns that we "...often insist on seeing them even when they aren't there, like constructing familiar shapes from Rorschach blots." For decades, the psychological and neurological foundations of these processes have been the subject of intense research. It is understood that pattern recognition and identification is largely based on human experiences and biases.

The blood trace configurations on the pillow and the corresponding opinions of certain experts in the Samuel Sheppard case are illustrative of the pattern recognition problem often referred to as pareidolia. This case is detailed in Chapter 10. With respect to the Sheppard case, one of the pathologists opined that he recognized a pattern in the pillow as having originated from a surgical instrument but paradoxically could not articulate what specific device it was. In this particular case, the opinion of the pathologist likely suffered from two problems; one is pareidolia itself, i.e., an unwitting desire to recognize patterns corresponding to objects that we are intimately conscious of – as a pathologist he would be prone to being very aware of surgical devices. Here too, the pathologist may have possessed a contextual bias since he knew that Sheppard was a medical doctor and would have access to and could have utilized the unknown surgical instrument to commit the murder. In the testimony of the pathologist, he could not name the device which underscores the shortcoming in this supposed scientific opinion – an unproven hypothesis that should have been recognized as flawed if a reasonable amount of skepticism had been applied. This skepticism could have been utilized in the form of peer review by a skilled criminalist or by a reasonable degree of self-examination. Blood trace configurations on fabric often present the examiner with poorly defined shapes that lend themselves to misinterpretation, especially to the poorly trained and less critical. Although a pathologist may be well educated in many areas of forensic science, it is unlikely that they have been adequately trained in the interpretation of blood trace configurations. The multitude of problems with blood traces on fabric are dealt with in more detail in Chapter 13; however, it should be recognized that issues arise from various factors including, but not limited to, folding/unfolding of the cloth, while the blood traces are still wet, texture, type of weave, fiber composition and degree of hydrophobicity. It is possible for crime scene examiners and even criminalists to imagine a still wet bloody knife as having been drawn across a fabric by the assailant based on the presence of a poorly defined bloody pattern that most likely originated from a previously bloodstained fabric becoming unfolded and producing gaps (with sharp edges). Another example of pareidolia is a case where an inexperienced examiner mis-characterized disturbed blood droplet deposits with a partial footwear pattern.

The pareidolia effect is clearly an integral component of the "belief engine" that Park has referred to and discussed in detail by Psychologist James Alcock (1995). The latter author has explained that new beliefs are created by our brains after processing new data detected by our senses. However, these beliefs, in general, are formulated or chosen based on beliefs that we already consider to be true at the cost of not considering what is actually true (Park 2000, 35; Alcock 1995, 15). Presumably, as discussed by Park, our behavior, in addition to our physical traits, is an aspect of genetics that evolved for the purpose of survival. This genetic capability allowed us to quickly recognize potential sources of food as well as the risk of predators, using pattern recognition skills for survival. We still possess this survival skill but unfortunately, in some modern contexts, it is prone to error because insufficient time has passed for our brains to evolve for modern pattern recognition needs – such as what we face daily in the forensic science milieu. Over the course of evolution, early humans could without hesitation imagine that they were about to be attacked by a predator, and if they were wrong, no harm no foul. Thus, an overactive imagination would help hominids in the ancient environment survive. In our context, the forensic science arena, an unbridled imagination can often lead to disaster that cannot always be remedied. Park has asserted that if we realize how readily we can tricked by the "belief engine", we are capable of utilizing more advanced sectors of the brain to provide improved strategies for coupling pattern recognition capabilities with observations involving nature – as opposed to the supernatural (p. 39).

We have previously discussed the importance of the active imagination associated with abductive reasoning (Chapter 1) and hypothesis development, but as pointed out in these discussions, this needs to be followed by and evaluated with the proper implementation of the verification stage of the scientific method.

7.3.5.2 Patterns Involving a Limited Number or Detail of Traces

When a pattern or configuration of airborne droplets is composed of a limited number or detail of traces, unavoidable ambiguities can be expected. As noted earlier in this chapter, Principle 8 states "A collection of a few seemingly related droplet deposits is not necessarily a pattern. It should never be treated as one." Given the inherent uncertainties that can arise in interpreting airborne droplet-like deposits consisting of only a few traces, one needs to avoid the temptation of over-interpretation. At times, the configuration will be uninterpretable as to cause or mechanism. This needs to be recognized and accepted otherwise over-interpretation will follow. The presence of too few airborne droplet deposits may not yield an interpretable pattern, and attempting to assign a single mechanism to its production risks leading to an inaccurate reconstruction.

A notable and well-publicized case where an early stage over-interpretation of a limited number of blood droplet deposits resulted in a misleading reconstruction that led to three criminal trials, with two convictions, was the David Camm triple murder case. This case and its complexities with respect to blood trace interpretations are described in detail in Chapter 11.

7.3.5.3 Chronological Sequencing It is often important for reconstructions to determine the order in which overlapping blood traces were deposited. As desirable as this capability is, it is often difficult to determine with a high degree of confidence. The interval of time between deposits is clearly a key factor in making such a determination; there is an inverse correlation between the duration of the interval and the success of chronological sequencing (Duggar 2000). This is because if the time interval is as short as a matter of minutes, then the blood deposits may blend together rather than have a discernable boundary, thus making sequencing impossible. On the other end of the time spectrum, a long interval would make determining the order of deposit vastly easier. There is a visual differentiation where there are distinct three-dimensional boundaries between the overlapping stains that allow for chronological sequencing. As an example, one could envision a blood pool that was completely dried, with a footwear outsole pattern in blood on top of it. The sequencing of these two blood deposits may be possible for a criminalist to determine through the use of oblique illumination. If the deposit of the outsole pattern had preceded the deposit of the blood pool, evidence of this might be detectible if the presence of the outsole pattern's features influenced the flow of the blood. Although it is possible to determine the chronological sequencing of different blood trace deposits, as the prior examples demonstrate, this often is not the case and a criminalist must be comfortable with making the conclusion that it is "indeterminate."

7.3.5.4 Effects Caused by Interaction of Blood and Target Surface The surface that a blood deposit forms on can be very important in the deposit's morphology and size, and for these reasons must be considered when rendering a blood trace configuration interpretation. There are four major features of a substrate that affect the blood

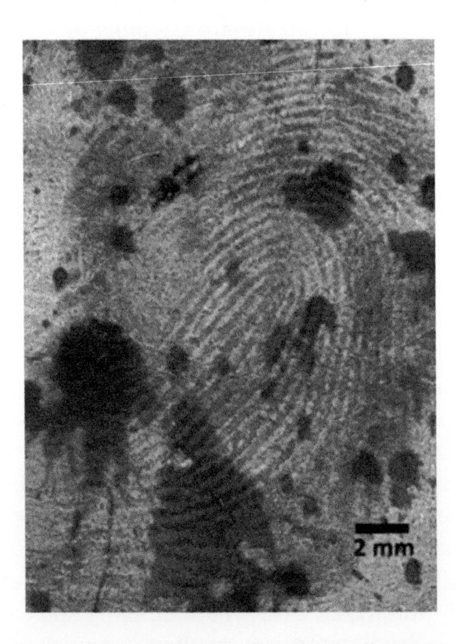

FIGURE 7.9 Overlapping blood trace consisting of a bloody fingerprint on top of spatter that had previously dried. The uninterrupted ridges can be clearly seen where they overlap the individual droplet traces. Photographed with oblique illumination at original magnification of 30 times. *Source: Photograph courtesy of Jeffrey Buszka.*

deposit formation: the surface texture, the sub-surface structure, the surface composition, and the target resilience. If a bloodstain pattern scientist intends on conducting a replication experiment, the target surface should be duplicated as closely as possible to ensure there are no deviations in the pattern due to differences in the substrate.

The surface texture plays an important role in the configuration and shape of the deposits produced from the impact of droplets. Very smooth, clean surfaces (e.g., glass and tile) produce deposits with featureless edges, most often devoid of scalloping and spines. In contrast, coarsely textured surfaces may produce prominent edge effects and spattering, as shown in Figures 7.10–7.12. It should be recognized that if blood drops into blood, then additional features would be present. For example, even though a single blood droplet impacting a smooth surface would have minimal edge effects, a second droplet impacting the first or falling into a pre-existing pool of blood would produce extensive spattering effects, as seen in Figure 6.26.

A second feature of importance when interpreting blood trace configurations is the sub-surface structure, specifically the porosity of the substrate. Blood droplets contacting porous surfaces can spread laterally or diffuse into the substrate, or both. Examples of porous surfaces are cloth, unfinished wood, and paper products including tissue, toweling, and cardboard. Non-porous surfaces include glass, tile, finished wood, and painted surfaces (i.e., walls).

The composition of a substrate material or possible surface coating is an important consideration for the interpretation of blood trace configurations (discussed in

(a)

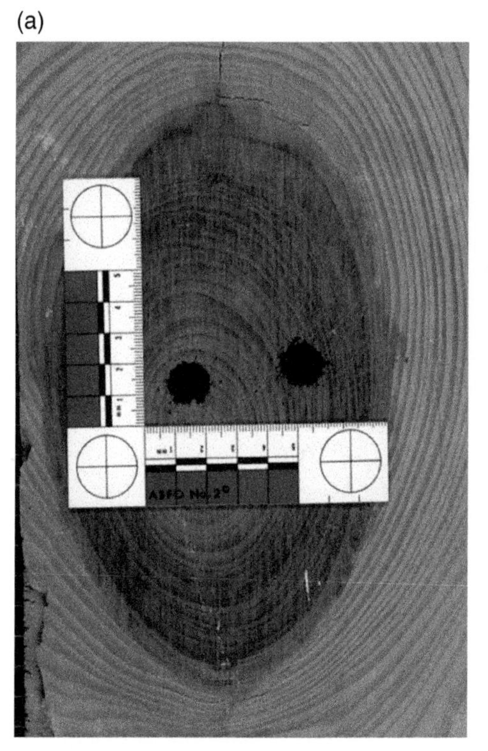

(b)

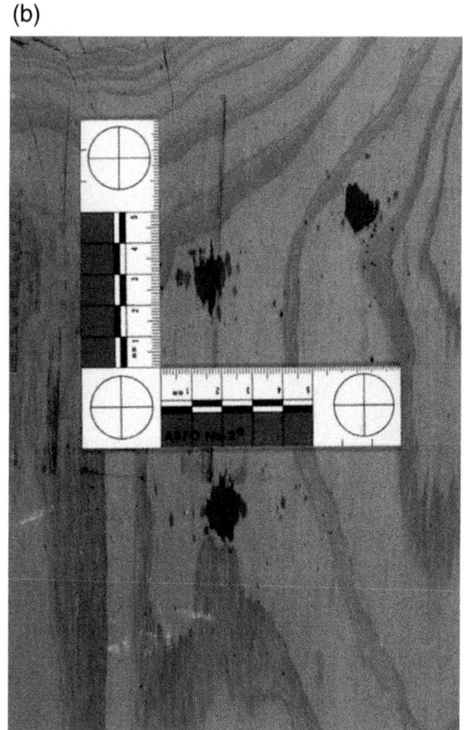

FIGURE 7.10 Very different trace morphology from impact with same wooden surface with different grain structure at same height of fall.

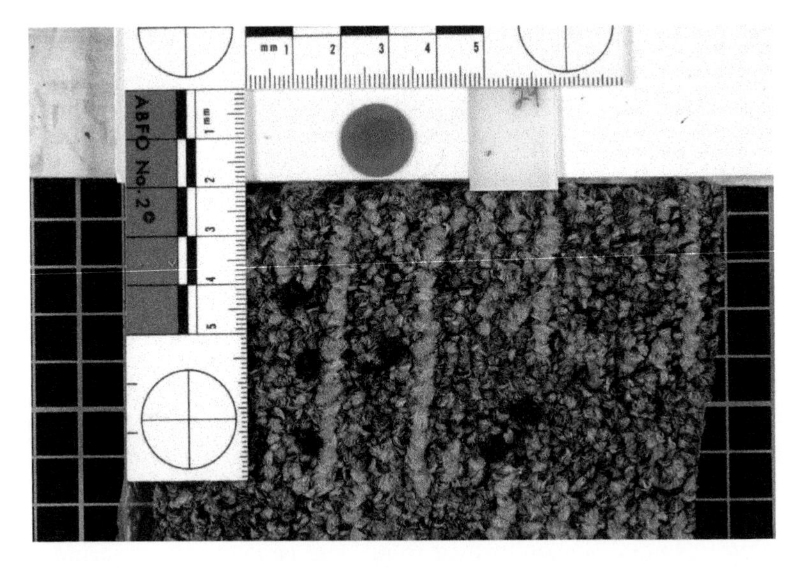

FIGURE 7.11 Same height of fall onto glass (top) and carpet (bottom). Carpet resilience is a major factor in the relatively small diameter of these traces.

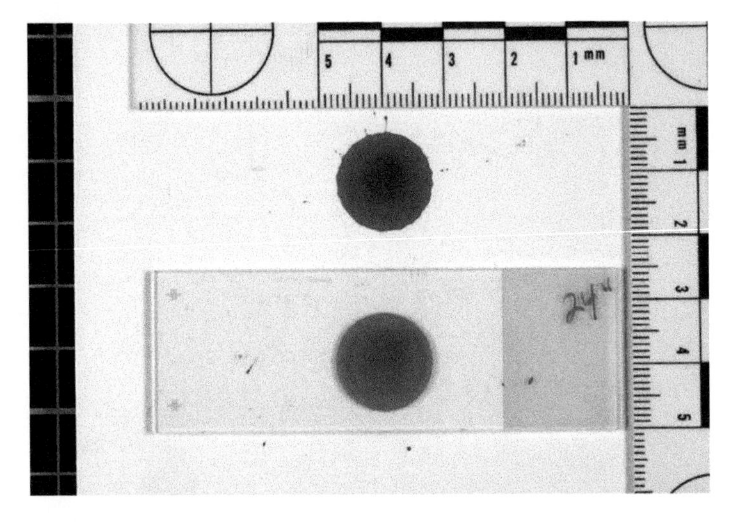

FIGURE 7.12 Glass v. ceramic tile. As commonly recognized, blood traces resulting from vertically falling drops onto a clean glass surface do not exhibit any crowning; in contrast, blood traces formed from vertically falling drops onto slightly textured glazed tile can exhibit crowning/edge effects (same height of fall for both traces).

Chapter 3). Of specific relevance is the hydrophobicity of the surface. An example is the contrast between a water-based liquid, such as blood, impacting normal porous paper and waxed paper. Waxed paper is impregnated with a hydrophobic material, and thus when it is impacted by blood, the droplet will initially spread from the impact and then rapidly contract due to repulsive effect from the hydrophobic wax-like material (Figure 3.7). When a blood droplet impacts porous paper, it may actually spread rather than contract. The interaction of aqueous drops with a hydrophobic surface is easily visualized when water droplets bead up on the surface of a newly waxed car finish or water-resistant garment. Consideration of the hydrophobicity of a substrate is important for the interpretation of blood traces on fabrics, the fibers of which may be coated with surface treatments such as water or stain repellants. A blood droplet on an article composed of such treated fibers may not soak into the body of the fabric and may dry before spreading in the fabric. Although the effect would not be as extreme, fabrics made with synthetic fibers (such as polyester, nylon, or acrylic) are more hydrophobic than those made with natural fibers (such as wool, cotton, and linen), and thus there are variations in the way that blood interacts with each of those textiles that must be considered when performing blood trace configuration interpretations.

The fourth major feature of a substrate that should be considered with blood trace configuration interpretations is the target resilience or "give." With respect to airborne droplets that impact the surface with considerable velocity, the target resilience is a factor to be considered. The target resilience refers to the physical resistance offered by the impacted surface. This can be important when blood drips onto a yielding surface, such as a plush or cut-pile carpet where the individual fibers are capable of recoil. Substrates that recoil lessen the impact force of the blood droplet with the surface resulting in a smaller blood traces. Most surfaces are unyielding, and thus, there is no diminution of the impact. It is not necessary to consider the substrate resilience with contact traces because there is no impact involved.

Fabrics pose additional challenges when trying to interpret airborne droplet deposits. If on preliminary inspection, it is observed that there are a large number of small traces arranged in a regular array on a woven textile, it should be investigated as to whether the blood trace configuration corresponds to the weave pattern itself. This would be an indication of contact rather than airborne droplet transfer. Contact with a bloody object can deposit small droplet-like traces on the high points of the fabric weave. On the other hand, airborne droplets impacting the low points (valleys or interstices) can be recognized as actual airborne droplets from a spatter event. An example is a situation where there are a large number of traces present on a fabric. If there are no small traces in the valleys and all of the small deposits are on the high points of the fabric, this would strongly suggest that the source was contact rather than airborne droplets. Consideration of the fabric substrate was important in the David Camm case (discussed in Chapter 11), where questions arose concerning the source of a limited number of small blood deposits on his t-shirt. There were conflicting opinions about whether the source of these small blood traces on the high points of the fabric weave were from contact with the daughter's bloody hair while removing the son from the vehicle or from spatter arising from the entry wound in her head. As mentioned earlier, these conflicting interpretations were due in large part to the minimal number of small deposits present on the t-shirt.

7.3.5.5 Configurations Observed after Application of Blood Presumptive and Enhancement Reagents

The application of blood enhancement and presumptive reagents, such as luminol and others discussed in Chapter 4, has the ability to reveal blood trace configurations (Figures 5.31 and 7.13b). However, one must be aware that their misuse or over-application will alter any blood trace deposit, thus making interpretations hazardous. The addition of these reagents runs the risk of changing the physical shape of a blood trace by spreading, diffusion, or flowing. For this reason, where possible, documentation of all blood configurations, especially airborne droplets, should precede attempts at chemical enhancement. Blood enhancement and presumptive reagents should only be used as a last resort to visualize

(a)

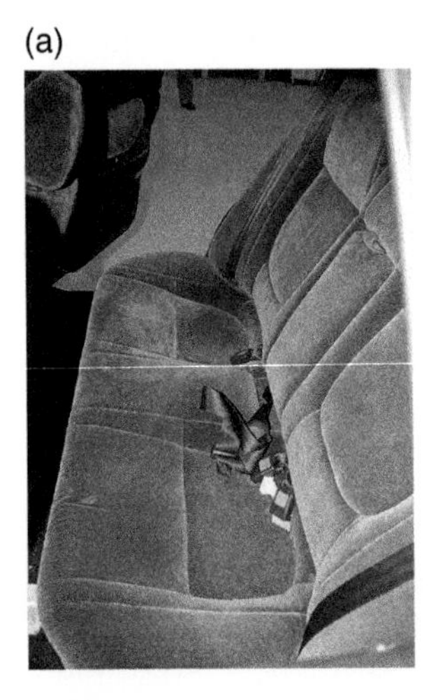

(b)

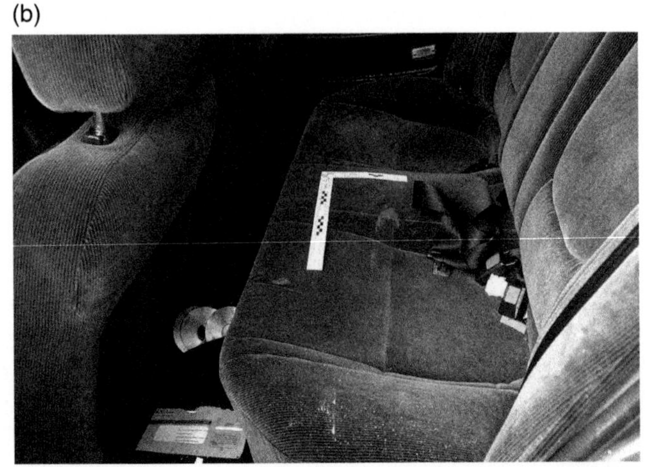

FIGURE 7.13 Photographs of a latent transfer blood trace on the rear seat of a vehicle, possibly from a baseball bat, before (a) and after (b) treatment with luminol. The chemiluminescence was documented photographically for 30s in essentially complete darkness; rear curtain flash was then triggered to illuminate the substrate and surrounding objects. *Source: Photographs courtesy of Norman Marin & Jeffrey Buszka.*

other types of blood traces, such as those created by cleanup activities, never to enhance airborne blood traces from a reagent enhanced trace. If blood trace configuration analysis is desired in a case, it is advisable to use other non-destructive non-contact visualization techniques, such as oblique illumination, infrared, or crossed polarized light photography (De Forest et al. 2009).

Despite not providing interpretable patterns and being destructive, blood enhancement reagents are useful for other purposes late in the scene investigation. They can show evidence of cleanup and also be used for locating and securing testable amounts of blood that otherwise might be overlooked, such as crusts of blood that have collected behind an electrical outlet cover, switch plate, or baseboard.

7.3.6 Problems with the Interpretation of Specific Blood Trace Configurations

Several specific blood trace configurations warrant specific discussion here because they are prone to flawed interpretations. Some of these are due to a lack of knowledge about the science, while others are due to popular misconceptions. Four of these are discussed separately below.

7.3.6.1 False Expectation of Airborne Blood Droplets from the First Wounding One common but incorrect assumption is that the first wounding action will produce airborne bloodspatter (Principle 4). As previously detailed in our discussion of the positional value of blood traces, the first impact does not result in the production of airborne droplets because there must be a prior accumulation of coalesced blood available to be spattered. This is true of actions such as gunshot wounds, stabbings, and bludgeonings. Theoretical considerations and properly designed experiments have proven this to be true.

A case example that demonstrates how this false expectation of extensive blood deposits from airborne blood droplets could potentially lead to an erroneous reconstruction was the murder of a driver in his vehicle by his passenger, detailed in Chapter 10.11.

7.3.6.2 Limitations in Determining the Origin with the Radial Spatter Configurations

When radial spatter blood trace configurations, or portions of them, are recognized at a crime scene, it is often desirable to attempt to determine a region or regions of origin for these. It is generally agreed-upon among knowledgeable experts that it is not possible or even necessary to determine a specific point of origin (see also the discussion in Chapter 6). There are two reasons for this. The first is that the origin may actually be an extended region rather than a point. This could be because successive blows could strike somewhat different areas in a mass of coalesced blood, or it may be the bludgeon used may have a broad area that would contact an area of the coalesced blood. Second, and perhaps even more importantly, it must be appreciated that there are errors or uncertainties inherent in the determination. The factors contributing to such errors do not seem to be properly appreciated. For this reason, there seems to be an illusory quest for precision. Such precision is neither possible nor even necessary. Attempts at extreme precision could even potentially lead to mis-interpretations and false reconstructions. A theoretical example of this is two bludgeonings that are close together may not be able to be spatially resolved using radial spatter configurations, thus questioning whether the victim had moved or been moved between the two actions.

There is a synergistic interplay between the geometric understanding of blood configurations and the use of DNA typing for source determination. For example, in the situation where there are two victims who produced possible overlapping radial spatter configurations, DNA can be used to help deconvolute the two configurations. Conversely, using the features of geometric blood trace configurations can provide a more economic and efficient sampling scheme or protocol.

The first step in calculating a region of origin is to identify droplet deposits with suitable geometries on nearby surfaces. In indoor scenes, these surfaces are normally walls, ceilings, or floors. Elliptically shaped single droplet deposits on such surfaces are generally the most useful. Once such traces have been identified and selected, their directionality and dimensions are determined. A discussion of this and the determination of the angle of impact and region of origin are in Chapter 6. Several well-formed blood droplet deposits are necessary for determining a region of origin.

In addition to the significant sources of uncertainty in the determination of the region of origin of a radial spatter configurations stemming from measurement errors in ascertaining the dimensions of the deposit and from the use of the trigonometric approximation, there are other sources of error related to the inability to assume a parabolic trajectory because of factors related to in-flight droplet oscillation and air viscosity, sometimes referred to as drag or air resistance (Chapter 6). Although they may provide a sense of security or comfort for the analyst, the use of computer programs does little to improve the situation. Computer-based measurements and calculations are wholly dependent on the quality of the input data. As the common saying goes "garbage in, garbage out."

7.3.6.3 Measurement Uncertainty and Significant Figures A topic related to the above discussion of determining the region of origin of a droplet deposit contributing to a radial configuration is that of significant figures. It is a fundamental error to report the result of a calculation to more places than is justified by the accuracy or degree of certainty of the input data used to make the calculation.

By virtue of their education and training, all scientists should have a clear understanding of measurement uncertainty as well as the concepts behind the terms precision and accuracy. It is important to remind oneself that all measurements have an associated uncertainty. The quality of the instruments used and the skill of the scientist or technician making the measurements will affect the degree of uncertainty. The use of the best instruments in the hands of the most skilled scientist will not eliminate uncertainty. The scientist making the measurements should have a clear understanding of the magnitude of the uncertainty of each of the measurements being made. In most situations, this degree of uncertainty should be reported along with the measured result. This is accomplished by reporting the results of a calculation or measurement with a +/− notation (i.e., range, standard deviation, or variance) or error bars for graphical presentations.

A hypothetical consideration of target shooting at a shooting range with a quality target rifle provides a useful example for understanding both precision and accuracy (Figure 7.14). If the target rifle is being used by a skilled shooter or, better yet, perhaps mounted in a machine rest where the skill of the shooter is no longer a factor, the tightness of the grouping or pattern of the impacts from several shots to the target is a measure of precision. If this grouping is quite small but is some distance from the bull's

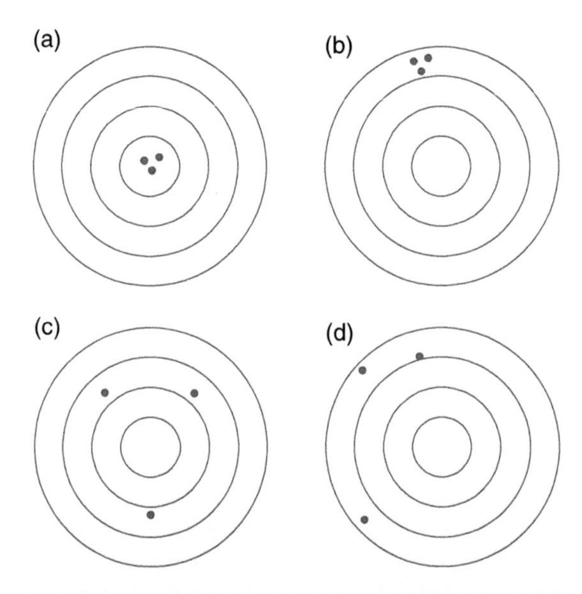

FIGURE 7.14 A concentric circle target (bull's eye) representation of the concepts of accuracy and precision where target (a) depicts a pattern displaying both a high degree of accuracy and a high degree of precision, target (b) depicts a pattern displaying poor accuracy but a high degree of precision, target (c) depicts a pattern displaying a high degree of accuracy (when averaged via vector addition) but poor precision, and target (d) depicts a pattern displaying both poor accuracy and poor precision.

eye, we would have an example of high precision but poor accuracy. This would be the anticipated situation in sighting-in a newly purchased target rifle. We would expect the high precision from a high-quality target rifle, but until the sights (open sight or telescopic sight) had been adjusted we would not necessarily expect good accuracy. The sighting-in operation of the target rifle would be complete when the tight grouping of the bullet impacts coincided closely with the bull's eye. The rifle then could be said to have both high precision and high accuracy.

7.3.6.4 "Height of Fall" Estimations Situations are sometimes encountered where it is desirable to attempt to determine the height from which a drop of blood fell to result in the production of a given deposit. This is not a trivial problem. Fortunately, this determination is not necessarily that frequent. The size of a deposit produced by a drop of blood falling vertically onto a surface, in addition to being influenced by the nature of the surface, is dependent upon its initial volume as well as the height from which it fell. It should be obvious that for a given height of vertical fall, a larger drop will produce a larger deposit. Similarly, up to a point, increasing dropping height will result in a larger deposit. The point where this correlation fails is the point where the dropping height is large enough for the droplet to approach its terminal velocity in air prior to impact. The terminal velocity for falling blood droplets depends strongly on the droplet size (measured as mass or diameter). For smaller droplets, the terminal velocity is lower and is reached in a shorter falling distance. Despite early assertions to the contrary, there is no such thing as a standard blood drop volume and this myth is discussed in Section 13.2.

Measurement of the diameter of a circular blood droplet deposit has the added complexity that the size may change prior to drying. If a blood droplet deposit is on a hydrophobic surface, such a certain plastics like polyolefins or waxed paper, the blood droplet will retract after initial impact (as discussed in Chapter 3). Conversely, when a blood droplet is deposited on a fabric, it could potentially expand after initial impact as the blood diffuses laterally into the textile material. This is one reason why it is important to perform height of fall experiments on exactly the same substrate as that bearing the questioned deposit.

In favorable circumstances, the volume of the drop that produced the deposit may be known or at least determinable. If, for example, the deposit is on a smooth non-porous surface, it may be possible to quantitatively recover the dried blood, weigh it, and calculate the volume of blood which would have been necessary to have made the blood trace. One would need to be able to make reasonable assumptions for the percent solids in whole blood and for the density of liquid blood to carry out this calculation. With necessary assumptions for these two numbers being made, a reasonable initial drop volume could be calculated. This is consistent with the use of Fermi estimates, which are commonly used in physics, engineering, and other disciplines to make good approximations using limited data. The density of liquid blood is somewhat greater than that of water at $1\,g/cm^3$. The percent solids in liquid blood will not be known exactly in a given case, and some variation can be expected; however, it is approximately 20%. It is understood that this will yield only a useful Fermi estimate. If one is unable to quantitatively recover the dried blood from the deposit, the situation becomes much more complex. Where the deposit is on a fabric, a piece of the same or similar fabric could be weighed and the areal density (mass per unit area) determined. Then, this could be used to calculate the mass of the evidence deposit alone through subtraction of the calculated contribution of the textile to the total weight. Such approximations may be useful in some instances. Another reported attempt at dealing with this overall problem on textured or porous surfaces is derived from correlating some measure of surface roughness with the number of spines produced for impacts from varied heights (Hulse-Smith et al. 2005). As noted earlier, no spines are produced by drops of blood that impact smooth surfaces such as glass. For surfaces that are not smooth, spines at the periphery of the deposit which appear to radiate from its center are produced by the impacts of energetic falling drops. The number of spines generally increases with dropping height. Studies of the number of spines on deposits produced by test drops on similar surfaces can be compared and correlated with those of the evidence deposit. The major problem with this approach is getting a reliable, meaningful, and reproducible measure of surface roughness of a difficult to characterize range of surface structures and morphologies. Specifically, the difficulty is finding a suitable test substrate that replicates the surface roughness of that bearing the evidence deposit. The present authors are not satisfied that this approach has been adequately researched to allow it to be applied to casework. Determining the dropping height when both the drop volume and the impact surface properties are unknown or poorly characterized remains a challenge. Again, fortunately, it is a relatively rare situation where a height of fall determination for a blood droplet deposit is necessary.

7.3.6.5 Crude Age Estimations of Dried Blood Traces Based on Appearance Although there have been instances of individuals attempting to determine the age of blood traces based on their appearance (e.g., color), reputable forensic scientists are aware of the complications associated with this practice. It may be useful for rough approximations to be used for investigative aids, definitive conclusions are not reliable and thus should not be offered in reports or court. The main issue comes from the fact that there are many factors that affect the perception of color, including the deposit's thickness, illumination, and the substrate material and color. Conditions of drying and storage (i.e., humidity, temperature) may be additional confounding influences. At best, only broad time estimates are possible. For example, bright red blood traces suggest that a time frame of hours or days, whereas a dark brown or black deposit points to a much older time since deposition such as months or years.

Attempting to make blood deposit age determinations from photographs presents even more pitfalls. This is because of the inability to assess factors such as deposit thickness, lighting intensity and color temperature, substrate albedo, etc. As previously stated, the dating of a blood trace from its color, especially when based only on its appearance in a photograph, should not be included in an expert scientific report or testimony.

There have been historical as well as current research attempts to use biochemical and chemical methods for assessing the age of a blood trace. However, as of this publication, there are no validated or accepted methods for more precise blood trace age determination.

7.3.7 Experimental Design

In the practice of the interpretation of blood trace configurations, occasions may arise where the conducting of experiments is indicated. Experimental design is a major challenge for any scientific investigation. A well-designed experiment must have a testable hypothesis (or hypotheses), be relevant to the question under study, and have controlled variables while considering other factors that could influence the results. There are two central problems with experimental design in the area of blood deposits trace configuration interpretations. The first is not accounting for competing mechanisms or hypotheses. The second is not accurately modeling the potential configuration of blood by employing naïvely designed experimental simulants. In cases where investigators want to demonstrate that a particular blood trace configuration could be created by a particular action or mechanism, it must be realized that this is not an experiment. It is conflating the goal of replication of a configuration with a well-conceived experimental design. These are discussed further below.

A replication that attempts to duplicate a particular blood deposit configuration as the end product of the investigation is not an experiment. It is a focus on duplicating a result and cannot be considered to be capable of proving that it is the explanation. In a replication, by merely getting it to "work" (i.e., achieving a result that appears to mimic the pattern being investigated), does not mean the event happened that way because there are other possible actions that could also get it to "work." Too often, when attempting to replicate an event, no concerted effort is made to consider alternative scenarios. Inappropriately, getting a similar pattern could prematurely and improperly end the inquiry. This could cause a naïve and/or inexperienced investigator to incorrectly focus on one pet theory at the expense of giving due attention to other possible explanations. This can be particularly problematic in situations where only part of the original blood deposit configuration may be present. An example of this is where the original mechanism for the presence of a dense configuration of airborne droplet traces on a wall near the floor was actually the result of blood dripping into pooled blood on the floor (secondary spatter), but due to subsequent cleanup, the original blood pool had been removed prior to the investigation. With only the pattern of airborne droplet traces on the wall, it could lead the investigator to explain the pattern by assuming that a bludgeoning occurred on the ground rather than the pattern arising from blood dripping from above into a blood pool. If a naïve investigator was to attempt to achieve a replication of a bludgeoning on the ground that might result in a similar pattern of airborne droplet deposits on the wall, he or she could miss other possible explanations and lead to an incorrect reconstruction. Awareness of the distinction between a simple replication and an experiment is necessary to avoid confusing a pattern produced by a mere replication with an explanation. This is not to say that replications should not be attempted, only that they should not be performed in isolation and instead be integrated with a scientific investigation.

Discussions on the limitations and misuse of simulants are not new in the field of bloodstain pattern analysis. In a 1990 book review in the Journal of Forensic Sciences, it was pointed out that "[t]here must be explicit recognition that simulations, although often necessary, cannot hope to deal with the variables and complexities that arise in a real situation. Simply attempting to simulate a particular phenomenon does not necessarily produce any more meaningful data and can provide a false sense of security" (De Forest 1990). Unfortunately, poor experimental design continues to be common in blood deposit configuration analysis with respect to employing naively designed simulations of anatomical structures or tissues. By definition, simulations will have limitations; however, these need to be recognized and accounted for when interpreting results. Some simulations will be totally unsuitable for particular investigations. The body is not a "bag of blood," as previously discussed. An example of a poorly designed simulation is the blood-soaked sponge that has historically been employed all too commonly by naive investigators when doing reconstruction experiments. The problem with simulants such as the blood-soaked sponge is that the distribution of blood in this situation will not replicate the distribution in tissue that is composed of finely divided capillaries and small blood vessels. By way of contrast, blood in polyurethane foam is coalesced in dramatically greater volumes compared to capillary rich tissue. The blood loading, the percentage of blood per unit volume, is vastly higher in such an object than in capillary supplied human tissue. Consequently, experiments using a blood-soaked foam do not accurately depict initial bludgeoning or shooting events involving portions of the human body where the blood is distributed in finely divided form in small vessels, such as local areas of the brain, tissue, or muscle. Other poor attempts at simulation design include blood-filled polystyrene foam mannequin heads (Eckert and James 1989) or blood-infused ballistic gelatin, which both neglect the concept that simulations should try to replicate the anatomical structure of the relevant region of the human body. "For the inexperienced, seemingly 'fancy' simulations often deflect attention away from failure to follow the scientific method. This must be guarded against. In addition, the design of simulations and the interpretation of the results obtained with such experiments are exceedingly complex and are proper roles for scientists, not investigators" (De Forest 1990).

7.4 Blood Trace Configuration Analysis as Part of a Holistic Approach to Reconstruction

Bloodstain pattern analysis cannot stand alone. It is clearly an integral part of some crime scene reconstructions, but it is not the only part. A holistic approach, as mentioned in Chapter 1, is the optimal way to extract the maximum amount of information from the physical evidence record. To only consider one aspect of that record is naive and could lead to missing vital information that could be critical to answering relevant questions that may arise during the course of the investigation.

There are two persistent myths in the bloodstain pattern analysis community that fail to recognize the importance of this holistic approach. The first myth is the idea that the analysis of blood trace configurations can be used as a sole aspect of a reconstruction. Significant information would be missed if this was the case. The second myth is that bloodstain pattern analysis is a stand-alone discipline that is not a part of crime scene reconstruction. If this is true, then what is the role of bloodstain pattern analysis? The absurdity of separating the interpretation of blood traces from reconstructions involving an array of other relevant traces cannot be understated. The case examples detailed in Chapter 12: Redefined Cases illustrates some of the dangers of confining the scientific investigative attention to these myopic perspectives. These cases demonstrate the fact that had the focus been exclusively on blood trace configurations, the cases would not have been solved or the correct conclusions would not have been reached. There is a great need for those persons performing blood trace configuration interpretations to be thoroughly conversant in other aspects of crime scene reconstruction. These cases also demonstrate the integral role blood trace configuration analysis has in crime scene reconstructions, and the fact that bloodstain pattern analysis is not a discipline that can stand alone.

References

Alcock, J.E. (May/June 1995). *The belief engine. Skeptical Inquirer,* 14–18.

Camm v. Faith (2019). *937 F.3d 1096.*

Daubert v. Merrell Dow Pharmaceuticals, Inc. (1993). *509 U.S. 579, 113 S. Ct. 2786, 125 L.* Ed. 2d 469.

De Forest, P.R. (1990). Book review: a review of interpretation of bloodstain evidence at crime scenes. Journal of Forensic Sciences 35 (6): 1491–1495.

De Forest, P.R., Bucht, R., Kammerman, F., Weinger, B., and Gunderson, L. (August 2009). *Blood on black - enhanced visualization of bloodstains on dark surfaces.* U.S. DOJ Award # 2006-DN-BX-K026. NCJRS Document # 227840.

Dror, I.E. (2020). Cognitive and human factors in expert decision making: six fallacies and the eight sources of bias. Analytical Chemistry 92 (12): 7998–8004.

Duggar, A.S. (2000). *Identification of initial and secondary stains in overlapping bloodspatter patterns.* Master's thesis. John Jay College of Criminal Justice, CUNY, New York.

Eckert, W.G. and James, S.H. (1989). Interpretation of Bloodstain Evidence at Crime Scenes. Elsevier Science Publishing Co.

Hulse-Smith, L., Mehdizadeh, N., and Chandra, S. (2005). Deducing drop size and impact velocity from circular bloodstains. Journal of Forensic Sciences 50 (1): 54–63. JFS2003224-10, doi:10.1520/JFS2003224. ISSN 0022-1198.

Kassin, S.M., Dror, I.E., and Kukuck, J. (2013). The forensic confirmation bias: problems, perspectives, and proposed solutions. Journal of Applied Research in Memory and Cognition 2: 42–52.

Kirk, P.L. (1953). Crime Investigation, 177–178. Interscience Publishers, Inc.

Kirk, P.L. (1955). *Affidavit of Paul Leland Kirk in STATE OF OHIO vs. SAMUEL H. SHEPPARD. State of Ohio, Cuyahoga County, Court of Common Pleas, Criminal Branch,* No. 64571, pp. 1–27.

Lentini, J.J. (2015). It's time to lead, follow, or get out of the way – criminalistics section. AAFS Academy News 45 (5): 5. and p. 28.

National Research Council (U.S.) (2004). Committee on Scientific Assessment of Bullet Lead Elemental Composition Comparison. Forensic Analysis: Weighing Bullet Lead Evidence. Washington, DC: National Academy Press.

Park, R. (2000). Voodoo Science: The Road from Foolishness to Fraud, 38. Oxford University Press.

Pizzola, P.A., Roth, S., and De Forest, P.R. (1986). Blood droplet dynamics – II. Journal of Forensic Sciences 31 (1): 50–64.

Thorwald, J. (1966). Crime and Science: The New Frontier in Criminology. Harcourt: Brace & World, Inc.

United States v. Davis. (8th Cir. 1996). *103 F.3d 660.*

Weinstock, R., Leong, G.B., and Silva, J.A. (2000). Ethics. In: Encyclopedia of Forensic Sciences, 706–712. Academic Press.

CHAPTER 8

Science and Pseudoscience

8.1 Science

Although a tremendous amount of knowledge has been gained since the scientific revolution during the Renaissance, it is important to acknowledge that the essence of the development of a scientist is not merely the memorization of some fraction of these facts. Rather, science is both a body of knowledge of interrelated concepts about the physical world and a way of extracting that knowledge. The latter is most commonly known as the scientific method, and having a deep understanding of this process is the hallmark of a scientist. A discussion of the scientific method and its critical role in crime scene investigation is detailed in Chapter 1.

8.1.1 The Need for a Generalist-Scientist in Crime Scene Investigation

To be most effective, a scientist at the crime scene should have broad scientific knowledge of many areas of forensic science combined with in-depth expertise in one or a limited number of specialty areas. The possession of both generalist scientific knowledge and in-depth expertise in one of a limited number of scientific specializations is not mutually exclusive. The derisive phrase "jack of all trades, master of none" is an unfair and misleading characterization of the generalist-scientist, as it ignorantly disparages and fails to recognize the critical importance of having this overarching knowledge-base when investigating a crime scene. There are many cases presented in this book, particularly in Chapter 11 ("Bad Cases"), that demonstrate the benefits of having a generalist-scientist for case resolutions. It is the generalist as described above who would be in the best position to decide if more specialized personnel are needed at a specific crime scene, or to examine specific items of evidence such as motor vehicles that have been impounded for additional detailed examination. The generalist concept

Blood Traces: Interpretation of Deposition and Distribution, First Edition. Peter R. De Forest, Peter A. Pizzola, and Brooke W. Kammrath.

that we have advocated for is similar to the idea supported by National Institute of Forensic Science of Australia & New Zealand referred to as the "Multi-disciplinary practitioner" (2019).

8.2 Pseudoscience

Pseudoscience consists of beliefs, practices, and claims that are neither grounded in science nor testable by the scientific method. Pseudoscience often masquerades as real science which makes it both alluring and potentially dangerous. Many examples of pseudoscience exist, with the bulk of them being relatively benign, such as astrology, phrenology, palmistry, and fortune telling. However, pseudoscience has no place in determinations made within our justice system, which should be grounded in the physical evidence and its scientific interpretation.

Properly understood events reconstructed from blood trace deposit configurations are based on tested and reproducible scientific principles. Bloodstain pattern analysis is not pseudoscience. However, there are some so-called experts, who some may call "enthusiasts," who can be characterized as pseudoscientists because of their nonscientific academic preparation and practice of bloodstain pattern analysis. Examples of this genre may include private investigators who fancy themselves as qualified experts with the notion that after having taken a 40-hour course or read a trade magazine or blog on investigations that they sufficiently and accurately interpret and reconstruct an event from the blood trace configurations. The distinction between pseudoscientists and enthusiasts may be difficult to discern in many instances.

8.2.1 The Pernicious Consequences with Respect to Reconstructions

There is a serious danger when a nonscientist attempts to reconstruct an event from a blood trace deposit configuration. Error is likely. By not understanding the fundamental science, the risk of an inaccurate conclusion and misleading opinion could lead to ills such as wrongful convictions or false exculpations. Two well-publicized case examples that demonstrate these perils are the David Camm and Joe Bryan cases, the first of which is detailed in Chapter 11 and the second was the subject of a two-part ProPublica-New York Times Magazine investigation by Pamela Colloff (2018a, b). In neither of these, can we know based on the physical evidence that the defendants were factually innocent; however, it is clear that the interpretations of the blood traces were erroneous and likely lead to convictions that were not justified by a scientific interpretation of the physical evidence.

8.2.2 Pseudoscience Characteristics

There are many indicators of pseudoscience that have been described by writers on the nature of science and scientific philosophy (Beyerstein 1996). Five characteristics of pseudoscience that Beyerstein discussed in detail and that the present authors subsequently have identified as being diagnostic in the field of forensic science are (i) isolation, (ii) nonfalsifiability, (iii) misuse of data, (iv) lack of replicability, and (v) claims of unusually high precision, sensitivity of detection, or accuracy of measurement.

8.2.2.1 Isolation Isolation of a laboratory, crime scene unit, or individual consultant may result in the practice of pseudoscience or some form of incompetence. Groups or individuals, especially those who have not been involved in mainstream forensic science before, can find themselves alienated to such an extent that they do not have a grasp on what is acceptable science and what is not. Many laboratories have found themselves in such a position when they are not actively engaged with other labs, professional development, science meetings, et cetera. An illustration of this is the Nassau County Police Lab ("Forensic Evidence Bureau") debacle that surfaced in the latter part of 2010 and ultimately resulted in the laboratory being shut down. It is in these kinds of settings when interacting with others that those that are practicing in such a fashion and environment have a good chance to find out that they as an individual or group are in serious need of improvement. Of course, nowadays, with regard to most disciplines, the individual examiner can go online (if they are reasonably selective) and have some notion of what is reasonable to do. Of course, online is only one avenue (although broad in scope); however, getting out physically and attending meetings, as well as attending scientific sessions, is an essential mechanism to not becoming isolated. With many of the forensic disciplines participating in the state technical working groups (TWGs) is greatly beneficial – reading the scientific working group (SWG) guidelines can often also be quite valuable. Groups or individuals have to realize too that allowing oneself to become identified solely with one particular association possesses its own set of hazards. For example, if an upcoming criminalist becomes a member of IABPA, they have to be cautious not to accept everything they read or hear as necessarily correct. Actually, any of us, who think of ourselves as scientists, must be critical thinkers (Browne and Keeley 2007) and evaluate all that is pertinent to our profession whether it is generated by the IAPBA, the AAFS, or other entities. An integral part of this enterprise, and the scientific method itself, is possessing the skill to ask the right questions of whatever issue we are presented with, either in casework or research.

8.2.2.2 Nonfalsifiability In the field of bloodstain pattern analysis, it is unfortunate that there have been made certain interpretations about blood trace configurations that are not able to be tested, and thus are nonfalsifiable. If an assertion is not even conceivably falsifiable, then it does not qualify as science and therefore can be classed as pseudoscience.

A case example of a pseudoscientific assertion made by a nonscientist "expert" was in the "Imagined Mist Pattern" case, detailed in Chapter 11. Another example of nonfalsifiability in cases is when there are claims that a defendant was present at a crime scene because of the presence of "bloodspatter" on a coarse knit sweater. In this case, the consultant who reached this opinion had access to only photographs that can be described as "overall" or "establishing" with no close-up photos. The blood trace deposits that were originally present had been cut out by a forensic biology for DNA analysis prior to examination by the consultant. Thus, the only source of evidence available for examination was the overall photographs taken prior to laboratory submission. There was no way to visualize the deposits on these kinds of photographs and describe them as spatter. It is possible that the blood traces were spatter to begin with but then were altered while still wet. The level of detail in the provided photographs did not exist to allow a qualified expert to opine that blood droplet deposits were present, thus making any conclusions incapable of being tested.

8.2.2.3 Misuse of Data An identifying characteristic of pseudoscience is the misuse of data to make unsupported and exaggerated claims. Part of this feature of pseudoscience identified by Beyerstein (1996, p. 31) is how old something is and honored because "...it has survived so long" and thus its worshippers mistakenly assume that it has strong or "rich" underpinnings. Forensic scientists must base their conclusions and reconstructions on the physical evidence without bias or embellishment.

The "Shooting of a Driver" and the "Dew Theory" cases detailed in Chapters 10 and 11 are prime examples where misused data led to erroneous and pseudoscientific reconstructions by police investigators and experts.

8.2.2.4 Lack of Replicability Reproducibility is an essential component of science. When individuals make claims that are unable to be replicated, those results cease to be scientific. Sadly, some nonscientist bloodstain pattern experts may fall victim to this especially when the only training is from a 40-hour course.

In the "Dew Theory" case (Chapter 11), the prosecution expert did not replicate the "fried-egg"-like appearance of the evidence deposits with his demonstrative exhibits, as there were readily observable differences between the diffusion evident in his presented exhibits and the actual observed separation of serum and cells of the evidence deposits.

8.2.2.5 Claims of Unusually High Precision, Sensitivity of Detection, or Accuracy of Measurement Nonscientist experts are particularly susceptible to making pseudoscientific claims of excessive precision, sensitivity, and/or accuracy due to their lack of knowledge concerning scientific concepts. These include a deficiency of understanding of these terms and fundamentals of measurement and significant figures. With computer programs affording the calculation of unlimited decimal places and unreasonably exact areas of origin (Chapter 6), this danger must be guarded against by criminalists performing reconstructions from blood trace deposit configurations.

8.2.3 Hallmarks of a Pseudoscientist

Even if the subject is a legitimate science, this does not guarantee that the individual performing the work is properly applying the science and not acting as a pseudoscientist. This "expert" may disagree that it is even a science, as if this somehow frees them from having an appropriate academic background. Such is the situation in bloodstain pattern interpretation where although there is universal agreement that blood behaves according to the laws of physics, many bloodstain pattern analysts either devalue the science or act in a pseudoscientific manner with respect to casework or research.

The practitioners or enthusiasts of any aspect of criminalistics would benefit from reading literary works like Beyerstein's "Distinguishing Science from Pseudoscience"

(1996) or Browne and Keeley's "Asking the Right Questions: A Guide to Critical Thinking" (2007). The need to self-examine one's own work or be skeptical of that of other individuals is equally essential in all areas of criminalistics, including the analysis of blood trace deposits. Of course, if one does not realize that they are attempting to practice criminalistics, then the chances of knowing that there is any reason whatsoever to study documents of this type is severely hampered. Beyerstein identified and discussed a number of "hallmarks" of the pseudoscientist. It should be kept in mind that not everything a practitioner does may fall under this label; it may be only one facet of what they are doing that is at fault. Some of the features that Beyerstein considered were impenetrability, bunker mentality, lack of formal training, ulterior motives, and the bandwagon effect and might be instructive to consider these in some detail. Unfortunately, in the experience of the present authors, many nonscientist practitioners of bloodstain pattern analysis quite often display these hallmarks of pseudoscientists. However, it is also true that some so-called scientists have also exhibited many of these characteristics.

8.2.3.1 Impenetrability Impenetrability is one attribute that has often manifested itself in the work of many forensic scientists and bloodstain pattern analysts. Impenetrability is "...an unshakable commitment to some questionable finding or hypothesis." Self-aggrandizement is another sign of the impenetrable one. In these instances, even when presented with overwhelming evidence of the wrongness of their belief, either with respect to case work, research, or both, they are unwilling to yield to even the most reasonable and justified criticism. This is especially true when the criticism directly contradicts their central ideas or findings. Beyerstein most succinctly stated "When the disputed work is a direct extension of the researcher's ideology or core belief system, biases are especially likely to interfere with objectivity." Modesty or humility is not commonly one of the attributes of "impenetrable" individuals and therefore genuine acceptance, or any level of admission at all, of having taken an incorrect or specious position is the antithesis of their personality. In the forensic field, all too often there may be a tendency or strong temptation to become overly enamored with oneself because of media attention often coupled with recognition from the agency the expert is employed by.

8.2.3.2 Ulterior Motives (Financial Gain/Recognition) Efforts must be made to eliminate ulterior motives. Having ulterior motives can have serious consequences for various aspects of forensic science. Blood trace configuration interpretation is not an exception. Ulterior motives, most prominently for financial gain or fame, are two very alluring temptations for forensic scientists. That is not to say that either the desire for financial gain or recognition is wrong. On the other hand, there are negative repercussions when these desires become too intense. One could argue that the workshop model of training has prospered from these ulterior motives. For some participants, the attraction of the workshop is the perception of increased income due to the addition of a "bloodstain pattern analysis specialty" with little intellectual or academic challenge or sacrifice in the form of significant investment of time. Additionally, those who attend these 40-hour courses may obtain an unjustified confidence in one's ability to effectively perform this type of work. For the workshop instructor and/or organizer, the financial gains are significant (as much as twenty thousand dollars for a week-long

40-hour course) as are the resulting prestige of being the grand "expert" teaching future so-called experts. They become increasingly successful financially while also obtaining more and more recognition – a dangerous and addicting elixir. This situation persists and has not been effectively challenged. Bloodstain pattern interpretation is very susceptible to this situation because few individuals see the need for a scientific academic background which opens the door for these 40-hour courses.

Prosecutors are also often susceptible to ulterior motives to hire outside consultants who can provide a desired opinion despite the lack of appropriate scientific credentials on the part of the proposed witness. The motivation for prosecutors is a conviction which potentially leads to reputation-boosting accolades and other rewards. Too often the prosecutor ignores the capabilities present in their local forensic science laboratories in lieu of obtaining an expert who will provide testimony consistent with the theory of the case. A prosecutor rarely possesses adequate knowledge to vet an external expert in forensic science, especially in crime scene reconstruction, and may not use available resources resident in local forensic science services. Prosecutors should heed the precaution of "buyer beware" and recognize that they are ill-equipped for assessing the qualifications and integrity of an individual purporting to be a scientific expert witness. For many prosecutors, the only qualification that they are concerned with is how often the consultant provided sufficient evidence for a conviction in their jurisdiction others.

8.2.3.3 Lack of Formal Science Education The need for a formal science education is stressed throughout this book and explicitly discussed in Chapter 9. One potential indication of a pseudoscientist may be centered on claims of being a scientist while lacking a formal undergraduate or graduate science degree. These pseudoscientists instead may cite a smattering of disparate college science or mathematics courses combined with a certificate from a 40-hour course. Science degrees are highly structured, hierarchical, and more than just a random assortment of seemingly science-related courses.

8.2.3.4 Unwillingness to Self-Correct One of the major strengths of true science is the intrinsic characteristic of self-correction. This is a natural process that comes with new discoveries or technologies for understanding the physical world. Although this is the nature of science, unfortunately in forensic science there is a perception that admitting error is fatal to the particular discipline as well as an individual's future as an expert witness. This needs to be countered.

In the subdisciplines of compositional bullet lead analysis and microscopic hair comparison, a new understanding of evidentiary significance has sadly mischaracterized them as "junk science." These are not pseudosciences! In most instances, the problem was inaccurate or exaggerated testimony going beyond the capabilities justified by the supporting science. Sadly, many lawyers paint these fields with a broad brush of "junk science" and as a result are unknowingly supporting the proposition of "throwing the baby out with the bathwater" rather than addressing the identified problems. In other fields of science, such as pharmaceutical, medical, and aviation sciences, self-correction is not a death knell to the future of the enterprise. Rather, serious errors are treated as sentinel events with learning experiences and corrective actions initiated (Doyle 2010). Forensic science is fundamentally a science, and thus self-correction should be welcomed and not feared, avoided, or concealed.

For forensic scientists, the fear of losing one's job or reputation makes admitting error difficult. However, when faced with new information or alternative interpretations, a scientist must be willing to self-correct. It is the obstinate refusal to self-correct in the face of an error that is characteristic of a pseudoscientist. This action was observed in the "Dew Theory case" (in Chapter 11), where the prosecution expert refused to correct his reconstruction despite evidence of his error.

Although it may be human nature for individuals to be unwilling to accept responsibility for errors with respect to a criminal proceeding, this is not an acceptable attitude for a scientist. There are different concerns over admitting fault for error by consulting and civil servant experts. A consultant may be in fear of a lawsuit or losing future business if they admit to any errors or wrongdoings. A civil servant is ordinarily indemnified against civil actions but may still have an unwillingness to self-correct for a number of reasons, not the least of which is ego. Some of this reluctance may also stem ironically from the accreditation perspective; if significant error is admitted, in addition to a possibly humiliating public outing, the likely result is a corrective action which may become public. As a field of forensic science, we need to clean up this practice and make self-correction not have such stigma. Accepting responsibility and doing the right thing by admitting error should not have significant adverse repercussions for the scientist.

Since forensic science is a science, when there is a significant difference of opinion, there should be a willingness to at least consider the other opinion. As Robert Park has stated rhetorically while commenting on the climate debate "If scientists all claim to believe in the scientific method, and if they all have access to the same data, how can there be such deep disagreements among them?" (Park, Skeptical Inquirer 2000). Some forensic scientists believe that it is perfectly acceptable to have deeply divided so-called scientific opinions and to express these in court without reservation. They seem to think that if forensic science is part of the adversarial system, then it is appropriate to argue our positions out in front of the Court. It is important to note here, unfortunately, that the opinion of a scientist is not necessarily a scientific opinion (Chapter 7). Two extreme examples of this are the David Camm and the "Dew Theory" cases which are discussed in Chapter 11. At a minimum, when there is a major issue between forensic experts, it would be advantageous for the court if these differences were resolved in advance of testimony, as was done in a New York City case by Judge Gerald Sheindlin (People of NY v. Castro 1989).

8.3 Bad Science

There is nothing about being a scientist that automatically ensures that the work will be done scientifically and guarantee accuracy. Bad science can be a product of a good scientist. Bad science is not necessarily the result of unethical actions or intent. Scientists are human and can make mistakes. Constant vigilance is needed to ensure that bad science does not pass through the technical or peer review process and into the court room. It is certainly possible that an individual is unaware that they are committing significant errors or are incompetent. Kruger and Dunning have pointed out that the incomepetence of some may lead to two different related problems – bad choices and overestimating their capabilities (1999). Since their incompetence greatly hinders their reasoning they don't recognize what they are doing. This situation can occur within a crime scene unit or forensic laboratory where the notion of constant vigilance – a key element of a quality assurance program is always essential.

The examiner cannot offer scientific opinions that blood trace configurations can establish the intent of an assailant or other mental concepts such as claiming that the victim was caught by surprise. This goes beyond forensic science, and this issue was raised in Chapter 7 ("Opinion of a scientist vs. scientific opinion"). Separating the scientific role from that of opinion and intent, is within the domain of the court or attorneys but not the scientist. It is bad science for an expert to conflate the two. Theories on intent or behavior cannot be hypothesis tested or falsified. As discussed in the case "A Vertical Crime Scene" (Chapter 10), the forensic scientist may be no more of an expert on these issues than a juror.

Another example of bad science would be to claim that there are cast-off patterns of blood on the back of a defendant's jacket based on enhancement of deposits of unknown origin with Amido Black. Anyone with suitable expertise in this area knows that Amido Black is a general protein stain not unique to blood. In a case example that illustrates the lack of selectivity of this enhancement reagent, one of the current authors used Amido Black on a floor to visualize latent cucumber prints. The cucumber patterns were not particularly helpful in that case, but did serve as a reminder that Amido Black will dye a wide array of material of questionable origin (animal and vegetable proteins). The recognition of this helps one to appreciate how potentially harmful the failure to understand the chemistry or biochemistry can be. Perhaps some of what we consider pseudoscience is in reality bad science and this should be considered as well.

8.4 Conclusions

In recent years, the scientific foundations of forensic science have been questioned (Balko and Carrington 2018). Claims have been made as to whether forensic science is a science. It is our contention that much of this criticism is misguided and due to the actions of pseudoscientists and/or rogue scientists. It is important for criminal justice stakeholders to be able to distinguish valid forensic science that has been used successfully for both inculpations and exculpations in criminal investigations, for more than a century, compared to the bad practices of a few bad scientists or pseudoscientists. It is important to put the flaws and shortcomings of forensic science in perspective with other criminal investigation methods (e.g., confessions, eye witness statements, etc...) and realize that forensic science is far and away the most reliable tool for informing decisions concerning guilt and innocence.

References

Balko, R. and Carrington, T. (2018). *Bad science puts innocent people in jail – and keeps them there: how discredited experts and fields of forensics keep sneaking into the courtroom. Washington Post* (21 March). https://www.washingtonpost.com/outlook/bad-science-puts-innocent-people-in-jail--and-keeps-them-there/2018/03/20/f1fffd08-263e-11e8-b79d-f3d931db7f68_story.html (accessed 1 October 2020).

Beyerstein, B.L. (1996). Distinguishing Science from Pseudoscience. British Columbia, Canada: Simon Fraser University http://www.dcscience.net/beyerstein_science_vs_pseudoscience.pdf.
Browne, N.B. and Keeley, S.M. (2007). Asking the Right Questions: A Guide to Critical Thinking, 8e. Upper Saddle River, NJ: Pearson Prentice Hall.
Colloff, P. (2018a). *Blood will tell: part I.* ProPublica. https://features.propublica.org/blood-spatter/mickey-bryan-murder-blood-spatter-forensic-evidence/ (accessed 1 October 2020).
Colloff, P. (2018b). *Blood will tell: part II.* ProPublica. https://features.propublica.org/blood-spatter/joe-bryan-conviction-blood-spatter-forensic-evidence/ (accessed 1 October 2020).
Doyle, J.M. (2010). Learning from error in American criminal justice. The Journal of Criminal Law and Criminology 100: 109–148.
Kruger, J. and Dunning, D. (1999). Unskilled and unaware of it: how difficulties in recognizing one's own incompetence lead to inflated self-assessments. Journal of Personality and Social Psychology 77: 1121–1133.
National Institute of Forensic Science (NIFS), Australia & New Zealand (2019). *A multi-disciplinary approach to crime scene management*, Australia New Zealand Policing Agency (ANZPAA), State of Victoria ©, version 1.0. https://www.anzpaa.org.au/forensic-science/our-work/products/publications (accessed 1 October 2020).
Park, R.L. (2000). Voodoo science and the belief gene. Skeptical Inquirer: 26–27.
People v. Castro, 545 N.Y.S.2d 985 (Sup. Ct. 1989).

CHAPTER 9

Modes of Practice and Practitioner Preparation and Qualification

9.1 Existing Modes of Crime Scene Investigation Practice

There are several different models of systems and personnel charged with the initial investigation of incident scenes and associated physical evidence. These run the gamut from full-time experienced scientists to nonscientist police officers that are investigating incident scenes on a part-time basis.

Although highly desirable, scientists work in crime scene units on a full-time basis in only relatively rare instances. A second model is where laboratory scientists are rotated to respond to and investigate crime scenes on a regular basis. In the latter example, the laboratory scientists can also serve as a case manager throughout the course of the particular case investigation and supervises and directs the laboratory physical evidence examinations. There is a significant advantage to this later model, specifically that the entire case would be overseen by a scientist, thus enabling a seamless continuity in the case investigation which would enable the totality of the event to be reconstructed based on the physical evidence. The investigation of blood trace depositions and configurations does not stand alone, and it is important to analyze it in context with all of the physical evidence. As stressed throughout this book, the importance and contributions of scientists at the crime scene are potentially extremely

Blood Traces: Interpretation of Deposition and Distribution, First Edition. Peter R. De Forest, Peter A. Pizzola, and Brooke W. Kammrath.

valuable (Chapters 1, 7 and 8). Scientists at the scene as opposed to poorly educated and narrowly trained bloodstain pattern experts can address a broad range of traces present at a scene, and not just blood configurations. Thus, a generalist-scientist at the scene presents the optimal model for effective crime scene investigation and reconstruction.

Another model system is composed of criminalists who are assigned to forensic labs as full time "bench" scientists who respond with some frequency to what are seen as major case crime scenes to assist police officers who work full time out of a crime scene unit. A "bench" scientist refers to a criminalist who has practical experience examining evidence on a laboratory examination table, colloquially called a "bench." In some instances, the crime scene response by these forensic scientists is not so frequent. This teamwork arrangement varies from agency to agency regarding how cooperative the situation is. A drawback to this model, as compared to the two earlier ones, is that the scientist only responds to the scene when called upon by a nonscientist police officer or attorney, often based on arbitrary criteria that may reflect concerns that are more political than scientific.

There are three primary nonscientist models for crime scene investigation response. One model is the crime scene units composed of full-time personnel (sworn or civilian) nonscientists. Another is where nonscientists are assigned to identification units (i.e., fingerprint, footwear, and other evidence analyses) and respond to crime scenes on a part-time basis. The least desirable model is where the first responding police officers handle the entire crime scene investigation. These all fall short of the ideal of a scientific approach to an investigation.

Outside consultants are often used both the prosecution and the defense counsel. With respect to the prosecution, consultants may be hired in many different jurisdictions because the requisite service may not exist within the jurisdiction. However, in many jurisdictions, although there is considerable expertise available locally from a county or state forensic laboratory, the prosecutor opts for the outside consultant whom they can be more certain will provide the opinion sought. It is usually the prosecutors who are in a position of possessing considerably wide discretion, with the authority to access special funds, to hire a consultant.

Like their government-based counterparts, the consultants may possess a wide spectrum of knowledge, skills, and ability. This continuum of expertise can range from exceptionally skilled to utterly unqualified. Unfortunately, the vetting of experts is inadequate or nonexistent. Some consultants may combine a woefully inadequate level of academic background and experience with exceptional theatrical skills and showmanship. As was discussed in Chapter 8, sadly, there is a good chance of an ulterior motive for helping the prosecutor in a current case, i.e., quid pro quo for being rehired for future cases. The more often the consultants align their opinions with the theory being espoused by the prosecutor, the more often the consultant will be rehired for future casework. From an adversarial perspective, that of many a prosecutor, this behavior is perfectly acceptable since they are convinced that they have charged the right suspect. From a forensic standpoint, this affair is a subtle form of a contingency fee: unscientific, immoral, and a violation of codes of ethics of a variety of forensic science associations, accrediting and certification boards.

There has been a widespread discussion of the distinction between civilian and sworn crime scene investigators. This should not be the crux of the issue and often diverts the focus of the main concern, which is the scientist vs nonscientist. Some sworn investigators may have a scientific background, and some civilian crime scene technicians will lack any science education. The focus should remain on the importance of scientists in crime scene investigations.

9.1.1 The Folly of Casting Technicians into the Roles of Scientists

It is too common for science to not be employed or misused in crime scene reconstruction, and too often, pseudoscience or bad science takes its place. A significant portion of casework illustrating this problem involves nonscientists conducting reconstructions (e.g., the David Camm case). Part of the reason for the prevalence of this situation is that forensic scientists have not played a significant role over the past forty or so years investigating crime scenes for physical evidence. Several factors have brought about this neglect. First, many forensic scientists are supervised by administrators who have a poor understanding of science and see no need for scientific expertise at crime scenes. Second, many forensic scientists themselves are not eager to leave the comfortable confines of the laboratory environment to conduct scientific examinations at crime scenes and do not encourage their own use by their agency. Certainly, many criminalists would not be happy to be called in the middle of the night to process a homicide scene. Third, laboratory staffing has been historically inadequate and so the task of crime scene analysis has been given to police officers even though ironically this places a greater economic burden on municipalities. The consequence is that both nonscientist administrators and narrowly educated scientists assume that science begins at the laboratory door. Without the "trappings" of science (i.e., test tubes, balances, laboratory setting, etc), people mistakenly think that scientific investigation is not occurring. This misunderstanding has continued to evolve into the present-day situation of crime scene analysis.

The void created by the lack of broadly educated scientists at crime scenes has led to the widespread use of evidence technicians for these purposes. Many technicians have ventured into crime scene reconstruction upon completion of a workshop or "school," eventually calling themselves "reconstructionists" – this is particularly true of so-called bloodstain pattern analysts. The preponderance of these examiners has not been adequately educated to handle the complexities involved in crime scene reconstruction. The general attitude that fosters this practice is that either science is not required or that crime scene reconstruction is not science, per se, but "...is the application of science to solve problems regarding the events in crime," (Bevel 2001, p. 161), and therefore whatever it is it can be handled by nonscientist analysts. Bevel has opined that "The use of the scientific method is not predicated on holding a science degree" (2001). While we would agree that it is possible to use some aspects of the scientific method without a science degree this is not sufficient, because the proper application of the scientific method also relies on rigor, integrity, and a knowledge of the sciences. We ALL may use a simplified version of the "scientific method" to informally analyze and solve everyday problems. Several important components may be missing from this simplified version. These would include skill with hypothesis development and testing, informed consideration and scientific evaluation of alternate hypotheses, scientific reasoning, experience with experimental design, a detailed understanding of applicable physical and chemical laws as well as knowledge of the properties of materials.

The scientist should adopt the stance of vigorously attempting to refute or disprove a favored hypothesis.

Not only do nonscientist practitioners not have an adequate scientific background to properly interpret results derived from natural laws, but they also lack the requisite scientific philosophy. Mere adherence to the scientific method is insufficient. The requisite scientific knowledge and philosophy are not acquired overnight by reading a recipe for the scientific method. Development of a scientist is an ongoing process that begins with years of hierarchical study as a science student and continues through possible graduate school and employment as a scientist. It is facilitated through a series of lectures and laboratory exercises and progress is validated through numerous examinations. Critical reasoning is refined gradually over a long duration. For many students, formal education is also insufficient. Graduate school helps immensely with this maturation process. Additionally, individuals have not developed into scientists until they have coupled the academic background with practical experience. Thus, it is unlikely that a satisfactory degree of a scientific ethos can be developed in a workshop milieu.

The respect for scientific philosophy and attitude should supersede the awe that the naïve hold for technology, i.e., the tools of the trade. Being a good scientist does not necessarily equate with knowing how to use technology. Tools or other gadgetry can be properly operated by skilled technicians. All too often, as technology advances there exists a strong temptation or an unwitting trend to supplant scientific reasoning with technical process. Because the results of the technology can be so impressive, the preponderance of time is taken up with the machines. Scientific reasoning falls to the wayside. Both science and technology are complementary in theory and practice. Technology cannot supplant science – if it does the final product is potentially meaningless or hazardous.The authors do acknowledge that it is certainly within the realm of possibility that a truly rare individual who has not acquired a science degree may be self-taught and have studied in detail physics, mathematics, the scientific philosophy, the scientific method, coupled with practical training under the auspices of an accomplished science mentor and become very skilled in the interpretation of blood trace configurations or other aspects of crime scene reconstruction. Nonetheless, the frequency of occurrence of the latter individual appears to be small. This would indeed be an unusual individual.

9.2 Preparations and Qualifications of Practitioners

What qualifications are required to conduct bloodstain pattern interpretation? We will consider this question from several perspectives – education, practical training (i.e., workshops and mentoring), along with casework experience.

Currently, there exists a broad continuum spanning the minimum and maximum education levels that crime scene examiners possess. This is similar to what is seen throughout the US and internationally regarding certain other forensic science practices. Police officers represent the vast preponderance of crime scene investigators, and many were assigned to their agency's crime scene unit. How they were assigned is another question that we may discuss. The real issue though, is what were the educational requirements to be assigned to the crime scene unit? For most police agencies, it is the basic educational requirement to be appointed a police officer, i.e., a high school diploma. A high school diploma is not an adequate educational preparation for meeting the scientifically challenging task of crime scene analysis. Some police departments are fortunate to have sworn personnel who possess science degrees. Progressive police agencies attempt to place these officers in either their crime scene units or laboratories. The NYPD in the 1990s and continuing into the next decade, for example, made a strong effort to assign it scientifically educated officers to the Police Lab's Firearms Analysis Section. Unfortunately, less academic emphasis was placed on the police officers assigned to the Crime Scene Unit. Clearly, there is a lack of recognition by police departments of the scientific challenges and complexity of crime scene investigations and blood trace deposit interpretations.

A number of contemporary authors in bloodstain pattern analysis are not of the opinion that a university diploma in science is a necessary educational qualification for experts in bloodstain pattern interpretation, and instead have suggested that college courses in physics, trigonometry and geometry would serve as alternatives (James et al. 2005). This is naïve, as it does not recognize the hierarchical nature of a science degree, where most courses require the completion of prerequisites, and higher level courses build upon the foundation provided by more fundamental classes. James, Kish, and Sutton also strongly recommend participating in a basic 40-hour course in bloodstain patterns. They advocate that "participation in laboratory experiments is crucial for understanding the dynamics of bloodstain pattern production and the mechanisms involved." Workshop attendees lacking science backgrounds who take isolated science courses are ill-equipped to properly design an experiment in conjunction with a case. It is unreasonable for an individual whose only "formal" training is via a workshop(s) to critically evaluate the efficacy of reconstruction models utilized for casework experimentation created by those with more skill (Eckert and James 1989; Kunz et al. 2015). Physical models of blood trace deposit mechanisms have serious limitations and are inherently difficult to design. Even in the best of circumstances, models will always be inexact and require compromises and caveats in their use and interpretation. Additionally, they are often misinterpreted and easily misused to provide false confidence about the certainty of a reconstruction. It is common for models to be designed which oversimplify assumptions, either with the action itself or the details of human anatomy. A gross oversimplification example is the use of blood-soaked sponges which have been used to simulate the human body with regard to gunshot wounding events. However, the body is not a "bag of blood," and the use of a sponge to mimic human tissue is absurd. There are a number of aspects to consider when designing, executing, and interpreting *ad hoc* experimentation. Some of these facets include the determination of when an experiment is actually needed, the experimental design, the number of replications, an evaluation of the conditions and variables, and choosing an appropriate model to approximate the human anatomy while understanding its limitations. Additionally, a qualified scientist understands that mere replication of a certain result or pattern is not sufficient for demonstrating a causal relationship, as there often exists alternative situations capable of yielding the same result. A result cannot be accepted unless alternative explanations are carefully considered and subsequently eliminated. Reconstruction experiments must not be conducted merely to impress the court. We concur with James *et al.* in their advocacy for the coupling of a formal education with years of experience and active membership in professional and scientific organizations.

Much of the practical training in the interpretation of blood trace configurations has and continues to be based on the workshop format. Herbert MacDonell was the main advocate of this form of training for several decades. In the early 1970s, he began training via workshops through his "Institute on the Physical Significance of Bloodstain Evidence." The IABPA has fostered this concept by requiring an "approved" 40 hour workshop for membership since the 1980s. This has had unintended pernicious effects, as pointed out in Chapter 2.

In-service training (IST) can be very beneficial to the acquisition of in-depth understanding of various aspects of blood trace configuration interpretation. Mentoring by a skilled and experienced criminalist can be an important part of this. The mentor does not have to be a co-worker. The mentor could be a local consultant who is available to assist and guide the trainee in a training program and with casework. For this reason, it is advantageous for the crime scene unit or laboratory to have an affiliation with a university or college so that these resources can be consulted and advice obtained from a faculty member who possesses pertinent knowledge and skills to act as a forensic science mentor. Additionally, forensic science associations and other scientific societies have services to pair a volunteer mentor, such as retirees and other senior criminalists, with developing scientists.

9.2.1 Education and Training

It is important to recognize that education and training are not synonymous. Although there may be some overlap, education and training have distinctly different purposes. Training is skill-based, task-oriented, and focuses on addressing particular problems. Training in the field of forensic science and specifically blood trace configuration analysis should be built on a solid scientific educational foundation as training does not provide the depth needed for scientific problem-solving. Initial training for those interested in the interpretation of blood trace configurations often consists of attendance at a 40-hour course that focuses on the basics of bloodstain patterns. Attendance at a 40-hour course alone should never be viewed as sufficient for creating experts in bloodstain pattern reconstruction, as it must be combined with education and experience for one to have the necessary knowledge, understanding, and expertise. Science education is concept-based and emphasizes the fundamental principles of nature and not focused tasks. In the field of bloodstain pattern analysis, training without the requisite scientific education is analogous to constructing a building on an inadequate and shaky foundation. People with training and/or experience without a proper scientific education may lack the insight and ability to know their limitations and guard against errors, which is a fundamental benefit to applying the scientific method.

The interpretation of blood trace configurations is an extraordinarily complex activity. It cannot be done algorithmically because no two scenes are the same and there are an infinite number of variables that can affect the totality of the scene, of which blood trace configurations are only one component. For one to have competence in the interpretation of blood trace configurations, it is important to possess a scientific education combined with training and extensive experience. Education without training or experience provides an individual with knowledge but no practical skills. It is important to remember that the purpose of a university education is not to prepare students for jobs where they are expected to "hit the ground running," and instead endeavors to provide the fundamental knowledge for them to adapt and apply to their appropriate training and use throughout their careers. Training without education or experience may create "experts" who can perform a particular task with no insight nor the ability to problem solve. Experience without education or training can be the most dangerous because that individual can have a misplaced confidence in themselves with no substantial knowledge or skills. These are comparable to the distinction between a scientist and a technician: a technician can be adept a particular procedure and perform valuable work, but would not be able to troubleshoot when the unexpected occurs. An individual lacking a scientific education can provide a useful function at a crime scene by recognizing and preserving potentially valuable blood traces, but is ultimately a technician who is ill-equipped to perform an interpretation of a reconstruction. It is the scientist with proper education, training and experience who is not only capable of identification, evaluation, and recognition, but who understands how to properly analyze and interpret the results as well as use the scientific method to define and solve an unforeseen problem. It has been said that you train people for *performance* while you educate people for *understanding*, both of which are critical for performing a valuable and adept reconstruction from blood trace configurations.

9.2.2 Experience

No amount of experience, in the absence of essential education and training, is sufficient for developing competency in the interpretation of blood trace configurations. If an individual lacks the fundamentals that are provided by suitable training built on rigorous science education, the value of these experiences cannot properly be predicted or assured. Poor experiences could have a negative effect and ultimately reinforce repeated errors throughout the individual's practice. This problem is conflated with an ignorance or naïveté concerning the concept of scientific skepticism. This scientific skepticism is essential in that it encourages critical evaluation of any opinion being considered.

Experience is an essential tool when it is built on a proper foundation of education and training. The following homicide case example effectively demonstrates this point. A defense expert was testifying at a pretrial hearing that the "expert" witness for the prosecution possessed an inadequate academic background and thus was not qualified to be opining on bloodstain pattern interpretations. During cross-examination of the defense expert, the prosecutor asked if he would feel differently if he knew that the prosecution witness had observed more than two thousand autopsies? The answer to this question was obviously no. Observing 2000 autopsies does not qualify anyone to be a bloodstain pattern expert. There are two things wrong with this question. First, witnessing an autopsy has very little relation to blood trace interpretations. Among the plethora of reasons why it is absurd to conflate the two, the behavior of blood is not the same when it comes from a live versus deceased person (i.e., changes in viscosity, surface tension, density, decomposition, clotting, etc...), and the activities that occur in a violent encounter are not mirrored in those of an autopsy. Second, the number of autopsies "attended" is meaningless. In an attempt to help the prosecutor understand the lack of adequate expertise and experience, the defense expert witness related to him that the current situation was analogous to a popular transmission repair commercial in which two incompetent mechanics were proud that they had a lot of practice – repairing the same transmission nineteen times. In another remarkable case, a prosecution witness that possessed no academic background claimed that "backspatter" is called "backspatter" because it is found on the back.

9.2.3 Mentoring

The importance of quality mentoring cannot be overemphasized in many fields of endeavor. This is particularly true in the area of the interpretation of the configuration of blood deposits and crime scene reconstruction. The value of mentoring is widely recognized and is an important component of the "Standards for a Bloodstain Pattern Analyst's Training Program" approved by the AAFS Standards Board (ASB), (ANSI/ASB, Std 032 2020). Unfortunately, the qualifications of the mentor as adopted as a standard by the ASB suffer from the same lack of requirements as the requirements for the trainee. Mentoring is more than just passing on knowledge conveyed solely by lecture presentations and readings. Mentoring consists of an extended period of one-on-one interactions of the trainee with an experienced expert which consists of an infinite number of opportunities for professional growth. These may take the form of formal mentoring programs or through unstructured relationships and interactions. Effective mentoring provides nearly continual informal constructive criticism, support, and encouragement over months or years. A mentee may benefit from interactions with more than one mentor thus gaining multiple perspectives and expertise from people with diverse experiences. The mentor–mentee relationship may span the entire career of the mentee, with the mentor serving as a lifelong trusted advisor.

9.2.4 Professional Development

Much knowledge can be acquired from attending scientific sessions at regional, national or international science meetings such as the American Academy of Forensic Science (AAFS), Northeastern Association of Forensic Science (NEAFS), California Association of Criminalists (CAC) and other regional societies. Attending scientific presentations helps in many ways – it assists the attendee to learn the "state of the art" and the structure of scientific research and experimentation. Much valuable information can also be passed onto both the novice examiner and the experienced criminalist in this manner. Moreover, the discussions and interchange in these sessions between the audience and the presenter is an important aspect of the overall peer review process.

9.2.5 Peer or Technical Review

The peer review process is a singularly important one. It has been articulated that scientific norms have evolved, utilizing the peer review process among other mechanisms, to minimize the chance of being led off-course by our human frailties, specifically our "self-serving" nature (Beyerstein 1996). This concept applies to all scientific enterprises and is just as critical in forensic science as it is in any other scientific or technical milieu. Clearly, this is of equal importance in the interpretation of blood trace deposits and other areas of crime scene reconstruction.

Peer review in science is the process by which any scientific conclusions or findings are evaluated by other scientists before they are generally accepted. In forensic science, in addition to the more general scholarship peer review based on publication and criticism, there is also the case-specific peer or technical review that assures that reports issued by a forensic science laboratory are relevant, reliable and supported by the data.

In forensic science examinations, the technical reviewer must determine if the analytical results support the conclusion(s) by reviewing all the relevant data that forms the basis for the opinion and conclusions in the report before it is issued. When peer reviewing the report of a reconstruction, the process is vastly more complex than is the case for many other laboratory examinations dealing solely with associative or identification evidence issues. In the latter cases, a "paper peer review" is often adequate. However for a reconstruction, a meaningful peer review requires that the reviewer have in-depth knowledge of details of the case. This would necessitate the full evaluation of the foundational documents supporting the reconstruction, including notes, reports, sketches, and photographs. The peer-review process should include a detailed evaluation of the expert's testimony. Although this is time consuming and not commonly done, it is important. Laboratory directors may receive positive or negative comments from prosecutors about a criminalist's testimony, the emphasis tends to be on the performance and communication ability rather than the scientific validity. This is best accomplished where the reviewer is physically present in the courtroom during the testimony, thus enabling the scientific merits of the testimony to be peer-reviewed. Additional peer review may occur in the adversarial process where an opposing forensic expert examines all case documents.

The qualifications of the case-specific peer reviewer are just as, if not more important than the expertise of the primary examiner. The qualifications of a technical reviewer are that they must possess expertise that is at a minimum reasonably similar to that of the person whose work they are attempting to review. In other words, for at least the specific case in question, with the questions that are being addressed by the basic examination, the reviewer must have a reasonable level of knowledge, skills and ability to competently evaluate the work of the other examiner. Moreover, the criminalist cannot be reluctant to tell the primary examiner that he/she has made a mistake, is overreaching or that more work is required to form an adequate opinion. In a large agency setting it is crucial that the primary examiner not lose sight of the fact that he/she retains primary responsibility for the examination and findings and not rely on the peer reviewer to discover any problems. Pursuant to the Supplemental Requirements of ASCLD/LAB-International program of accreditation the "...perceived responsibility for the scientific findings..." should not be shifted to the technical reviewer from the examiner (2006). The peer reviewer must also take the obligation very seriously, and not "rubber stamp" the work of the examiner. Learning how important this process is, as well as its component parts, is an integral part of the qualifications of the peer reviewer.

The second common form of peer review, as with other sciences, is in the submission of a paper for acceptance for journal or book publication. Over time, this then has a second peer-review, whereby the published journal article is either accepted by the scientific community and incorporated into practice, modified to satisfy shortcomings, or rejected entirely and potentially withdrawn from publication. Last, the process of presenting research and other findings at meetings of professional science associations offers ample opportunities for peer review, both in their acceptance for presentation and the response from attending scientists.

An introduction to the concept of the peer review process is often found in the university at the undergraduate level, or at least at the graduate level, where students are taught concepts of critical thinking and evaluation. These concepts are essential for students to learn to not merely accept some publication as accurate simply because it is widely distributed, perhaps by the government, or in a journal format.

9.2.6 Certification & Qualification Standards

The minimum education requirement standards for the field of bloodstain pattern analysis are severely inadequate. As of 2018, the International Association for Identification (IAI) only requires a high school education combined with 240-hours of study in related fields (e.g., Crime Scene Investigation, Crime Scene/Forensic Photography, Evidence Recovery, Blood Detection Techniques, Medico Legal Death Investigation, Forensic Science and Technology), which must include an approved forty (40) hour workshop in bloodstain pattern analysis followed by three years of experience (https://www.theiai.org/certifications/bloodstain/requirements.php). Neither SWGSTAIN nor the NIST OSAC has yet to improve upon this situation. This demonstrates a fundamental lack of appreciation for a science education and its importance for the interpretation of blood trace configurations, which is not unexpected considering that many committee members do not have degrees in science. It is illogical for both SWGSTAIN and OSAC to have the term "scientific" as part of their names but not require even an undergraduate degree in science to establish expertise in their disciplines. As a baseline, an undergraduate science degree should be mandatory for all forensic scientists.

There are serious concerns about the qualifications of those who do bloodstain pattern analysis. This was the topic of a hearing of the Texas Forensic Science Commission on 22 January 2018. These are not novel concerns, as they were also raised in the 2009 National Academy of Sciences (NAS) report (National Research Council of the National Academies 2009). This is ironic, because the formation of the NIST OSACs was stimulated by the findings of the NAS report, yet it has not led to the development of higher standards for education and certification requirements. This is one of the biggest disappointments of the NIST OSACs. The determinations of the Texas Forensic Science Commission were that to be allowed to practice bloodstain pattern analysis, one must be certified by an "appropriate" organization (e.g., the IAI) and do yearly proficiency testing. There are two fundamental problems with these standards. First is the lack of sufficient standards for certifications, as discussed earlier. Second, the current bloodstain pattern proficiency tests (such as those given by Collaborative Testing Services (CTS) are simplistic and do not adequately represent the scientific problem solving and the higher-level reasoning required for a reconstruction. An over-simplification is reflected in Part 1 of the "pattern description" test. The participants are provided with images of "patterns" and are given (approximately 13) pattern types to choose from, reflecting various mechanisms of deposition (CTS Report 2019). Part 2 is more challenging in which the participants are provided with images that may have more than one mechanism of deposition with no selections to choose from. It is clearly stated by CTS that Part 2 of the test is "...not a reconstruction of a scenario, but simply a test of pattern recognition and description." Recently, CTS has introduced an on-site crime scene proficiency test that is customizable based on the customer's needs which can include reconstruction components (Hockensmith 2020). This type of program certainly has the potential for much more realistic and challenging proficiency tests.

To date, no organization has adequately addressed the insufficiently rigorous qualifications with regard to the need for fundamental scientific knowledge and standards for those who do crime scene and bloodstain pattern reconstructions.

The two studies done by Taylor et al. (2016a, b) regarding reliability appear to be of preliminary value in establishing an anticipated error rate for bloodstain pattern classification. However, it must be questioned if the authors were correct when they assumed that the errors made were not a function of the competence of some of the examiners, since they were all "experienced" bloodstain pattern analysts. From a scientist's perspective, the lack of science-based higher education requirements could have contributed to the high number of erroneous responses. The panelists were invited to participate based on activity in bloodstain pattern analysis casework for at least five years, been qualified and testified in court as an expert on bloodstain pattern analysis (BPA), completed a minimum of 80 hours of training, coupled with their standing (in the BPA community) and their experience. There was no educational requirement, which is unfortunately common practice which has been reinforced by the recent publication of the ASB standard (AAFS Standards Board ANSI/ASB 032-2020 2020). Another contribution to the high error rate could have been relatively small size of the substrate (40 cm × 40 cm) bearing the blood traces. This inhibits the interpretation of the blood traces when compared to an actual crime scene where the configuration is viewed in context with the entirety of the scene. As we have written throughout this book, blood traces do not exist in isolation and a holistic approach to their interpretation is required. Anecdotally, this limitation in the dimensions of the object being studied is a similar problem experienced by proficiency tests participants in the present and past, as discussed above. We understand the difficulties in conducting such practitioner error rate studies, but the lack of context coupled with the limited size of the configuration on the substrate must be considered significant impediments to calculating a meaningful error rate for this subject. Other concerns regarding the accuracy of these two related studies concern the fact that neither were blind, thus this could have contributed to an increased number of inconclusive or multiple answers.

References

AAFS Standards Board, ANSI/ASB 032-2020, 1st ed (2020). *Standards for a Bloodstain Pattern Analyst's Training Program.* http//www.abstandardsboard.org/published-documents/bloodstain-pattern-analysis-published%20documents.

ASCLD/LAB-International® (2006). *Supplemental requirements for the accreditation of forensic science testing laboratories.* (Clause 5.9.4.1, Note 2) p. 21 (accessed 1 October 2020).

Bevel, T. (2001). Applying the scientific method to crime scene reconstruction. Journal of Forensic Identification 51: 150–162.

Beyerstein, B.L. (1996). Distinguishing Science from Pseudoscience. British Columbia, Canada: Simon Fraser University http://www.dcscience.net/beyerstein_science_vs_pseudoscience.pdf.

Collaborative Testing Service – Forensic Testing Program (2019). *Test No. 19-5601: Bloodstain Pattern Analysis,* 113.

Eckert, W.G. and James, S.H. (1989). Interpretation of Bloodstain Evidence at Crime Scenes, vol. 141–150, 289–297. New York: Elsevier Science Publishing Company.

Hockensmith, R. (2020). *Forensic Program Coordinator, Collaborative Testing Services*, Inc. personal communique.

International Association for Identification (IAI) (2018). http://www.theiai.org/certifications/bloodstain/requirements.php (accessed 1 December 2019).

James, S.H., Kish, P.E., and Sutton, T.P. (2005). Principles of Bloodstain Pattern Analysis: Theory and Practice, 9. Boca Raton, FL: CRC Press.

Kunz, S.N., Brandtner, H., and Meyer, H.J. (2015). Characteristics of backspatter on the firearm and shooting hand—an experimental analysis of close-range gunshots. Journal of Forensic Sciences 60 (1): 166–170.

National Research Council of the National Academies. Committee on Identifying the Needs of the Forensic Science Community. Committee on Science, Technology, and Law Policy and Global Affairs. Committee on Applied and Theoretical Statistics Division on Engineering and Physical Sciences (2009). Strengthening Forensic Science in the United States: A Path Forward. Washington, DC: The National Academies Press.

Taylor, M.C., Laber, T.L., Kish, P.E. et al. (2016a). The reliability of pattern classification in bloodstain pattern analysis, part 1: bloodstain patterns on rigid non-absorbent surfaces. Journal of Forensic Sciences 61: 922–927.

Taylor, M.C., Laber, T.L., Kish, P.E. et al. (2016b). The reliability of pattern classification in bloodstain pattern analysis-part 2: bloodstain patterns on fabric surfaces. Journal of Forensic Sciences 61: 1461–1466.

CHAPTER 10

Interesting and Illustrative Cases

A variety of cases will be considered in this chapter; most often to illustrate some point about where the investigation went awry. It should be noted that the discussions of these cases do not reflect the totality of the physical evidence record in each investigation but instead highlight the important and sometimes overlooked or controversial physical evidence. In some of these cases, we have provided additional references for fuller discussion of the case and physical evidence.

THE CASES:

1. The Sam Sheppard Case
2. Knife in the Gift Bag
3. The Farhan Nasser Case
4. Passive Documentation
5. British Island Case
6. Absence of Evidence is not Evidence of Absence
7. Triple Homicide
8. The O.J. Simpson Case
9. A Vertical Crime Scene
10. Tissue Spatter from a Large Caliber Gunshot
11. Shooting of a Driver
12. A Contested Fratricide

Blood Traces: Interpretation of Deposition and Distribution, First Edition. Peter R. De Forest, Peter A. Pizzola, and Brooke W. Kammrath.

10.1 The Sam Sheppard Case

10.1.1 Case Scenario/Background Information

On 4 July 1954, a woman named Marilyn Sheppard was found murdered in the bedroom of her home in a suburb of Cleveland, Ohio (EngagedScholarship @ Cleveland State University n.d.). She had been bludgeoned to death. There was a considerable quantity of blood spatter traces in the bedroom. Her husband, Dr. Samuel Sheppard, an osteopathic surgeon, rapidly became the primary suspect following a local newspaper blitz asserting his guilt. He was charged with the homicide and was tried and convicted in December of that same year. He was sentenced to life in prison. The case attracted a great deal of national media attention. As happens in some of these highly publicized cases, public opinion was sharply divided on the issue of Dr. Sheppard's guilt. This is a curious but common phenomenon in that members of the public who have very little knowledge or expertise in arriving at such conclusions have very firm opinions. In any case, this homicide offers useful examples for students and practitioners in the field of reconstructions derived from the configuration of blood trace deposits at a crime scene.

Following the conviction, the defense attorney, William Corrigan, pursued avenues for post-conviction review of the case. He contacted and retained Dr. Paul L. Kirk, a well-known criminalist, to advise on the case. Dr. Kirk was a Professor of Biochemistry and Criminalistics at the University of California at Berkeley and had published many papers and text books on the subjects of microchemical analysis and criminalistics. His book "Crime Investigation" is a well-regarded classic and had just been published the previous year. Ultimately, as part of the appeal process, Attorney Corrigan asked Dr. Kirk to examine the scene of the homicide, which had been fairly well preserved. In January 1955, about a month after the conviction, Dr. Kirk traveled to Cleveland and spent two days (23 January–24 January 1955) thoroughly studying the homicide scene and other evidence taken from the scene at the prosecutor's office. At the conclusion of his work, he submitted an affidavit about the physical evidence aspects of the case at the request of defense counsel. This affidavit has become an important document in the field of blood trace configuration-based reconstructions and has been cited by several authors. The affidavit criticized the original crime scene work and detailed Dr. Kirk's reconstruction in this case.

10.1.2 The Physical Evidence and Its Interpretation

There are other examples of blood traces and non-blood physical evidence that were considered in this case analysis but will not be delved into in this book. This discussion will focus on the blood trace deposits and configurations located in the bedroom. For further information on all evidence items, there are a plethora of scholarly resources in printed books and online, such as the Cleveland State University website http://engagedscholarship.csuohio.edu/sheppard/, which contains the "Sam Sheppard Collection."

The crime scene sketch by Kirk and photographs in Figures 10.1–10.5 show the blood trace deposit evidence from the crime scene bedroom.

We have already discussed pareidolia in this book in detail (Chapter 7), but it is of significant importance with respect to the first trial of Dr. Sheppard. The Cuyahoga

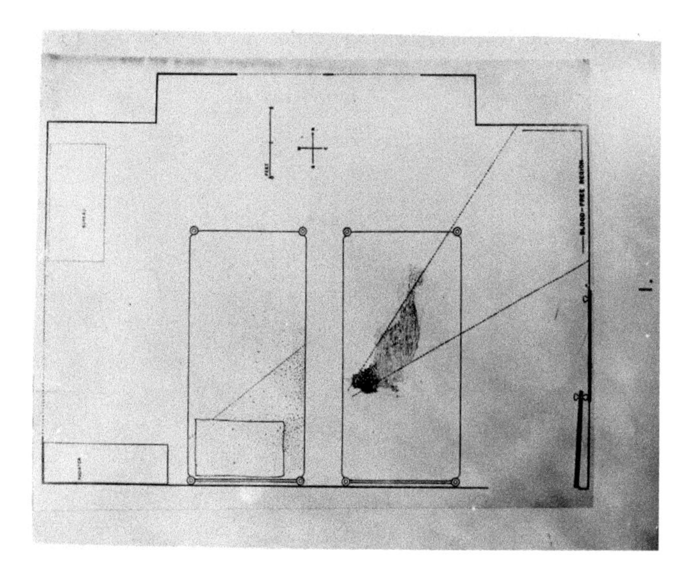

FIGURE 10.1 Paul Kirk's diagram showing the configuration of blood traces on the walls and the void area. Kirk's Affidavit: "The bedroom in which the murdered body of Marilyn Sheppard was found is shown in approximate scale diagram in accompanying photograph No. 1. The diagram represents the condition at the time it was examined by the undersigned. The two twin beds and bureau, shown in the drawing, are in the same position as indicated in prosecution photographs. The drawing omits the rocking chair in the northeast corner of the room, which carried no visible blood or other significant evidence, and the small telephone stand between the two beds which did not figure in testimony, or in this investigation." *Source: "From the Sam Sheppard Case Collection at the Cleveland-Marshall College of Law, Cleveland State University."*

County Coroner, Samuel Gerber, "saw" the imprint of a "surgical instrument" in the blood pattern on a pillow, although he was unable to identify the kind of instrument. This testimony may have influenced the jury in reaching their guilty verdict but was criticized in Dr. Kirk's affidavit and ridiculed in many quarters.

Although scientific publications dealing with the interpretation of the configurations of blood traces preceded the Sheppard case and Dr. Kirk's affidavit, the latter document represented an early illustration of the types of higher-order deductions that can be derived from blood trace configurations present at a crime scene. The affidavit described airborne droplet deposits on the walls all around the room except for an area devoid of any deposits (Figures 10.1 and 10.2). This led to the interpretation that the assailant's body had intercepted airborne droplets in-flight creating a pie-shaped sector of what would have been a complete 360-degree radial configuration creating a void. This allowed the position occupied by the assailant during the bludgeoning to be determined. It also illustrated that the murderer had remained in essentially one location during the time the blows were being administered. Complementing the configuration of the deposits on the walls were deposits on the adjacent twin bed (Figure 10.3). Dr. Kirk studied the droplet configurations on the closet (wardrobe) door and concluded that there were two larger droplet traces that were unlike the other airborne droplet deposits he had ascribed to the bludgeoning (Figure 10.4). He theorized that these might have been cast off from the assailant's hand, perhaps from a wound.

FIGURE 10.2 The strings illustrate the cone of the void pattern on the wall, which measure approximately 2 feet at the edge of the bed. Kirk used this to indicate the attacker's position as being close to the edge of the bed. *Source: "From the Sam Sheppard Case Collection at the Cleveland-Marshall College of Law, Cleveland State University."*

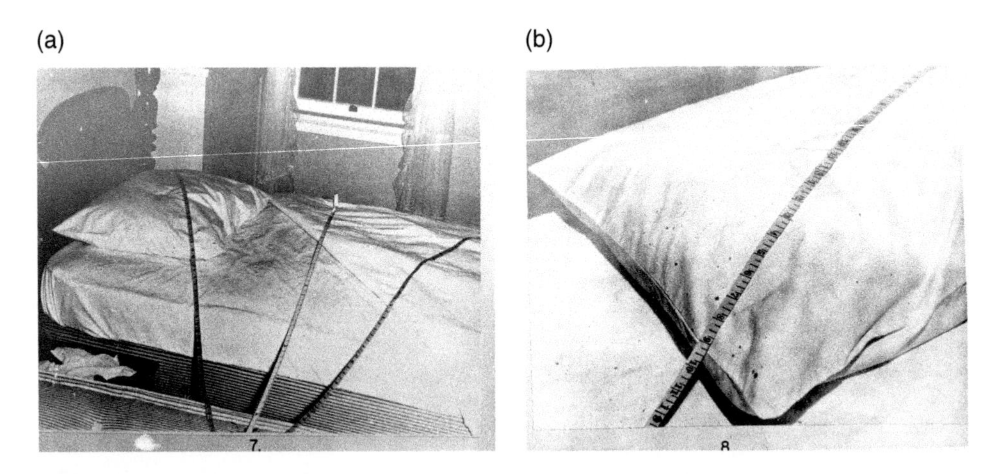

FIGURE 10.3 Paul Kirk's photographs of the adjacent twin bed (a) and a closer-up view of the pillow (b). The measuring tapes are used to give a crude delineation of the radial distribution of airborne blood deposits from the site or origin of the bludgeoning. *Source: "From the Sam Sheppard Case Collection at the Cleveland-Marshall College of Law, Cleveland State University."*

10.1.3 Conclusions

Although an extensive investigation took place shortly after the event, it would appear that it was naïvely done with a lack of recognition for the distribution of blood trace deposits. This was not recognized until the post-conviction scene analysis and

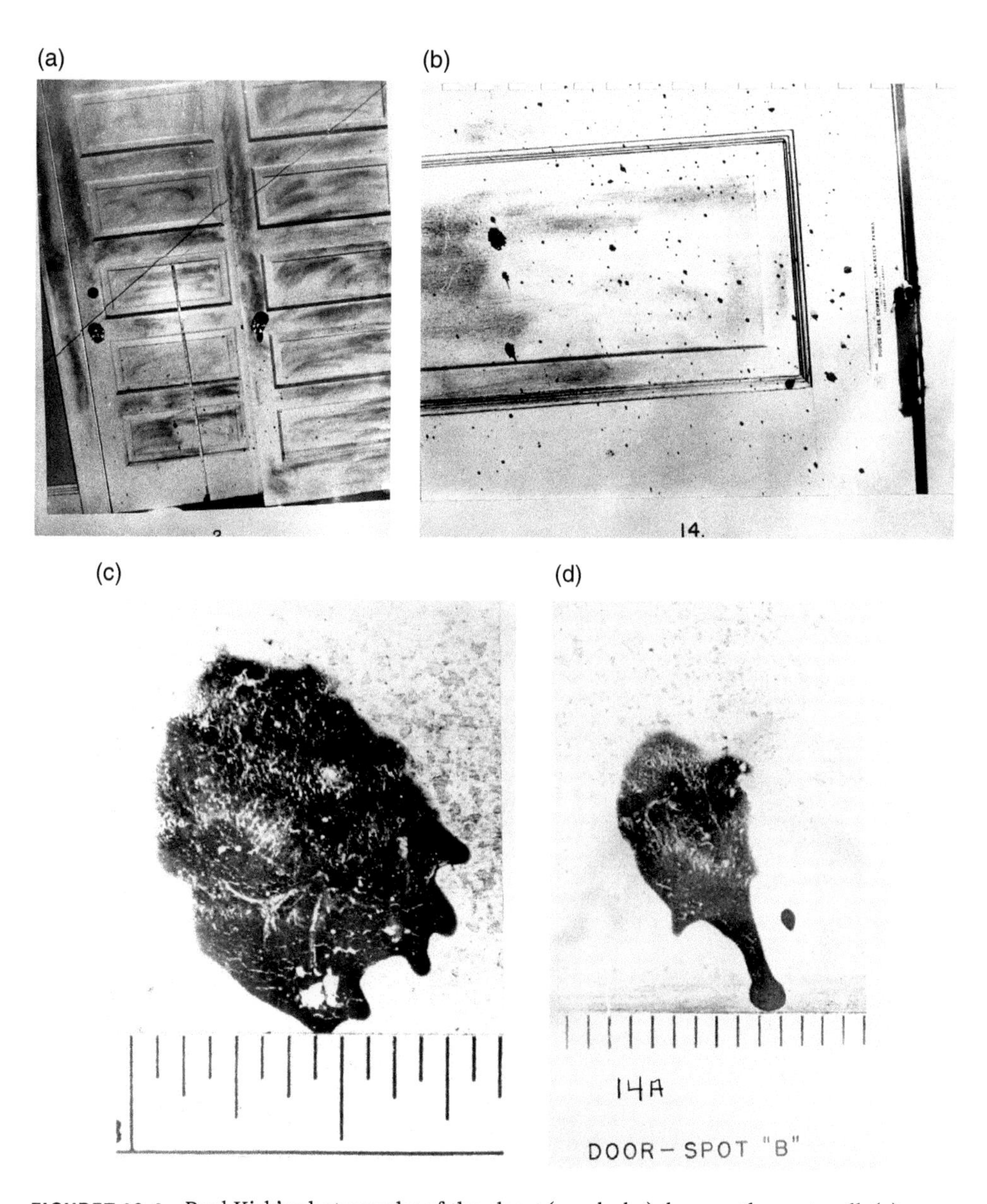

FIGURET 10.4 Paul Kirk's photographs of the closet (wardrobe) door on the east wall: (a) on the left side of the photograph is the closet door with evidence of dusting with fingerprint powder and blood traces visible on the lower panels, and on the right side of the photograph is the exterior face of the entry room door which has been opened fully, (b) midrange photograph of the lower panel with blood trace deposits, (c) close-up of the largest blood deposit, and (d) close-up of the second largest deposit. *Source: "From the Sam Sheppard Case Collection at the Cleveland-Marshall College of Law, Cleveland State University."*

documentation completed by Dr. Kirk in 1955. Although blood was collected for subsequent laboratory testing, there was no apparent appreciation of the potential information that could be gained from the interpretation of the geometric configurations of the blood traces.

FIGURE 10.5 Police photograph from the initial investigation showing the east wall with the closet door and the exterior face of the entry room door, prior to dusting for fingerprints. *Source: "From the Sam Sheppard Case Collection at the Cleveland-Marshall College of Law, Cleveland State University."*

The influence of this case on the field of bloodstain pattern analysis is widely known; however, it was the source of a mischaracterization. Dr. Kirk's Affidavit, specifically the term "high velocity" to describe the speed of the blood droplets and how far they traveled, was misinterpreted by Herbert MacDonell to describe the action used to create the blood trace deposits. This was then used to create confusing and misleading terminology (high-, medium-, and low-velocity spatter) that continues to be used despite being discredited.

10.1.4 Lessons

This case is important historically because it was an early case where blood trace evidence lead to a detailed crime scene reconstruction. This case took place before the value of the configuration of blood trace evidence was widely recognized. There are several things that Dr. Kirk identified and emphasized in this case, that heretofore had not been used for reconstruction. The lack of awareness prior to Kirk's involvement is evident from the dusting of fingerprint powder on the closet door with no apparent recognition of the significance of the blood traces. Dr. Kirk's investigation in this case continues to serve as an example on how physical evidence, and specifically blood traces, can be utilized in crime scene reconstructions.

10.2 Knife in the Gift Bag

10.2.1 Case Scenario/Background Information

This case illustrates the dangerous and all-to-frequent tendency of the criminal investigators to prematurely lock in on a certain hypothesis or theory. This often closes off other avenues of investigation.

The victim was found naked in the master bedroom with very substantial blood trace deposits on the walls and floor. He had suffered one stab wound to the femoral artery in his groin, which was unltimately fatal. There was no bloodspatter in the living room or the hallway directly adjacent to the living room which led to the master bedroom. A knife, presumably the weapon used, was found in a gift bag in the living room. An unidentifiable bloody fingerprint was found on the knife (the victim could not be excluded as the source of the print). The investigating detectives concluded that the victim stabbed himself in the living room, dropped the knife in the bag, and then moved to the bedroom where he collapsed and died.

10.2.2 The Physical Evidence and Its Interpretation

The present authors have too frequently found that some criminal investigators will prematurely adopt an overall theory early in an investigation. In some of these situations, there seems to be a natural temptation to jump to an early accident or suicide conclusion since the determination of noncriminal act allows the investigation to be "closed" or "cleared" more readily. Then, when contrary physical evidence is found, the criminal investigators may be very reluctant to change their minds. Part of this reluctance may be due to being humiliated by the error, but this should not prevent investigators from revising theories based on the physical evidence, as demanded by the iterative process of the scientific method. Self-correction is a valuable characteristic of a scientist (Chapter 7), a practice that is sadly not necessarily shared by nonscientist investigators.

The finding of the bloodstained knife in the gift bag in the living room approximately 20 feet from where the victim was found with no intervening blood deposits near the living room should have raised doubts about the contention that victim's wound was self-inflicted. The physical evidence is better explained by the knife being placed in the gift bag following the homicide. This is further supported by the following two more detailed considerations.

Blood in the femoral artery is under pressure, thus wounds to the femoral artery result in significant amounts of blood being projected from the wound immediately and followed by continued substantial blood loss. The lack of blood traces in the living room and the adjacent hallway, either in the form of single droplet configuration or as distribution patterns, coupled with the fact that the victim wore no clothing that would tend to retain blood, demonstrates that the victim was not wounded in the living room.

This lack of significant blood traces in the living room stands in stark contrast to the amount and distribution of blood deposits found on the floor and wall by the body of the victim, as well as on and near the bed. The blood trace configurations adjacent to where the victim was found on the south hallway wall adjacent to the bedroom and floor resulted from impacts of blood drops traveling from the east toward the west (leading from the direction of the bedroom toward the living room). For example, when facing the south wall in the hall adjacent to where the victim was found, these blood deposits were produced from drops impacting the wall from the left to the right (i.e., moving away from the bedroom, not toward it). The directionality of these blood traces were likely due to the blood being projected from the wound site. If the victim had been wounded in the living room, the general direction of the blood deposits would have been the opposite (from the living room toward the bedroom). It is apparent, based on the blood traces on the victim's bed and on the floor in that area, that the victim was stabbed while positioned on the bed or in close proximity to the bed. In addition to the drip patterns on the bed itself, a pool of blood is present on the floor corresponding to the drip area. A large volume of blood dripped onto the floor in that location producing considerable secondary bloodspatter on the surrounding floor and bedding, allowing for the conclusion that there was little or no movement of the victim.

10.2.3 Conclusions

Despite the overwhelming physical evidence supporting a conclusion of a homicide, no charges were ever filed. It is unknown how vigorous of an investigation took place after this was incorrectly assigned as a suicide.

10.2.4 Lessons

There is no convincing explanation that would support the detective's conclusion that the naked victim knifed himself in the groin compromising the femoral artery, went into the living room to leave the knife with a bloody fingerprint in a gift bag, and then return to the bedroom without leaving a blood trail. This demonstrates a premature narrowing of the inquiry by nonscientist detectives without consideration of the blood traces leading to an erroneous suicide determination.

10.3 The Farhan Nassar Case

10.3.1 Case Scenario/Background Information

In 1979, the proprietor of a neighborhood deli was found lying unconscious on the floor, by a customer, in the rear of the store in a pool of blood at 6:30 a.m. Police officers with emergency medical training responded to the scene and treated the injured party who was transported to a local hospital where he was treated in the emergency and operating rooms. The emergency room physician informed detectives that he had observed seven separate lacerations of the scalp, fractures of the skull, and bone fragments in the brain. The physician went on to state that it was his opinion that the victim had probably been struck by a blunt object on the right side of his head causing the fracture and also stated that it was a very remote chance that the injuries were caused by the victim falling while accessing an upper shelf. The victim died approximately three weeks later. At first, the incident was treated as a serious assault by the police department but subsequently was concluded to be an accident by the senior investigators. In addition to detectives, crime scene unit and forensic lab personnel (one of the authors) responded to the scene.

10.3.2 The Physical Evidence and Its Interpretation

The important physical evidence was as follows:

1. Apparent cast-off blood trace configurations on the underside of shelving (14–18 inches above the floor).
2. Large clumps of blood-encrusted hair (50–100 head hairs) deposited on the underside of the same shelving
3. Blood spatter on a vertical column with an origin of about 6–12 inches above the floor
4. Large clumps of blood-encrusted hair deposited on the top of shelving several inches in from the relatively sharp edge, with no evidence of contact between the victim and the edge.
5. Blood spatter several feet high and 4–6 feet away from where the blood spatter was found on the aforementioned column.
6. Cash register, that was 40 feet from where the victim was found, with its drawer fully open with no bills present and currency on the adjacent floor.
7. Partial bloody print (deemed of no value for comparison purposes) on the cash register's outer surface and there was no evidence that the victim could have traveled 40 feet to the register and then back again, another 40 feet, to where he was found without leaving some sign or indication of this movement.
8. Foreign material found in the victim's wound site by a surgeon. This will be discussed in more detail in this case presentation.

Certain detectives and ultimately the deputy medical examiner erroneously and tenaciously held to the conclusion that the incident was the result of an accident. They hypothesized that since the victim was short, he had to climb the wooden shelves to stock the higher ones and then fell. They concluded that the victim's head hit the edge of the shelving before impacting the floor. They believed that the blood spatter on the column originated from the head impacting the floor. Forensic lab personnel immediately notified detectives that this hypothesis was clearly contradicted by the physical evidence but was unfortunately disregarded. There was no spatter on the floor emanating from the pool of blood under the victim's head nor any blood traces at the lowest portion of the base of the center island. The physical evidence made it abundantly clear that the origin of the spatter on the column was significantly above the floor level. Furthermore, there was no indication that the victim's head had struck the edge of the shelving. Rather, it was obvious that two large groups of hair had either fallen or been projected onto the shelving surface and a third very large group had been projected along with blood (likely from a cast-off) onto the underside of a shelf (Figures 10.6–10.9).

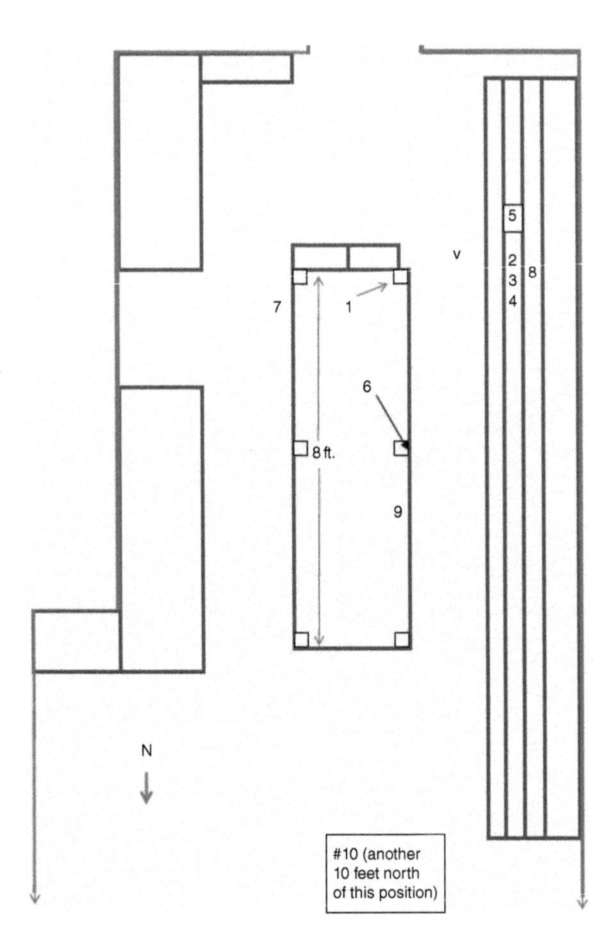

FIGURE 10.6 Crime scene sketch of the store. Legend:

#1 spatter pattern on vertical column and lowest shelf of center island.
#2–#4 clumps of hair on top of second shelf from floor.
#5 bloodspatter (cast-off) and large clump of hair adhering to underside of shelf (cut-out).
#6 bloodspatter on vertical column of island.
#7 blood and hair on dented can lying in cardboard box positioned partially on floor and lower shelf of island.
#8 blood on third shelf from floor.
#9 bloodspatter on can.
#10 bloody ridge pattern on cash register.

When confronted with the presence of the partial bloody print on the cash register surface, and the certainty that it could not have been deposited by the victim, the detectives inexplicably dismissed this evidence and stated that the initial uniformed emergency medical personnel deposited the friction ridge pattern when they "went through the cash register." However, the responding author interviewed the aforementioned emergency medical personnel who denied contacting the cash register.

The scientific investigation was complicated by the fact that the victim survived for several weeks in the hospital after surgery on head wounds. The surgeon's report was ambiguously written but noted the presence of "blue plastic" in one of the head wounds.

FIGURE 10.7 A picture of the shelving showing blood traces and hair bundles on the upper surface of the second shelf. *Source: Yonkers Police Department.*

FIGURE 10.8 The directional blood trace deposits on the shelving pillar with stringing reconstruction that demonstrates the position of the area of origin. This was approximately 6 inches above the floor level. *Source: Yonkers Police Department.*

FIGURE 10.9 The undersurface of the shelving showing the blood droplet trace deposits and blood-encrusted hair bundles. *Source: Yonkers Police Department.*

After the victim expired, a deputy medical examiner discussed the case with the detectives, read the surgeon's report, and visited the incident scene, most notably without input from laboratory personnel. He ruled that the cause of death was an accident.

The forensic lab personnel continued their investigation and independently decided to examine the "blue plastic" that had not been submitted for examination. Standard macroscopic visual and stereomicroscopic examination clearly demonstrated that this material (two small pieces) should have been more accurately described as irregular fragments of apparent black vinyl (PVC) electrical tape. Lab personnel hypothesized that the tape could have been wrapped around the unknown weapon used for the assault. Subsequently, the identity of the material was confirmed spectrophotometrically. The Chief of Detectives was furious with lab personnel for taking the initiative and conducting the examination which was clearly at odds with the surgeon's description of the material. It is also important to note that no sign of electrical tape was found near the victim or anywhere else in the store. Scientific personnel were ordered back to the scene and re-examined it for the presence of tape, and as before, found no sign of it. The fragment of black vinyl electrical tape from the wound and subsequently identified by the forensic laboratory suggested that the bludgeon-like weapon had been wrapped with this tape. This type of weapon is not one of opportunity in that that it is not readily available for a lay perpetrator, which implies that it was brought to the scene and indicates this may have been a planned and organized attack.

10.3.3 Conclusions

Almost a year later, one of the authors who had been involved with the initial investigation was approached by a private investigator (a former detective lieutenant from the same agency) about the incident and asked for his opinion. Up to that point in time, the private investigator was not aware of the author's involvement in the case. The private investigator requested a referral to an independent criminalist to act as a consultant to re-examine the case. An independent criminalist (a second author of this book) was hired and met with the private investigator, the chief medical examiner and the responding criminalist to discuss the entire case. The consulting criminalist recommended that scientific personnel revisit the scene and use the "string method" to back-project into three-dimensional space to more closely approximate where the impact occurred that produced the bloodspatter on the column; that is, did the spatter result from the victim striking the floor (as claimed by detectives) or from somewhat above the floor? As can be seen from the accompanying photographs, it is clear that the spatter originated from above floor level. There is not even a scintilla of physical evidence to indicate that any spatter resulted from an impact with the floor. Ultimately, the initial criminalist and the consulting criminalist presented a summary of the totality of the physical evidence in the case to the chief medical examiner and when provided with the entire physical evidence record, he overruled the deputy medical examiner's findings and changed the cause of death to homicide.

10.3.4 Lessons

There are very few cases that can be solved solely with blood trace configuration interpretation, including this one. This case is one in which resolution of the relevant questions and the understanding of the event was based on the totality of the traces which included the complementary blood trace configurations. The finding of the black plastic material in the wound by the emergency surgeon was further confirmation contradicting the accident theory. The locations where the large bundles of hair were found also strongly dispels the notion of an accidental cause and supports the hypothesis of being projected from a bludgeon.

This complex case clearly illustrates the need for the skilled and experienced criminalists up-front at crime scenes and the integration of contextual information by improved communication among other sources, such as with medical examiner and surgeon in this case.

Scientists should not be constrained by the possible narrow focus of detectives and/or prosecutors and should arrive at their own conclusions independently based on the physical evidence. As is commonly the case, the interpretations from the blood trace configurations in this case are not obvious, even though they appear to be so when they are written up and explained after the fact. Instead, they are the result of the thought process of an experienced scientific investigator.

This case also very well demonstrates that investigators and medical examiner staff must not "lock" themselves into premature decisions. The impulsive assignment of accidental cause, which was not corrected until over a year later, greatly reduced the likelihood that this case would be reviewed and solved.

10.4 Passive Documentation

10.4.1 Case Scenario/Background Information

On a bitterly cold and windy night, a crowd gathered outside a local bar. During a brief struggle, a young man was stabbed to death on the sidewalk and in the street. Complications immediately arose during the ensuing investigation because of the sheer number of possible witnesses which the investigators and the police department were ill prepared to handle. The crime scene was processed by one of the authors in conjunction with a crime scene unit that did not work the graveyard shift, because of labor issues, and had to be recalled from off-duty status. The lack of a rapid response to the scene by the crime scene unit certainly did not help the resolution of the affair. However, one of the present authors did respond to the scene immediately, and upon arrival, found that the scene was not being adequately secured. Possessing command authority, he promptly took steps to correct the situation.

The prosecutor's office and the police department feuded over other more general aspects of the police investigation. Unfortunately, much of the police department's investigative functions were curtailed or obstructed by the prosecutor's office because of a lack of trust.

Within several days of the incident, the FBI observed individuals under surveillance discard large garbage bags in a parking lot in an adjoining county. The FBI investigators mistakenly thought that the discarded material were gambling records and collected what turned out to be bloodstained clothing. The FBI laboratory processed the clothing for DNA. The key item of evidence turned out to be a sweater believed to have been worn by the suspect. Blood traces were found on the cuffs of both sleeves. Ultimately, it was found that the DNA profile of the blood on one cuff was consistent with that of the victim, while that on the other was matched to the suspect.

10.4.2 The Physical Evidence and Its Interpretation

The prosecutor's office hired a freelance bloodspatter "expert" to interpret the deposits on the cuffs of the sweater. The retention of the outside expert was done without vetting by the forensic science resources available locally. The crime scene and blood trace interpretation capabilities existed within the local forensic science laboratories. Of critical importance, the FBI laboratory had already taken cuttings from the area where the alleged blood traces had been deposited, thus the sweater itself was no longer available for laboratory examination. Furthermore, no close-up or laboratory examination quality photos had been taken of the stained areas. Additionally, the sweater was of a coarse knit. The author was requested by the prosecutor's office to review the findings of the freelance expert. The prosecutor made it clear that it would be advantageous to have two experts testify and agree to the same outcome. It was explained to them beforehand that there was no guarantee of that. This author found that the photography was incomplete as only overall photographs of the sweater had been taken. As a further complication, the coarseness of the knit made it impossible to differentiate between airborne droplets and contact traces. As discussed in Chapter 13, contact traces, depending on the nature of the contact, can also mimic droplet-like deposits. Based on the examined photographs, it was determined that there was no way of distinguishing among the possible methods of blood trace deposition, specifically whether they had resulted from contact with a bloody surface or impact by airborne blood droplets.

10.4.3 Conclusions

In a conversation with the prosecutors, the substance of the anticipated testimony of the nonscientist freelance expert was made apparent to the aforementioned author. They stated that he would testify that the blood deposits demonstrated that the clothing was worn when the incident occurred. This author was adamant in explaining to the prosecutors that this finding could not be supported scientifically because there simply was not enough detail in the photographs to claim the presence of bloodspatter. He advised them to present the facts to the court that DNA profiles consistent with both parties was present on the respective sleeve cuffs, but not submit testimony of the mechanism of deposit. This author was not called to testify presumably due to his contrary opinion. Additionally and regrettably, he did not prepare a report of the examination of the photographs because he did not think that there was much to report – merely that there was insufficient information to reach any relevant or meaningful conclusions. Naïvely, he thought the prosecutors understood and respected his criticism of the outside expert.

Several days later, newspaper accounts of this case reported that the outside expert testified that the sweater was worn during the incident. For the abovementioned author, this was an unsupportable opinion that was both alarming and disappointing. Not knowing exactly what was testified to by the expert, the author referred the matter to a certain scientific association for further investigation and recommended they obtain a copy of the transcripts. No further contact was made between the association and the author. It was noted in a different newspaper article that the expert was hired just prior to the advent of the trial at a significant cost and based his conclusions solely on the photographs.

Following the conclusion of the trial, two of the authors realized that both had worked on this case unbeknownst to the other. Working as a scientific consultant for the defense, the other author examined the sweater prior to cuttings being taken by the FBI. He informally opined that no conclusions would be possible from the blood traces to support the assertion that the sweater was worn during the occurrence.

10.4.4 Lessons

The fact that the initial government examination and sampling of the clothing item took place in the FBI laboratory rather than in the county laboratory illustrates the failure to coordinate the physical evidence analysis on the part of the local prosecutor's and federal investigator's offices. This lack of coordination irreparably hampered the conclusions drawn from the investigation of the blood traces on the garment. Further, there is a responsibility of the prosecutor's office to be aware of locally available resources, and not "shop around" for favorable opinions offered by alternate experts.

Extreme caution should be exercised by expert consultants in interpreting blood deposit configurations from photographs. This is particularly problematic for blood deposits on textiles, where there are numerous factors that will affect the deposit's morphology and configuration (see Chapter 13).

This case demonstrates an overemphasis by investigators and attorneys on obtaining a DNA profile rather than evaluating the blood trace deposition. There needs to be an appreciation of both in order to maximize the information obtained from the physical evidence. The destruction of the blood traces for DNA extraction should only have been conducted subsequent to a thorough documentation and physical examination of the blood deposits.

In retrospect, the government criminalist recognized that he should have issued a report regarding the examination of the photographs explaining that there was insufficient scientific evidence to support conclusions about the deposit mechanism of the blood traces. This documentation of this opinion would have been discoverable by the defense and potentially would have prevented the prosecutors from putting a witness on the stand presenting erroneous testimony. This concept is further discussed with respect to the British Island Case. It appeared that the prosecutors failed to notify the defense of the author's admonition. Assuming this was the case, in the opinion of the authors, the prosecutors had at least a moral obligation, if not a legal one, to notify the defense of the government expert's opinion. Additionally, the government criminalist could have discussed the matter with the outside expert with the hope of a consensus being reached. Unfortunately, there was no guidance provided under criminal law, at least not in the state involved, to protect the expert faced with this situation. Government employees who take such proactive steps may face serious disciplinary action.

What is the responsibility of the scientist to stay on top of how the evidence is conveyed in trial? Currently, this is clearly an unsettled issue. The present authors feel that it is the responsibility of the scientist to communicate an appropriate and valid interpretation of the physical evidence for the triers of fact. There are a plethora of apparent wrongful convictions where bad forensic science is pointed to as a cause, where the reality may be that it is the distortion of the scientific findings and conclusions by attorneys. It needs to be appreciated that practical application of the principle of scientists maintaining responsibility for the interpretation of the physical traces would be very difficult in current practice. Unfortunately, the need for such a principle is not widely recognized and appreciated.

10.5 The British Island Holiday Case

10.5.1 Case Scenario/Background Information

In the summer of the mid-1990s, at approximately 3:45 a.m., a female's body was found stabbed and lying partially naked in the middle of an infrequently traveled street through a park on a popular British commonwealth island destination in the Atlantic. She had been vacationing with a friend and was last seen at a nightclub accepting a ride from two men on a moped at 3 a.m. In addition to more than 30 stab wounds and other cuts and bruises, she had been sexually assaulted both vaginally and anally. Investigators arrived at the scene at approximately 4 a.m. A light rain shower occurred between 6 and 6:14 a.m. Two men were rapidly developed as suspects. According to the police, the victim had accepted a motorcycle ride from one of the suspects, to be referred to as suspect 1, and that the second suspect (suspect 2) sat behind her. Suspect 1 claimed that they stopped at the park where he had consensual sexual intercourse with the victim and afterward walked to the bay water to wash himself. When he came back, he found suspect 2 punching and stabbing the victim. Suspect 1 also claimed that when he and suspect 2 left the crime scene, the latter threw the knife into a bay from a bridge crossing. Suspect 1 provided the police with the location of the knife which was recovered. No useful traces were found on the knife.

Prior to the results of the DNA analysis from the sexual assault kit being known, suspect 1 was allowed to plead guilty to accessory after the fact to murder in exchange for incriminating suspect 2 as the assailant. As part of the plea deal, suspect 1 received a five-year sentence. Subsequently, the DNA analysis came back and demonstrated a full profile consistent with suspect 1 as the sole source.

Two years after the plea deal, in response to the DNA evidence, suspect 1 was charged with murder. Britain's Privy Council rejected this attempt to prosecute suspect 1 again. Suspect 2's murder trial also took place in the late 1990s, and in the end of that same year the judge directed the jury to acquit suspect 2 due to a lack of evidence. In subsequent years, attempts were made to issue charges against both suspects, but these failed.

10.5.2 The Physical Evidence and Its Interpretation

There were three locations of apparent activity at the crime scene where the body was found. The first, marked "1," in the grass 25 feet from the body, was the location where the victim's denim skirt was found and contained significant amounts of what appeared to be blood deposits. The second, marked "2," in the grass 32 feet from the body, showed disturbance in the ground and was the location where one of the victim's shoes and her panties were recovered. The panties were cut and had minimal amounts of contact blood transfers. The third was the final location of the victim in the roadway.

A local government analyst employed by the Crown offered a scientifically unsupportable opinion regarding toolmark evidence. He astonishingly claimed and testified that he could definitively associate the knife that was found in the bay alongside the park with puncture defects in the victim's clothing. This knife was shown to be similar to knives found in suspect 2's residence by the bunter marks (brand stamping) on the blades, which was testified to by a toolmark examiner from Miami, FL. Regarding the claimed association of the knife slits in the clothing to the knife recovered in the bay, there is no scientific basis for this conclusion. No knowledgeable forensic scientist would ever conclude that a specific knife could be "matched" to cuts in fabric. This is just not possible due to the total lack of any individualizing details. Additionally, even assuming the knife found in the bay actually originated from the set found in suspect 2's kitchen, it is not possible to determine who removed it from his house, when and whether it was used in this stabbing murder.

In addition to the autopsy pathologist, the Crown retained a private pathologist and criminalist who supported the prosecution's theory that the victim was transported to her final position by more than one person. Apparently, the prosecution was attempting to prove that both suspects had committed the murder and carried the victim to the street where she was eventually found. This was ostensibly motivated by a desire to include suspect 2 in the act of the homicide, while reconciling the clear DNA exclusion of suspect 2 and single profile inclusion of suspect 1. The Crown's private pathologist was explicit that another person was involved, as discussed below.

Two of the three authors of this book were hired by the defense attorney prior to the trial to assist in evaluating the examinations and conclusions of the Crown's pathologists, the Crown criminalist, and the local government analyst.

On 10 December 1997, the pathologist retained by the British Crown stated

> *The absence of dirt or soil on the soles of [the victim's] feet in the photographs, the absence of blood in the roadway extending from the adjoining area of grass, the directionality of the dried blood on the body, and the absence of drag marks indicate that she did not walk or drag herself onto the roadway and that she was not pulled there; these findings do indicate that her body was lifted and carried there by more than one person. The absence of defense wounds also indicates that there was more than one assailant.*
>
> It is my opinion, based on the information available to me, that [suspect 1] did actively participate in the sexual assault and murder of [the victim] together with [suspect 2].

On 6 November 1996 he stated

> *There was blood staining of the denim skirt such that it indicated the bleeding occurred after it was cut off from [the victim].*
>
> *Flowing type blood stain patterns extended downward from the neck and shoulders over the front of the chest and abdomen which indicate that she had been seated and that the upper part of her body was upright when these stab and cut wounds were inflicted. Grass and soil fragments adherent to the buttocks and nowhere else, apparent in the scene photographs, indicate that she was seated in the grass after her skirt and panties had been cut off and that her flowing blood then stained the grass at that site.*
>
> *There were no scrape marks on [the victim's] back, buttocks or heels and no blood drops on the roadway between the grass and the body which, together with the blood flow patterns on the skin, demonstrate that she was not dragged to the middle of the road and that she did not move herself there. Rather, these findings indicate that she was carried there face up by more than one person. The large amount of blood on the roadway near the neck area indicates that she was also stabbed after she was placed there.*
>
> *There were no defensive-type wounds which would have most likely occurred in these circumstances if she had been able to struggle and to defend herself. Rather, there were multiple fresh bruises on the body, including contusions on both upper arms typical for marks caused by squeezing while being tightly held.*
>
> *It is my opinion, to a reasonable degree of medical certainty, that the cause of [the victim's] death was bleeding from multiple cut and stab wounds that were inflicted while she was seated in the grass while she was being held by the arms; ...that her body was lifted and carried from the grass by more than one person to the middle of the roadway where she was dropped still barely alive; and that*

the entire circumstances of her death, including the cutting and tearing of the clothing, the pattern of fresh bruises to her upper arms and the lack of defense wounds are evidence that none of the sexual contact was consensual and that more than one person was involved.

The statements made by the Crown's pathologist included three meaningful misrepresentations that should be dispelled:

1. There were no heavy grass deposits on the part of her upper thighs and buttocks that would be expected if she were sitting on the grass. The photographs indicate that the grass deposits were just below the small of the victim's back, which are more consistent with vaginal intercourse with the assailant on top with added weight than they would be with the victim sitting on the grass.
2. No scientist can reasonably conclude that two individuals were required to murder or carry the victim to her final position. Contrary to the opinion of the Crown pathologist, one person of average strength could carry the victim.
3. There is no scientific basis for concluding that suspect 2 had sexual contact and murdered the victim together with suspect 1. Only one male DNA profile was found in the vaginal tract of the victim, and it proved that suspect 1 was the sole source and excluded suspect 2.

The claims of the Crown's private pathologist that both suspects were responsible for the murder while perhaps logical is not based on physical evidence. Certainly, it would be entirely reasonable for a police investigator to strongly suspect that both defendants committed the murder because both picked her up on the moped. However, police suspicions should not be included in an ostensibly scientific report and presented as science.

A third pathologist utilized by the government, based on autopsy and scene photographs, stated that "A very faint linear scratch-type injury is on the right palm. In addition, an incised wound with surrounding blood extravasation is on the left palm. These are consistent with defense-type wounds." She also noted the presence of a faint impression on the victim's left thigh consisting of a "...series of letters that are barely discernible." She attempted to enhance the pattern photographically but was unsuccessful and recommended further examination.

It is unknown why the Crown's private pathologist ignored the findings of the third pathologist regarding the claim that the incised wound on the left palm was consistent with being defensive. Additionally, the third pathologist's observation of the lettered pattern on the victim's left thigh was not explored further, despite its potential as significant evidence.

The private criminalist retained by the Crown claimed that the lack of vertical droplets within 2 feet of the victim's body suggested she was carried to the middle of the street. This claim is not supported, as there are alternative explanations for the failure to detect these blood trace deposits. First, there was a large pool of blood near the victim's head which could have obscured or dissolved any small deposits that resulted from blood falling vertically as the victim walked. Since the Crown private criminalist did not have the benefit of examining the crime scene shortly after the incident, he would not have known if any small stains were present. He cannot know how well the pavement was searched. The rough coarse surface of the asphalt makes it difficult to detect small blood trace deposits. Further, it is evident that at the time the photographs of the road were taken, the asphalt still possessed considerable areas of moisture due to the light rain between 6 and 06:14 a.m. Some moisture is apparent on the asphalt immediately adjacent to the victim's feet. The Crown private criminalist should have considered the possibility that some of the smaller stains could have been missed during the investigation or washed away to prevent their detection.

The Crown private criminalist also asserted that there was "no excessive soil or vegetation on the soles of her feet or on her toes" in support of the contention that the victim had not walked to her final position in the street. It is unclear what was meant by the term "excessive," as there is no way that this is quantifiable. In addition, if the victim walked from the grassy area to where she was found, contact with the pavement could have removed some vegetation and soil.

Additionally, the Crown's private criminalist stated that a detailed examination of the road between where she was found and yellow marker "1" was conducted. There is no way the quality and reliability of the scene investigation could have been known based on the available evidence at this time. He further claimed that "This indicated the victim was carried from location "1" to her final position in a relatively rapid fashion, such that there was insufficient time for the blood to drip onto the road surface."

There are directional blood trace present on the victim's inner left thigh that stand in stark contrast to the assertion that there was no blood evidence indicating that she was vertical. There are several elongated blood droplet deposits on her upper left thigh that likely resulted from the impact of nearly vertically falling drops while she was in a standing or kneeling position. There is no reason to assume that these came early in the attack and could be consistent with having been deposited while she was walking to the roadway. Clearly, it is impossible to know at what point in the sequence of events these were deposited. We can conclude that they were not deposited at the beginning of the attack while she was wearing clothing. It is also safe to conclude that they were not deposited while she was in her final horizontal position in the roadway.

10.5.3 Conclusions

Although the prosecution experts placed a lot of value of the blood deposits, the interpretations were flawed. The blood configurations were seized upon by prosecution experts as evidence of the two men participating in the crime; however, the conclusions and interpretations were open to serious questioning and alternative explanations. This likely influenced the judge's decision to instruct the jury that there was insufficient evidence to conclude that suspect 2 was guilty of the homicide.

10.5.4 Lessons

An important lesson from this case is the need for all scientific experts to consider all possibilities and alternative hypotheses. This is a part of the scientific method, in that one should actively try to disprove a hypothesis before accepting it. This is one valuable method for minimizing bias. In this case, the Crown's expert opinions appear to have been tailor-made to what the prosecutors needed to make their case against suspect 2. If the Crown experts had made use of the scientific method and tried to actively falsify their hypotheses, then they could have recognized the failure in their interpretations to consider other scenarios that could have explained the physical evidence.

Obviously, the government analyst's toolmark conclusions were patently absurd and should not have been admitted in court. The Crown's private criminalist, who clearly was aware of the state of the art, should have emphatically stated in his report that the government analyst's examination and conclusion were not based on valid science, i.e., it lacked even a scintilla of a scientific underpinning. In fairness, perhaps the Crown's private criminalist may have verbally informed the prosecution of this, however, there is no evidence to support this. All criminalists should affirmatively deter prosecutors from using pseudoscience. Criminalists should feel that they have an obligation to not only verbally communicate these concepts but to also include them in a report. This will make them exceedingly difficult to ignore because of disclosure requirements which will enable the defense to have an opportunity to view such opinions. This should not be a dissuading concern of a true scientist. Additionally, some criminalists or other scientists, faced with such scenarios, may be reluctant to write a report making such bold statements out of concern that their views will be rejected by the prosecutor, and that they will not be utilized in the present case or in the future. However, it is the opinion of the authors that all scientific experts must be willing to take the chance of losing favor with prosecutors to minimize occurrences of wrongful conviction. It must be recognized the obligations of a defense expert relative to disclosure differ from scientists working for the prosecution.

Based on a recommendation to the defense attorney by one of these authors, the Miami toolmark examiner could be used to undercut the pseudoscientific testimony of the government analyst concerning the association of a particular knife with regard to damage to the fabric. In cross-examination, the toolmark expert agreed that it would be impossible and improper to claim this type of association. This showed how a qualified opposing expert can be utilized to counter absurd claims by other "expert" witnesses.

There was a blatant lack of respect for the physical evidence by the Crown. Had the prosecutor waited to make the plea deal until after the DNA typing results were made available, the case resolution would have potentially been more accurate. To this day, although this is not a concern for scientists, the public outcry about this case remains because of concerns that there was no real justice for the victim.

10.6 Absence of Evidence is Not Evidence of Absence

10.6.1 Case Scenario/Background Information

In the late 1980s, a 17-year-old young adult claimed that he awoke to find his mother stabbed to death and his father stabbed, beaten and unconscious in their suburban residence. He notified the police. The son was rapidly developed as a suspect, interrogated and confessed to the double homicide. The validity of the confession would subsequently come under intense scrutiny but not before he was convicted of the crimes and sentenced to two consecutive terms of 25 years to life in prison. In the early 2000s, the defendant was granted an appeal which ultimately led to the overturning of the conviction. Ultimately, after the man spent 17 years in prison, the State's Attorney General's office dismissed the charges based on an investigation which determined that there was insufficient evidence for another trial. The man subsequently sued the County and State, eventually, settling with each for a sum equal to more than $13 million.

10.6.2 The Physical Evidence and Its Interpretation

Two of the authors were involved in the case, with one hired by the defense during the initial trial and the other by the state during the civil proceedings. While the criminal case was under appeal, a private forensic expert was hired by the defense, and he continued through to the civil case. The civil appeal case was settled out of court; thus, no expert testimony was required from either side. However, prior to the settlement, the plaintiff's civil attorney provided a description of the opinions on which the private forensic expert was expected to testify as part of disclosure.

A considerable portion of the expert's opinion, as detailed by the plaintiff's attorney, focused on a lack of biological traces on either of the weapons identified in the confession (both collected from the crime scene), a failure to detect blood between the master bedroom and the office (which is where the two bodies were found), and the lack of biological traces in the traps and drains throughout the house. In other words, it was indicated the expert hypothesized that since there is no physical evidence, two or more perpetrators committed the murder as opposed to simply one. This is unjustified and in the realm of dangerous pseudoscience. One must be extraordinarily cautious when reaching conclusions when there is no evidence. The common maxim "absence of evidence is not evidence of absence" applies here. Weapons can be washed leaving no detectable traces of blood, biological material, or material evidence. Depending on how long water is run, the same is true of traps and waste lines. Relevant traces could have gone unrecognized at the scene. Additionally, a statement was made implying that two perpetrators makes it perhaps easier to kill two people at nearly or the same time, thus explaining the lack of blood traces between the bedroom and office. However, the more individuals involved also means the greater the opportunity for traces of various types to be created and transferred to other locations in the residence. These statements should not be construed to mean that the plaintiff's expert is incorrect, simply that it is a very precarious situation when one bases a conclusion on the absence of expected supportive traces.

Another claim that was made by the plaintiff's attorney regarding the expert's opinion was that the defendant would not have been able to generate sufficient force to produce the blood traces in the office where the father was murdered given the amount of space available. This assertion appears quite strained. There was nothing remarkable about the blood traces located on the desk (impact spatter) or ceiling/wall (cast-off) to support this hypothesis. The amount of force necessary to fatally stab or bludgeon an individual depends on a number of factors, including but not limited to the sharpness of the knife, the location of the stab wound(s), the type and characteristics of the bludgeon, physical abilities of the assailant, etc. There is no method for measuring the amount of force that could be generated in the provided space. Depending on the instrument, a minimal force may be sufficient for generating these injuries. Additionally, someone had sufficient space and force to commit this assault. No information was provided about the defendant that would exclude him from being able to commit these crimes. Again, we must emphasize that the expert for the plaintiff may be correct in the conclusion, but the opinion is not supported adequately by physical evidence.

There was also contention over the weapon or weapons used in the murder of the mother. Specifically, the plaintiff's expert proposed that the bloodstain patterns and the location of evidence indicated that she was attacked with both a blunt object and a knife while she was lying in her bed. The specific blood traces were claims of the presence of a "knife pattern" in blood on the bedding of the victim. These are disputable and were never satisfactorily proven. A triangular pattern that appears like a knife blade can be generated by other objects, as discussed in Chapter 13. Thus, concluding a triangular pattern must have come from a knife blade may be explained as pareidolia (Chapter 7).

In addition, there was an exceedingly subtle pattern, which may have originated from footwear that was observed on the bedding by one of the experts hired by the state for the civil proceeding years later. There is no record that this possible footwear outsole pattern was recognized by anyone other than the state's expert, which occurred years after the confession. Had this been properly recognized in a timely manner, it could have been compared to the footwear outsole pattern of the defendant's shoes to provide valuable associative or exclusion information. It is interesting to note that the presence of a possible footwear pattern is consistent with the narrative of the defendant's disputed confession. This illustrates the danger of reaching conclusions based on the absence of evidence because it may just not have been initially recognized.

10.6.3 Conclusions

The recognized and documented physical evidence was insufficient to resolve the ultimate question of guilt or innocence. This may have been a contributing factor to the adjudication by the court to dismiss the charges. However, the conclusions of the plaintiff's expert, excluding the son from having been the perpetrator, as outlined in the attorney's representation, are an unsupported overreach.

10.6.4 Lessons

Generally, the previously mentioned maxim and title for this case, "the absence of evidence is not evidence of absence," must be carefully considered in reconstructions. In this case, the lack of recognized traces was used improperly in an effort to eliminate the possibility of a single assailant. There are alternative explanations for the lack of traces, and these should have been seriously considered and not dismissed.

When performing the examination of the bedding, the present author who was working for the State in the civil trial, received excessive interference by the plaintiff's attorney. The plaintiff's expert was performing a sequential enhancement of the bedding, and the present author was prevented from exercising unencumbered documentation of the evidence. During sequential enhancement, it is necessary to document the physical evidence at each stage since it cannot be known a priori what effect the process will have on the resulting configuration (i.e., enhancement or risk of diminishing it). Ultimately, a criminalist should anticipate and prepare for this situation, and the examination protocol needs to be discussed and agreed upon in advance between the attorneys representing both parties. If this were done, time-consuming negotiations that hinder a complete documentation would not need to be conducted during a physical evidence examination.

10.7 Triple Homicide

10.7.1 Case Scenario/Background Information

In the summer of 2001, a triple homicide took place in the first floor apartment of a mother of three children following a party the night before with significant alcohol consumption. She had been fatally stabbed sometime after the party had ended. Her nine-year-old son had escaped out the back window of the childrens' rear bedroom of the apartment. He had been stabbed numerous times but nevertheless was able to reach his grandparents' residence next door and alert them to what had taken place. The grandparents notified the police who responded to the crime scene. When the police entered the scene of the assault, they found the mother's boyfriend lying prone on top of her in the childrens' bedroom; he was alive but had approximately 12 wounds on his abdomen and chest, while the female victim had expired with numerous stab wounds to her chest, back, and neck. Additionally, the bodies of the two other children (eight-year-old boy and four-year-old girl) were found in the upper bunk of a bunkbed. They died from numerous stab wounds and had apparently been killed where they lay. The oldest son who had escaped had occupied the lower bunk where he apparently had been assaulted.

The boyfriend was transported to a local hospital where he was treated. His wounds were photographed, and medical personnel characterized them as superficial due to their limited depth of penetration (one to one and a half inches). The police interviewed the boyfriend, and he claimed to have discovered the female victim standing on the bunkbed ladder stabbing her children in their beds. In addition, he asserted that when he tried to stop her, she in turn stabbed him in the chest and abdomen, but he was able to gain control of the knife and stab her in self-defense.

10.7.2 The Physical Evidence and Its Interpretation

The crime scene was examined in detail over the course of several days by one of the present authors, assisted by two skilled and experienced evidence technicians. One of the most telling items of physical evidence was found very shortly after the childrens' bedroom began to be examined. An open bloody folding knife was found on the floor to the left of the deceased mother's body. The knife had numerous apparent blood traces on various areas including the blade and the handle along with patent friction ridge patterns. The blade had a blood traces that extending about 1 inch from the tip with a well-defined demarcation separating it from blood residues further up the blade. A photograph of the knife blade is shown in Figure 10.10. Since the senior criminalist

FIGURE 10.10 A photograph of the knife blade showing the blood demarcation approximately 1 inch from the tip. *Source: Yonkers Police Department.*

that examined the crime scene also had responded to the hospital to view boyfriend's wounds, he hypothesized that the distinct blood configuration on the tip of the knife might correspond to the defendant's superficial wounds.

Based on this preliminary hypothesis, the knife was carefully documented under laboratory conditions, sampling was guided by the blood trace configuration on the blade, and then DNA analysis was used for typing of the selected blood deposits. The senior criminalist requested that both sides of the blade on the tip of the knife, corresponding to the distinct blood configuration, be separately analyzed from the remainder of the residues by the forensic laboratory scientist assigned to this DNA analysis. The laboratory detected blood containing a full profile consistent only with the DNA of the defendant from the tip of the knife. DNA of the victims were found in other parts of the knife. Moreover, only friction ridge patterns of the defendant were found on the knife.

As the autopsy of the four-year-old daughter was conducted, the deputy medical examiner detected black nylon fibers in a wound site. However, the victim wore no nylon clothing nor was there any nylon found in proximity to her. The only black nylon found in the childrens' bedroom, or anywhere else in the apartment for that matter, was the bra worn by the mother with obvious defects corresponding to stab wounds in that immediate locus of her body. The forensic laboratory criminalist compared the nylon from the wound site of the daughter to the nylon from the mother's bra and found them to be consistent, i.e., the bra could not be excluded as the source of the nylon fiber segments found in the wound site.

A thorough reconstruction was conducted by the senior criminalist that took many months to complete; however, only the most salient issues are discussed here. Of particular interest was whether the physical evidence supported or refuted the defendant's account. Clearly, the physical evidence did not support his contention of the sequence of events. It was clear that he was the last person stabbed with the knife based on his sole DNA profile being found on the tip end of the knife blade. The appearance, distribution, and depth of the defendant's wounds also strongly suggested that they were self-inflicted. Based on the single-source DNA profile, the blood of the prior victims had to have been removed in some manner (e.g., wiped off) up to the point of the well-defined demarcation prior to the defendant creating the self-inflicted wounds or during the creation of the self-inflicted wounds. Alternatively, a combination of both actions could have resulted in the removal of the various victims' blood from the tip of the knife blade. The transfer of the black nylon to the wound site clearly demonstrates that the knife had contacted black nylon prior to being used to stab youngest victim. Given the closed set of possible fibers, the mostly likely source of the black nylon is the mother's bra.

10.7.3 Conclusions

Based on a consideration of the totality of the physical evidence, the mother was stabbed early in the sequence of events, clearly prior to the stabbing of the four-year-old daughter whose wounds contained the black nylon fibers mentioned above. The boyfriend was the last one stabbed, based on the sharp demarcation of blood deposits on the knife blade. This compellingly refutes the account of events given by the boyfriend. This contributed to the conviction of the boyfriend of first-degree murder with a sentence of life in prison.

10.7.4 Lessons

Some current thinking concerning the prevention of bias (Chapter 7), had it been heeded, may have precluded the author from responding to both the hospital and the crime scene. This was justified because of the necessity of scientifically evaluating a complex case in context which enabled the recognition of the interrelationships of the potential evidence. Furthermore, this author was in a position to ensure that the knife was promptly documented (producing examination quality photographs), examined under laboratory conditions, and prepared for submission to the local government lab for serology/DNA analysis. This latter phase was critical to the scientific investigation of the case because of the singular nature of the deposit at the tip of knife blade. It was essential that the significance of the blood trace configuration be clearly articulated to the DNA laboratory analyst. Well-intentioned but naïvely promoted current thinking about controlling the flow of information may have precluded the communication between the aforementioned senior criminalist and the DNA laboratory analyst. However, failure to do this could have led to destruction of evidence. If the DNA laboratory analyst had not been made aware of the need for informed selective sampling, it is possible that the criminalist would have only examined a portion of the unusual deposit severely limiting key information that could have been learned from the physical evidence. This case exemplifies the benefit of having a case manager with generalist knowledge who is able to evaluate the totality of the traces produced by the event for an accurate understanding and reconstruction of what had taken place.

This case also demonstrates the value of having an experienced scientist directing the recognition of evidence at a crime scene, and the integration of scene observations and laboratory analysis.

10.8 The O.J. Simpson Case

10.8.1 Case Scenario/Background Information

This case includes a description and an analysis of some of the forensic evidence in one of the most highly publicized criminal cases of the twentieth century. The victims of the double homicide, Nicole Brown Simpson and Ronald Goldman, were found deceased on 12 June 1994. Their bodies were found outside Nicole Simpson's residence, on the ground level of 875 South Bundy Drive, in Beverly Hills, California. Both victims were found on the east side (near the north corner) of the residence in proximity to stairs and a patio on the Bundy Drive side of the premises (Figure 10.11 shows a floor plan of the residence). Nicole Simpson's body was located on the ground next to the lowest stair tread (Figure 10.12). Nicole Simpson's main injuries were a brain contusion to the right side of her head and a very deep incised wound that extended from side to side and from the front of the neck to the spinal column. Ron Goldman's body was found in a partially enclosed or fenced-in area, about 4 × 6 feet, approximately 5–10 feet north of Nicole Simpson (Figure 10.13). He had two parallel slashes across the neck as well as stab wounds to the chest, abdomen, and left thigh, among numerous others. Reportedly, Nicole Simpson had lunch with her mother at Mezzaluna Trattoria restaurant earlier that day, and a pair of her mother's eyeglasses were left at the restaurant. Goldman was in the process of returning them to Nicole, in a white mailing envelope, when they were both assaulted.

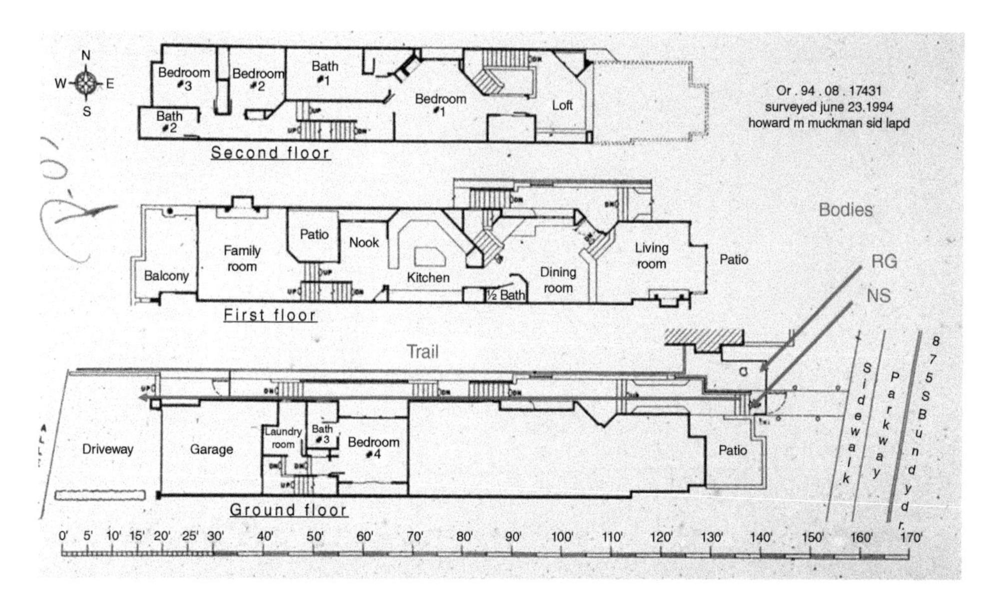

FIGURE 10.11 Plan view of the three levels of the Bundy Drive residence. The bodies of Nicole Simpson (NS) and Ronald Goldman (RG) are indicated, as is the location and direction of the blood trail.

FIGURE 10.12 Crime scene photograph of Nicole Simpson's body, as found. The large pool of blood on the first step and floor indicate that her throat was slit above the bottom step. On the second step are traces of airborne blood droplets and dynamic transfer traces, possibly made by her hand or another object dragging across the stair. Of particular note are the circular blood droplets on her back, and the envelope containing the eye glasses in the upper right area of the image. The controversial blood droplet deposit is visible in this image on the envelope.

FIGURE 10.13 Crime scene photograph of Ronald Goldman's body, as found. Of particular note is the envelope containing the eye glasses in the lower central area of the image, with the controversial blood droplet deposit visible in this image.

Orenthal James (O.J.) Simpson, a celebrity actor and former football player, was the ex-husband of Nicole Simpson who was charged with the double homicide. There were a significant number of diverse physical traces analyzed throughout the investigation of this case that served to associate O.J. Simpson with the homicides. The analysis presented here will not attempt to include a discussion of all of them and instead will focus on the examination of those involving blood evidence: the blood trails, blood traces on the envelope containing the eyeglasses, the blood on and in the defendant's Ford Bronco SUV, the blood evidence on defendant's socks and subsequent testing for ethylenediaminetetraacetic acid (EDTA), and the separated pair of gloves found individually at the homicide scene and at the defendant's residence.

10.8.2 The Physical Evidence and Its Interpretation

10.8.2.1 Trails of Blood Droplets and Footwear This case consisted of three significant blood trails: two at the crime scene located at Bundy Drive and the third at O.J. Simpson's home on North Rockingham Avenue. These will be discussed separately.

The two trails of blood at the homicide scene delineated a common path that was on the street level of the exterior north side of the premises and consisted of dripped blood deposits and footwear patterns (Figures 10.11, 10.14–10.16). This lead from the Bundy Drive side of the residence, where the bodies were found, to the driveway to the location where the perpetrator's vehicle was believed to have been parked. The footwear outsole patterns had a distinctive design that was exclusive Bruno Magli shoes.

FIGURE 10.14 The bloody trail leading away from the bodies, with the footwear outsole patterns visible in the center of the pathway. The scale rule in the middle of the image is positioned next to a left footwear outsole pattern.

(a)

(b)

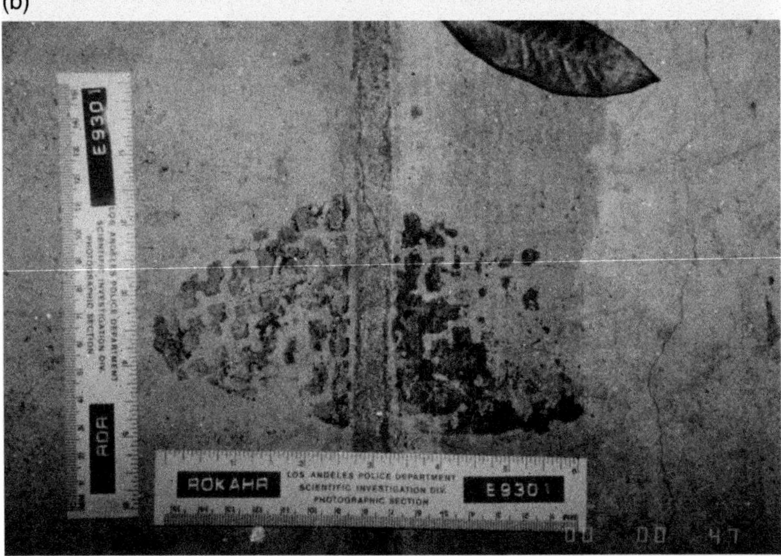

FIGURE 10.15 (a and b) Close-up photographs of the bloody footwear outsole patterns showing detail of the tread design.

The size was determined to be a US 15, which is substantially larger than the average shoe size for men (which is a US 10.5).

The trail of blood at the North Rockingham Avenue crime scene started from Simpson's white Ford Bronco SUV parked on the street at the curb in front of the house, led up the driveway, into Simpson's home via the west doorway, and continued into his house (Figures 10.17–10.21). What appeared to be blood traces were found in the sink basin and the shower drain in the bathroom adjacent to the bedroom where the socks

FIGURE 10.16 Blood droplet deposit on the driveway apron, at the end of the Bundy trail. The DNA profile agreed with that of O.J. Simpson. The defense claimed that the blood collected from this deposit had been tampered with in the forensic laboratory and replaced with one from O.J. Simpson's known blood.

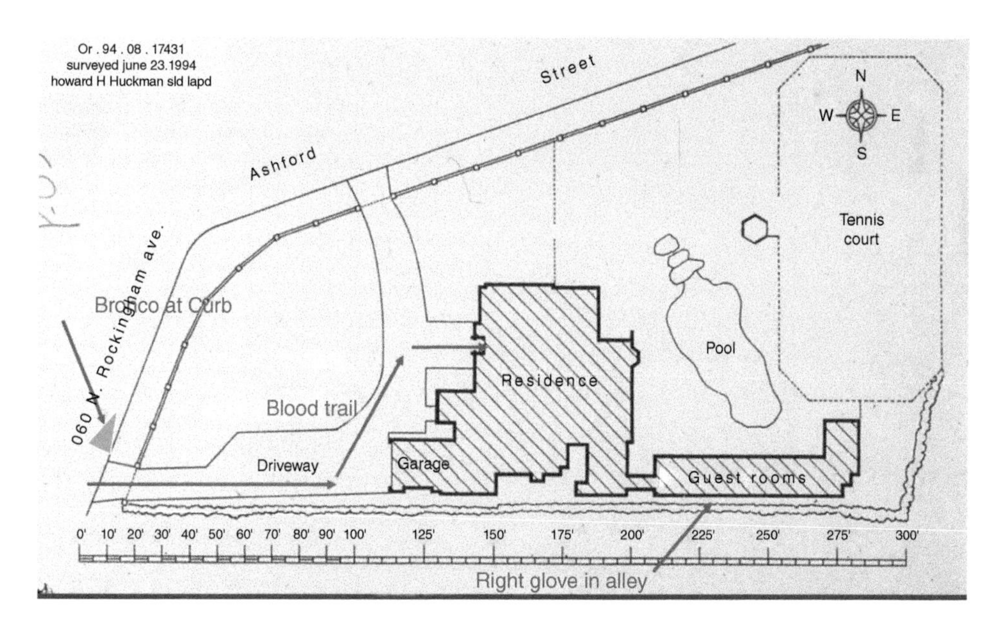

FIGURE 10.17 Plan view of the North Rockingham Avenue residence. The position where the right glove and Bronco were found is indicated, as is the location and direction of the blood trail.

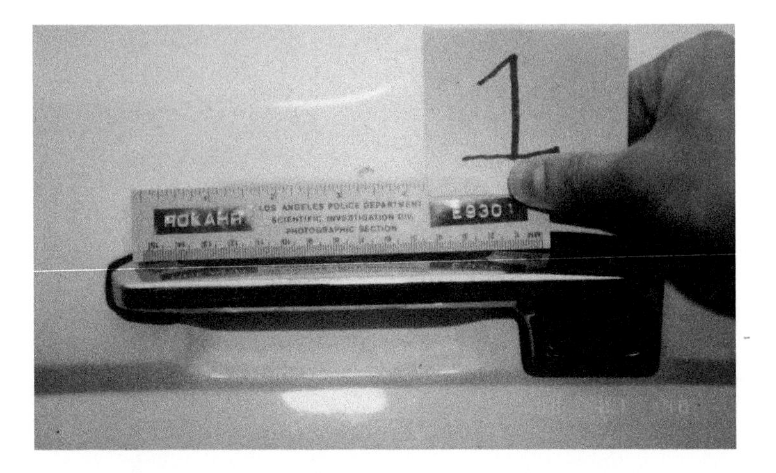

FIGURE 10.18 Photograph of a blood deposit on the driver's side door of O.J. Simpson's Ford Bronco, parked on the street outside of his residence on North Rockingham Avenue.

FIGURE 10.19 A view of the driveway of the North Rockingham Avenue residence of O.J. Simpson, taken from the street. The white markers indicate the locations of blood droplet deposits.

FIGURE 10.20 A blood droplet deposit on the street pavement in front of the Rockingham residence.

FIGURE 10.21 Blood droplet deposits on the driveway outside of O.J. Simpson's North Rockingham Avenue residence.

FIGURE 10.22 A swab of the master bedroom basin drain showing a presumptive positive test result for blood using phenolphthalin.

were found (Figures 10.22 and 10.23). Preliminary examinations for blood via Kastle Meyer (phenolphthalin) testing of residues in the sink and shower drain were positive. This positive blood testing was ruled inadmissible during trial, since the results for blood were merely presumptive and considered prejudicial by the court. The implications of this will be discussed in the lessons section of this case.

It was apparent that both blood droplet trails originated from a bleeding wound in contrast to weapon dripping with blood because of the consistency in the droplets with no sign of a reduction of the flow as the trails continued (discussed in Chapter 5). The DNA analyses of the blood droplets from the trails found at both scenes matched the DNA profile of O.J. Simpson.

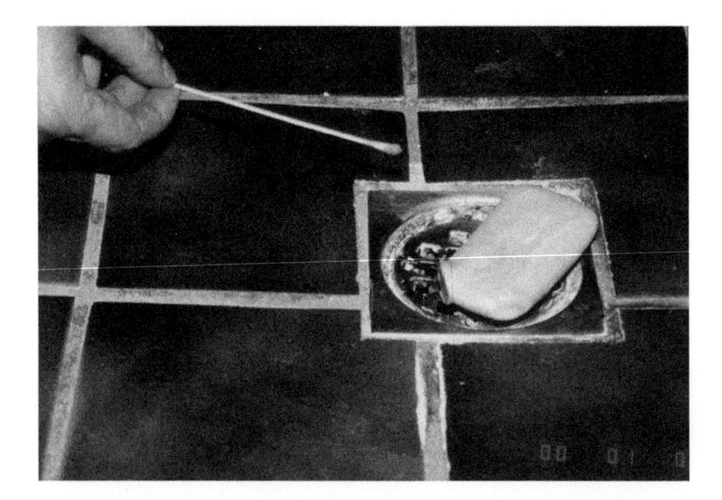

FIGURE 10.23 A swab of the master bedroom shower drain showing a presumptive positive test result for blood using phenolphthalin. The presence of the bar of soap on the floor of the shower with what appears to be a fresh impact from presumably falling is interesting. Although not scientific, an attorney could argue that whoever used the shower was in a hurry and thus did not take the time to pick up the soap.

The presumptive positive test results for blood from the bathroom would have been something that the jury would have expected to hear about because of its probative value which is context-dependent (relative to the blood trails). Since the jury was not permitted to hear this evidence, it is reasonable that the jurors assumed that no blood traces were found in either the bathroom or the bedroom with the exception of that on the socks. Although this is more of a commonsense idea, rather than one of science, a reasonable person would find it logical that the deposits in the bathroom were a continuation of the trail that began outside the house. Thus, it would seem that the presumptive positive test results for blood, again because of context, was not prejudicial to the defense. One could reasonably argue, however, that the failure to allow the jury to hear this evidence was prejudicial to the prosecution's case.

The defense made claims that the swatches of blood that were collected from both scenes were tampered with, thus the association to O.J. Simpson was unreliable. The basis for this argument was the transfer pattern of blood from the swatches onto the bindle, as seen in Figure 10.24. The swatches from the Bundy scene were described as being dry prior to being put in the bindle, thus finding transfer traces from them suggested that they had been tampered with at the laboratory. However, the swatch-drying process was conceptually flawed. The swatch collection and drying process had been developed during the era of doing protein type and polymorphic protein typing. Swatches were used at that time due to the convenience of teasing a yarn from the swatch which could then be directly put into a slot in the gel in preparation for the electrophoresis. The swatch collection process began using a swatch moistened with saline or distilled water to forcibly rub on the blood trace to quantitatively collect the entire deposit. If not all of the deposit was collected with a single swatch, multiple swatches would be collected and put together in a plastic envelope. These swatches would be moist at this time. At the laboratory, the moist swatches were taken from the bag and put into an 8-mm test tube and left to dry. If several swatches were collected

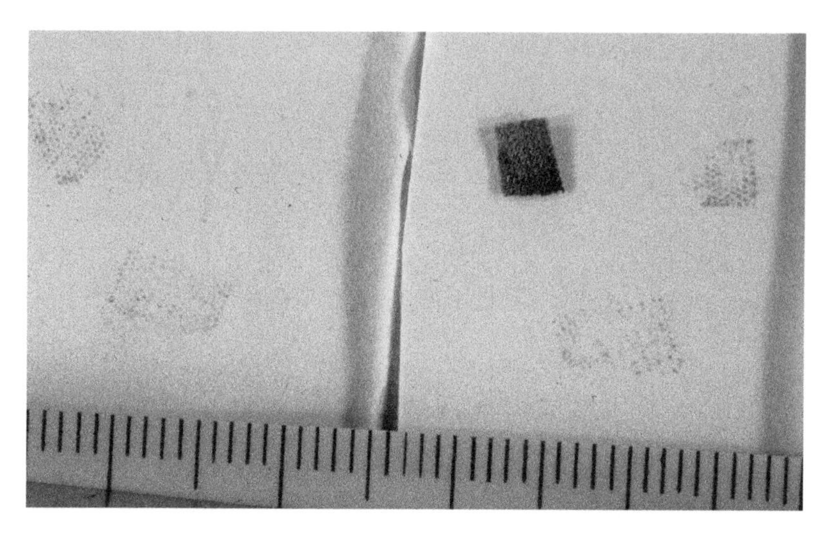

FIGURE 10.24 Inside of cut open bindle used to contain swatches used for blood collection from a droplet deposit (shown in Figure 10.16) from the driveway of the Bundy homicide scene, after "drying" in a single small test tube in the laboratory.

from a single blood trace, they would be dried in a single test tube, thus complete drying would not occur. In a later experiment designed to test the drying process, it was shown that swatches were still wet after 48 hours. It must be noted that the blood trace collected from the apron of the Bundy Driveway took eight swatches for collection. Thus, it is likely that these swatches were not in fact tampered with and instead were not actually dry at the time they were put into the bindle.

Further, the blood samples that were taken at the Rockingham scene (which was the first one attended to by the scientific investigators) contained cultured bacteria and had begun to decompose during the time interval between collection and laboratory submission. There are two factors which contributed the degradation of the blood samples. First, the refrigerator in the evidence collection van was not working, thus the samples were not kept at a cold temperature to prevent bacterial growth. Second, the wet swatches were sealed in plastic envelopes, which is acceptable for temporary transport (two hours or less) but not for longer-term collection because it promotes the growth of microorganisms which will destroy or alter evidence. The investigators did not consider this, and rather than going from the first scene back to the laboratory, they went to the second scene (which has its own conceptual problems, as one team should not process two potentially related scenes). This resulted in a longer delay in getting the blood evidence back to the lab where it could be refrigerated and processed.

10.8.2.2 The Blood on and in the Bronco There were several areas on and in O.J. Simpson's Ford Bronco where blood traces were recognized, documented, and collected for laboratory analysis. The most well known of these is a contact blood trace on the outside of the vehicle, specifically on driver's side door handle (Figure 10.18). Reverse dot blot polymerase chain reaction (PCR) DNA analysis of this blood trace showed evidence of this being a three-person mixture, with markers for both victims and O.J. Simpson included. No other DNA analysis (e.g., restriction fragment length polymorphism, RFLP) was completed. There were also a numbers of locations of blood traces inside the vehicle: on the carpet, console, steering wheel, and door. On the carpeting by where the clutch would normally be in a car with a manual transmission, there was an ill-defined bloody left footwear outsole imprint. However, because of the carpet's texture, the specific outsole pattern could not be determined. In addition, there was blood smeared all over the console between the two front seats. This console was composed of a pigskin-textured plastic polyolefin material. Similar to the blood trace by the door handle, the blood inside the Bronco was analyzed by PCR and concluded to be from a three-person mixture with likely contributions from both victims and O.J. Simpson.

10.8.2.3 The Socks and EDTA Testing

Black dress socks were found on the bedroom floor at Simpson's residence which were approximately ¾ inside out (Figures 10.25–10.29). The fabric making up the socks is problematic with respect to blood trace interpretation. There are three major problems with this fabric in this regard. First, the very dark color of the socks obscures dried blood. It becomes very difficult to visualize blood deposits on this material. Illumination is very critical (Chapter 4). Second, the structure of the fabric is not conducive to the production of blood traces which retain their original shapes. Wet blood traces can be expected to diffuse into this type of material distorting the shape the deposit assumed when it first formed (Chapter 13). Third, the elastic nature of the knit fabric and the inelastic nature of dried blood crusts result in the flaking off of blood when the socks are manipulated.

The socks were first examined by the defense's pathologist and his assistant pursuant to a court order. To allow this, the seals on the packaging were broken before the evidence was examined by the LAPD criminalists. The mere handling of these stretchy socks by a non-trained criminalist risks alteration and loss of the blood traces deposits. It is not understood why the defense examined them prior to the LAPD laboratory scientists nor why a pathologist would examine the socks at all. Although a pathologist is able to provide meaningful analysis of the cause of death, it is unclear what

FIGURE 10.25 Black dress socks found on the bedroom floor at Simpson's residence.

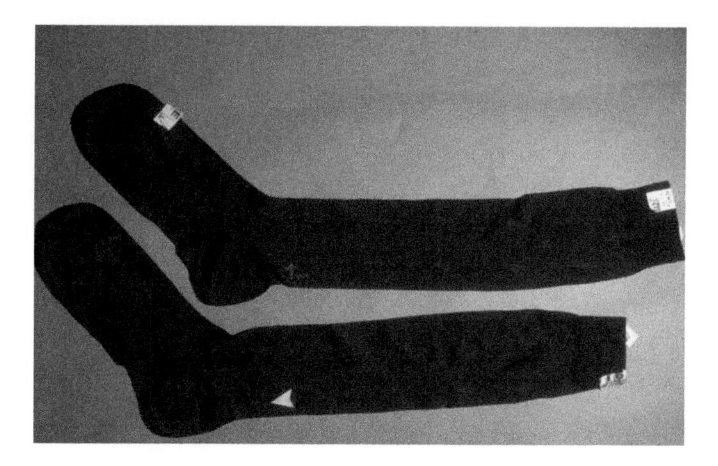

FIGURE 10.26 Evidence photographs of the black dress socks found on the bedroom floor at Simpson's residence.

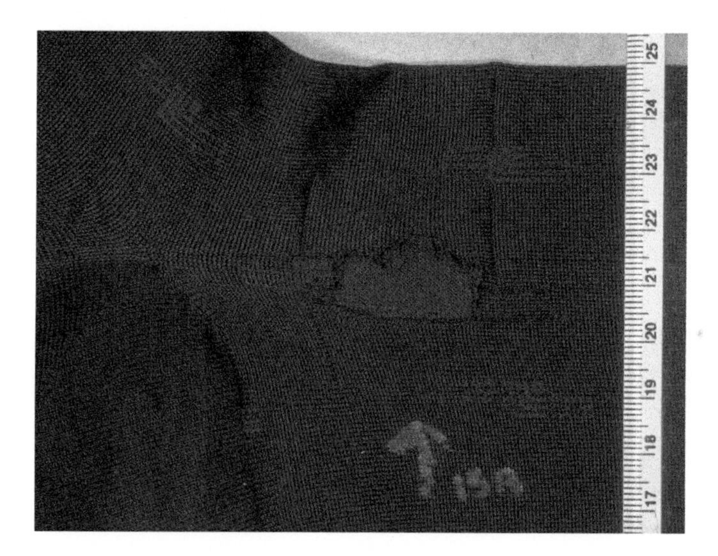

FIGURE 10.27 Close-up photograph of the black dress socks found on the bedroom floor at Simpson's residence, showing a portion of the socks cut out for defense DNA and EDTA analysis.

a pathologist could contribute to the examination of socks. This kind of examination needs to be done by an experienced criminalist.

Subsequently, an additional prosecution expert, California Department of Justice Crime Laboratory criminalist Gary Sims, did examine the socks. Blood traces were not visible with the naked eye and required a stereomicroscopic examination for their detection. Sims found numerous microscopic flakes of blood on the surface designated #3 (one of the two inside surfaces of the sock), which could have originated from dried blood traces from any area on the sock when handled and stretched by the pathologist. The pathologist could have unwittingly transferred blood traces between surfaces

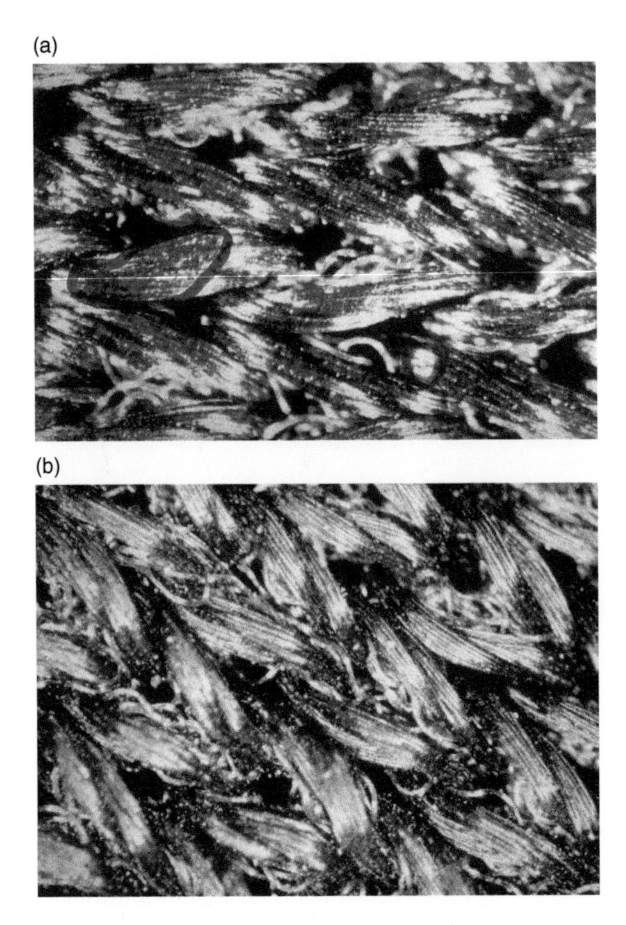

FIGURE 10.28 Photomicrographs of the black dress socks found on the bedroom floor at Simpson's residence, as viewed with a stereomicroscope. In image A, the fibers of the black socks appear white (an example was circled in black marker during the testimony of a defense expert) which are not parallel to the camera's image plane. This is caused by oblique illumination creating a selective overexposure for the fibers which are oriented roughly 45° to the camera's film plane resulting in a false white appearance in the photograph of these fibers. The fibers running parallel to the film plane are not overexposed and thus appear red. This led to the misleading interpretation that the blood is only on the surface of the fibers (an example area outlined in blue marker during the testimony of a defense expert) as opposed to soaking into the textile. Image B shows the same area of the socks as shown in A, with orthogonal illumination; however, the upper surface is selectively overexposed to enable visualization of the blood deep into the textile.

of the socks during this ill-advised and premature handling. Alternatively, the blood traces on the inner surface of the sock could have been transferred when the damp bloodstain surface #2 made contact with surface #3 after being removed from the foot, or from the hand of the wearer when removing the socks from his foot. The manner and mechanism for the deposition of these blood traces cannot be known. Ultimately, a blood deposit on one of the socks yielded a 15-probe RFLP match to Nicole Brown Simpson. RFLP was a highly discriminating technology during that era; however, it

FIGURE 10.29 Image C shows an alleged "ball of blood" (circled in black marker during the testimony of a defense expert) on surface #3 viewed through the cut-out shown in Figure 10.27.

was very time-consuming, thus resulting in lengthy turn-around times, and in addition required relatively large amounts of sample.

To counter this highly incriminating and normally incontrovertible evidence, the defense claimed the blood was applied to the sock while it was off the foot and thus planted. One defense criminalist stated that the blood deposit would have dried faster if the socks were being worn and therefore could not have transferred liquid blood to the other inner surface following the removal of the socks. This is not necessarily true. If the wearer were sweating, the socks would be expected to be still damp when they were removed. The blood traces, particularly the inside surfaces of such, would not dry rapidly under such conditions. There are two additional issues with the blood on the sock which are interrelated and based on the manner and mechanism proposed for its "planting." The defense claimed that the evidence socks were removed from the packaging in the laboratory, and a finger was dipped into the known Nicole Brown Simpson blood tube, and rubbed (or "swiped") on the outside of the sock (surface #1). This blood was then mysteriously forced through the fabric's weave and deposited onto surface #3 as wet beads of blood that were not absorbed into the fabric, thus drying as "balls." The application of the blood to the sock (surface #1) via a rubbing action was illustrated using a photograph purported to illustrate blood confined to the surface yarns of the textile, as opposed to blood having soaked into the sock (Figure 10.28a). The photograph in question was incorrectly interpreted and misleading as presented, because it was selectively overexposed. A subsequent examination by one of the authors of the same location of the sock, with different illumination, showed that the blood was not a surface phenomenon but had soaked completely into the fabric (Figure 10.28b).

Additionally, the claimed "balls of blood" on surface #3, as viewed through an evidence cut out (Figure 10.27) for DNA sampling, were used to assert that a "wet transfer" to the surface designated #3 took place. One "ball of blood" documented in this area was photographed with a stereomicroscope and is shown in Figure 10.29.

Unfortunately, there were no contemporaneously taken notes provided on discovery to the prosecution and made available to this author specifically detailing the "balls of blood." Without the proper documentation of evidence by taking contemporaneous notes of the examination at the time of photography, the integrity of this observation and its interpretation is in question. The use of stereomicroscopes, also known as dissecting or sample preparation microscopes, for the documentation of the balls of blood deserves attention. Stereomicroscopes are low-magnification microscopes that provide realistic three-dimensional non-inverted images. This enables sample manipulation and facilitates preliminary examinations. They are not intended for detailed analysis at high magnifications. This microscopical analysis of the socks was done at or beyond the resolution limits of a stereomicroscope (e.g., the ability to see fine details), which leaves uncertainty about the actual morphology of the blood trace designated as a "ball."

A wet transfer on the other sock was documented in bench notes made by Criminalist Gary Sims. This observation was confirmed by the fibers within a yarn being cemented together with blood. This matter was not contested by either side, likely due to the results of the DNA analysis indicating that it was from a single source: O.J. Simpson.

A particularly controversial item during the trial was the detection of EDTA in the blood recovered from O.J. Simpson's sock. This blood sample was identified as being a match to Nicole Simpson, and to mitigate its significance, the defense alleged that the blood was planted based on the presence of EDTA. EDTA has a variety of industrial, medical, and cosmetic uses. It is commonly known for being a preservative added to blood collection vacutainer tubes; however, it is also used in the dying process during the manufacturing of textiles, in the soft drink industry, and in personal care products such as shampoos and moisturizers. In this case, the FBI's reported concentration of EDTA that was detected in the blood traces on the socks was less than 1% of the expected concentration of EDTA in blood from a purple-top vacutainer tube. Thus, it is illogical to think that the source of the EDTA in this blood sample was due to it having originated from a vacutainer tube, unless it was diluted. Further, a dilution is entirely inconsistent with the concentration of the blood components that were observed. In addition, if the blood was diluted a 100-fold, it would have been next to impossible to see evidence of deposits on the dark socks. Therefore, one might conclude that if the intent was to fabricate and implicate O.J. Simpson in these murders, dilution of the blood by a detective would not make sense because they would want the observer to readily see the traces. A forensic toxicologist who testified for the defense, Dr. Frederich Reiders, initially reported and testified that the presence of EDTA indicated that the blood could have been planted; however, he later changed his testimony and acknowledged that FBI scientist Roger Martz found an equivalent level of EDTA (parts per million) in his own blood as was found on the sock (Walraven 1995a, b).

10.8.2.4 The Envelope The envelope ostensibly containing the eyeglasses that Ron Goldman brought to the scene was observed following the homicides and photographed at the boundary between the soil and the imitation paving stones near the body of Nicole Simpson (Figures 10.12 and 10.13). Blood was found on the envelope, and of apparent value was a blood droplet deposit, which when tested with reverse dot blot DNA technology, was consistent with having come from O.J. Simpson (Figure 10.30).

The envelope was removed from its original position at the crime scene in connection with removal of the bodies of the victims from the scene by the coroner's

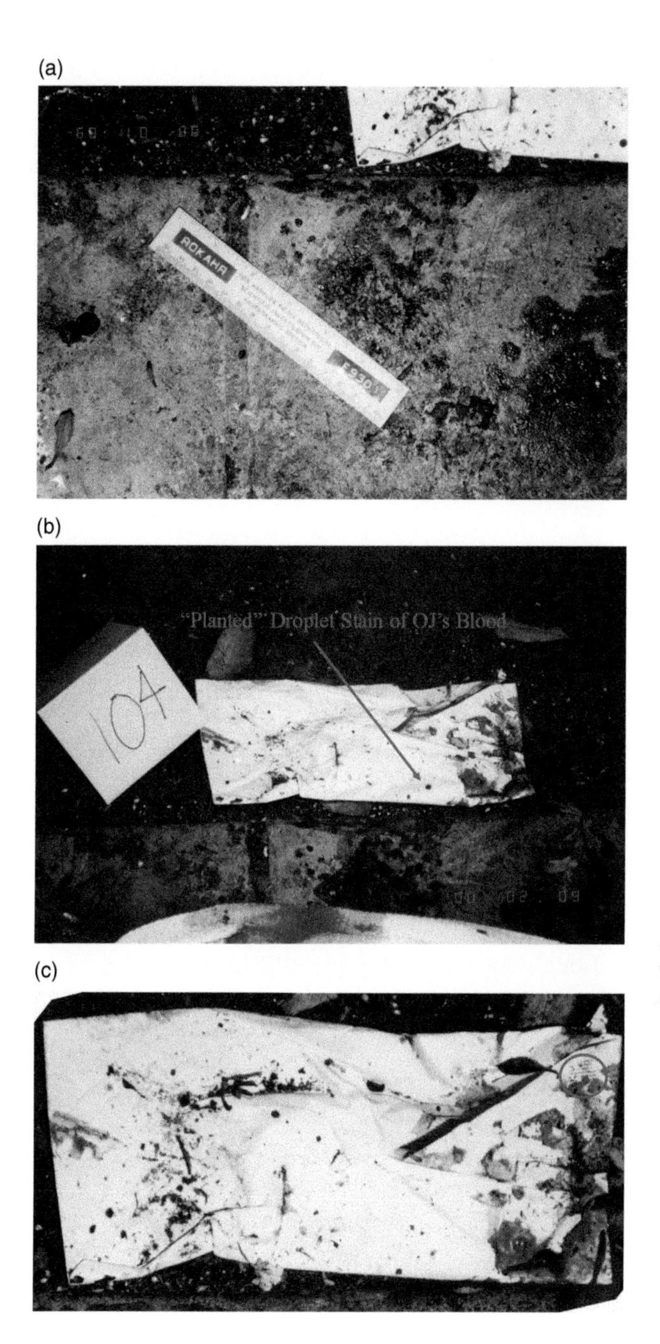

FIGURE 10.30 Photographs of the envelope before (a) and after (b) it was picked up and repositioned. An arrow pointing to the allegedly "planted" blood droplet is shown in (b), and the parallel line pattern circled is (c).

office personnel. Following this, instead of being documented and secured separately as evidence at that time, the envelope was apparently put back to what was thought was its original location. The defense exploited the differences in the positions of the envelope seen in the photographs in order to make a claim that the blood trace on the envelope was planted. However, the alleged added deposit was clearly present, but partially obscured by glare, when the envelope was first photographed in its original position. This refutes the defense claim that the bloodstain was added after the envelope was moved. Considering the totality of the photographs, the appearance of the blood droplet before and after the move is the same. Although it is true that in a number of intermediary photographs, the blood droplet is significantly washed out because of improper use of the camera mounted flash, there is no doubt that the droplet deposit is clearly present in other premove frames.

Another blood trace on the envelope was a transfer pattern of parallel lines, circled in Figure 10.30c. The defense made claims that this could have been a footwear outsole pattern from another perpetrator; however, the dimensions and characteristics of this pattern are more likely to be an impression from a twill fabric and not a footwear outsole pattern. Additionally, no footwear outsole pattern with this parallel pattern was detected at the crime scene at the time of the initial investigation.

10.8.2.5 The Hat and Gloves A black knit hat was found in the garden plantings near both victims. The hat contained two hairs that were consistent with O.J. Simpson as well as two different kinds of carpet fibers that were consistent with two types of carpet fibers from Simpson's Bronco (Allied Signal fibers). Of note, the FBI examiner had performed research about the rarity of these fibers, which were used to estimate statistics for their frequency of occurrence which were not very common. This evidence was not presented to the jury because the defense was successful in having the court rule that the hat and the hair evidence were inadmissible due to the failure by the prosecution to disclosed the FBI lab's findings to the defense in a timely fashion.

A left-handed black Aris Light leather glove with a cashmere lining, size extra-large, was also recovered near the two victims on the ground level of the Bundy crime scene. This glove had blood traces on it from both victims and O.J. Simpson, with no traces of EDTA. Although not part of the scientific investigation, Nicole Brown Simpson bought these same type of gloves as a present for her then husband, O.J. Simpson, from Bloomingdales in New York City in 1990, and there is documentation of him wearing them from 1990 to 1994.

A right-handed black Aris Light leather glove, ostensibly the companion to the glove discussed above, was recovered at O.J. Simpson's North Rockingham Avenue residence, in a narrow passage way behind an annex to the main residence that had guest rooms (Figures 10.17 and 10.31). DNA analysis established that blood on this glove was a three-component mixture consistent with having originated from O.J. Simpson, Nicole Brown Simpson, and Ron Goldman. Although the defense claimed that this glove was planted by Detective Mark Furhman, there is no evidence to support this claim.

It is noteworthy that O.J. Simpson's left hand had a substantial cut, the presumed source of his blood found at several locations during the investigation. Since the left glove was found without any corresponding defects at the primary crime scene, it is logical that the left glove came off first, assuming Simpson had worn it, and was actually present at the crime scene. Then, O.J. was cut on the left hand and fled the scene; subsequently removing or dropping the right glove at the Rockingham scene.

In perhaps one of the most memorable moments of the trial, O.J. Simpson put on the gloves in front of the court. This was an ill-conceived experiment initiated by the prosecution and ultimately used by the defense to support claims of innocence.

FIGURE 10.31 The right black leather glove recovered at O.J. Simpson's North Rockingham Avenue residence, in a narrow passageway behind an annex to the main residence that had guest rooms.

However, it is not wholly accurate that the gloves did not fit. First, Simpson donned latex gloves prior to putting on the leather gloves found at the respective scenes. When worn, latex gloves form a web between the fingers (approximately a 0.5–1 inch above the base of the fingers) which could prevent a hand from being fully inserted into the leather gloves. Another explanation is that the gloves had shrunk from successive wetting followed by drying. Furthermore, in this courtroom, so-called experiment, O.J. Simpson was in effective control, presenting to the jury his biased demonstration that they did not fit. Simpson was free, during the experiment, to flex the muscles in his fingers and palm, for maximum expansion, to suggest a snug fit.

10.8.3 Conclusions

The physical evidence in this case was substantial, and in most cases, would likely be considered more than sufficient for associating the defendant with the murders. However, there were numerous errors in this case which contributed to confusion and were taken advantage of to cast doubt about the physical evidence. Most of the errors, made by both the prosecution and defense, refer to concepts discussed throughout this book. A list of major errors is bulleted below. Many lessons were brought to light as a result of these errors, which inspired considerable improvements to necessary appreciation of nuances applicable to crime scene investigations, laboratory analyses, and issues of evidence admissibility.

An example of this is the positive presumptive blood test findings within the Rockingham master bathroom which were excluded by Judge Ito. Although context-free presumptive blood test results are commonly not admitted into evidence in most trials due to the conflict between their probative value and prejudicial effects. In this case, it can be argued that it should have been admitted because its inclusion would have completed an understanding of the blood trail. To exclude this evidence denies the finder of facts the opportunity of evaluating these findings in the context of the totality of the case. By not having this evidence presented at trial, it misleadingly gives the finders-of-fact the impression that there was no continuation of the evidence trail into the O.J. Simpson home. In post-trial polling and interviews of jurors, some jurors stated that they expected to see blood evidence in the Rockingham residence, and its absence contributed to their non-guilty verdict. Consequently, the exclusion of the blood evidence gave an incomplete picture and was misleading.

Major errors in the OJ Simpson murder investigation:

- Scientific investigators did not have full authority at the Bundy scene. The police investigators had nearly completed their analysis of this scene, which included photographic documentation and fingerprinting, prior to arrival of the laboratory team arriving from the Rockingham residence. The scientific team were ultimately relegated to evidence collectors rather than afforded the autonomy to do a proper scientific investigation.
- Access to the scene was not limited to essential personnel (Chapter 1), with an unnecessary number of police officers and detectives allowed to enter the crime scene.
- Incredibly, a blanket from the interior of the residence was used as a body covering at the scene (Chapter 1). This raised the spectre of contamination of the traces recovered from the victims. Traces must be recognized, documented, and collected at the scene from the bodies and the surrounding areas prior to their removal.
- The bodies of the victims were prematurely removed, as they were did not remain at the scene for the scientific investigation. Any area that may be influenced by removal of the body must be thoroughly scientifically investigated, which includes the documentation and collection of traces, prior to arrival of the medical examiner.
- There was a failure to recognize and collect relevant blood traces at the scene. Of particular value were circular blood traces on the back of Nicole Brown Simpson that may have fallen vertically from the assailant (Figure 10.12). The opportunity to analyze these was lost and gone forever after the body was removed, either being obscured in the body bag or when the body was washed at autopsy.
- The forensic mantra, "separate teams for separate scenes," was not adhered to.
- Some of the defense experts failed to properly document their scientific examinations, which included a lack of recording observations.
- Some of the evidence (e.g., the socks) was examined by defense personnel before analysis by criminalists from the forensic laboratory as a result of an ill-conceived court order. As a corollary, the evidence seals were broken and packaging was opened before proper laboratory analysis. There was excessive handling of this evidence prior to proper laboratory analysis, which affected the integrity of the traces.
- Physical evidence occurs in context. This was not properly considered with regard to using the physical traces in the event reconstruction.
- A generalist-scientist, or "Maestro," was not used to maintain scientific oversight over the physical evidence and crime scene (Chapter 1). This prevented the totality of the evidence to be considered as a whole. Instead there was a myopic item or test focus which detracted from the holistic approach necessary for a comprehensive reconstruction.

10.8.4 Lessons

Several lessons can be drawn from errors made in this case that are specific to this particular case and can be appreciated from the preceding "Conclusions" section discussed above. There are lessons that can be drawn from errors that clearly compromised traces. There are other errors, equally important, in that they leave the mere possibility of compromise open. Both need to be stringently guarded against.

There are additional lessons that are more general in nature that will be discussed here.

Ideally, experienced scientists need to be present in a decision-making role at the scene. A teamwork approach with synergism between the criminal and scientific investigators is optimal, as discussed in Chapter 7.

There is a strong inclination on the part of first responders and others who feel that they need to protect the privacy of a victim by placing coverings over the body. This is a serious mistake and should be avoided. It can alter blood deposit configurations on the body and the clothing, and in addition, can compromise other traces by depositing them or removing them. Similar problems exist with the nearly universal practice of using body bags. An argument can be made for having experienced trace scientists examine, document, and secure traces on the body before it is placed in a body bag. Whoever has the responsibility to remove the body from the incident scene, such as the forensic pathologist, medical legal investigator, or other medical examiner personnel, must be trained in the fundamentals of trace evidence so that they have at least a modicum of appreciation of the failure to properly consider the trace evidence. However, they must seek approval from the trace evidence scientist before taking any action.

Some cases, such as the above one, involve more than one scene. There may be primary scenes, secondary scenes, and automobiles, for example. It is important to recognize that different teams should not be physically present at more than one scene, although it is fundamentally important that there be scientific oversight over all of the scene activity. Modern technology allows information to be easily conveyed to and exchanged with the scientist in charge.

Access to evidence from scenes should be provided to scientists, with appropriate qualifications, representing the defense. However, this should be done in coordination with scientists representing the state. So-called "viewing" of evidence by attorneys, prosecutors, forensic pathologists, and police "brass" should only take place in a laboratory environment, following the completion of scientific examinations.

10.9 A Vertical Crime Scene

10.9.1 Case Scenario/Background Information

A "robbery in progress" call brought four uniformed police officers to a 10-story converted loft building in a major city. The call originated from an apartment that occupied the entire tenth floor of the building. The female caller said that she and her male partner had been bound by the partner's adult son at gunpoint and that she had escaped her bindings to make the call. The son was apparently motivated by fear that his wealthy father's young girlfriend was going to "cheat him out of his inheritance." When police arrived, the defendant went to the roof in an attempt to flee and retreated back into the building upon seeing additional officers covering the fire escape. This effectively trapped the suspect in the stairwell. The officers who entered the lobby of the building could not gain entry to any of the floors because the building was secured by locks and alarms. The elevator could not be operated without a pass code or a key. An occupant of one of the lower floors of the building, on learning that police were present, descended the inner stairwell and opened a door from the lobby to the stairwell from the inside. One of the officers entered the stairwell and began an ascent to the tenth floor, but the door closed and locked before the other officers could enter the stairwell. The other officers then turned their attention to gaining access to the elevator. In the meantime, the officer who became separated from the others continued his ascent until he sustained what was ultimately a fatal gunshot wound. A 25 ACP FMC bullet fired downward had passed through the space between his shirt collar and his body armor and then entered his upper back traveling downward through left lung, the diaphragm, and lodged in a kidney. The wounded officer retreated down the stairs coughing up large quantities of blood along the way. He collapsed and expired from exsanguination on the second-level landing.

After an extensive search of the building, the son, bleeding from minor gunshot wounds, was found hiding in the cluttered basement of the building with the officer's 38 SPL revolver and a 25 ACP semiautomatic in his possession. He was arrested and charged with the murder of the police officer. At trial, the defense claimed that the officer had fired on the defendant first, and that the defendant only fired in self-defense.

10.9.2 The Physical Evidence and Its Interpretation

A re-examination of the secured scene by one of the authors, a consulting criminalist and a crime scene scientist employed by the medical examiner's office, a day or so following the initial examination by the crime scene personnel yielded data that allowed a detailed shooting scene reconstruction. The stairwell contained numerous relevant traces in addition the blood deposits (Figures 10.32–10.34). These included projectiles and projectile fragments, ejected cartridge cases, impact marks, wood fragments, paint, fibers and fabrics, and plaster or gypsum wall traces. Much of this had been documented and collected in the previous investigation. Additional relevant traces recognized during the later investigation by the authors included projectile, paint, and wood fragments. There were two shooting events, and based on physical traces, the location of the shooter and the wounded subject in each case can be known.

Several well-defined bullet impact sites, many of which contained flattened copper disks, were found on the tenth-level landing and on stair treads just above and below this level. The flattened copper disks were recognized as the remains of the nose portions of 25-caliber, full-metal jacketed, bullets that had been rapidly decelerated on impact to painted concrete or steel. These observations taken together made it clear that the 25-caliber semiautomatic had been fired almost vertically downward from a restricted area near the roof entrance to the stairwell. Although single impact points do not define a trajectory, when the geometry of the 11-level stairwell is taken into account, the possible points from which the shots could have been fired are extremely limited.

FIGURE 10.32 A image showing the stairwell of the tenth floor landing and stair treads above (designated 10-1 for the first step, 10-2 for the second step, etc...). The rosin paper and duct tape were applied to the stair treads prior to the event to protect them from construction-related activities for renovations taking place in an apartment. The impacts and fragments of vertically fired 25-automatic projectiles were recognized and recovered between the ninth and tenth floor levels. Important here, as is the situation in other large or extended scenes, is the use of labels to document specific locations and orientations (as shown with the yellow cards designating the floor and tread number in addition to the north-pointing arrow) in photographs to allow for a detailed understanding of complex relationships among traces.

FIGURE 10.33 An image showing blood traces on the stair treads between the seventh and eighth floors. From the blood pattern, it can be known that the officer was descending the stairwell, after being shot, coughing up blood on the way down.

There were deposits of aspirated blood found on the walls of the stairwell at various points from the area of the ninth level and below. A trail of large quantities of dripped blood was present on the landings and stair treads from the ninth level to the second level. A horizontally projected configuration of aspirated blood was observed on the wall just above stair tread height in the area of the sixth level. The other configurations

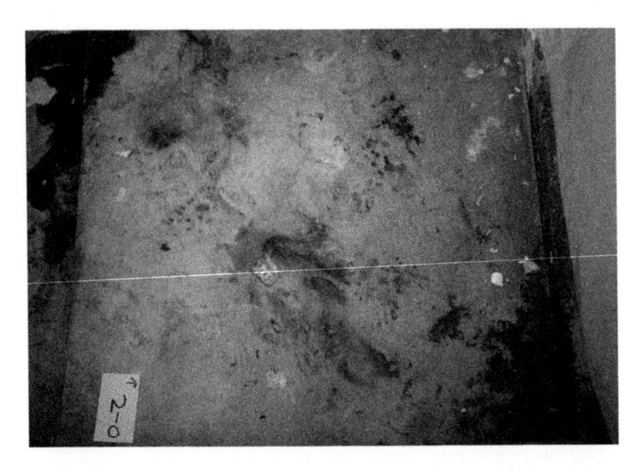

FIGURE 10.34 An image showing blood traces on the second floor landing where the officer ultimately bled out. On the left side of the image, there is a portion of a blood pool.

of aspirated blood were present on the walls at heights consistent with being expelled from the mouth of the victim in a standing position.

In addition to the 25 ACP bullet impacts described above, evidence of another set of bullet impacts was observed in a lower area in the stairwell. All were consistent with having resulted from lead bullets fired in an upward direction from the officer's 38-SPL revolver. The approximate trajectories for these could be traced back to a restricted area of the stairs just above the sixth level. This was in the same area where the low configuration of projected aspirated blood on the wall was located. The physical evidence made it clear that there were two and only two sets of gunshots in the stairwell. One set was fired upward from the location just described. The other was fired downward from the portion of the stairwell near the roof mentioned earlier.

Since all the shots fired originated from only two well-defined areas in the stairwell, and since the two weapons used were grossly different from each other, it was easy to demonstrate that all the shots associated with the roof area origin were from the 25 ACP and all of the shots originating from the **Level #6** area were from the officer's revolver. By examining the bullets and the clothing of the defendant for trace evidence (Figures 10.35 and 10.36), it was possible to show that one bullet from the revolver had traveled upward and entered the lower side of an oak stair railing on **Level #7**. This bullet exited the railing and penetrated a nylon jogging suit jacket and wounded the defendant in the left flank and the inner aspect of the left forearm. Paint chips and wood fibers similar to those comprising the railing were found on the recovered bullet and on the defendant's jacket in the vicinity of the bullet holes. In addition, fibers consistent with having originated from the jacket were found associated with the exit hole in the top of the oak railing.

The scene and laboratory examinations of the physical evidence were combined to reconstruct some significant details the shooting. From the stairwell geometry and from the details described above, it was concluded that the officer was fatally wounded as a result of a shot fired downward from near the roof level while he was on the stairs between levels nine and ten. It was also concluded that the defendant's wounds were acquired while the front of his torso was against or leaning over the railing on **Level #7**. When these conclusions were made available to the defense, they argued that during

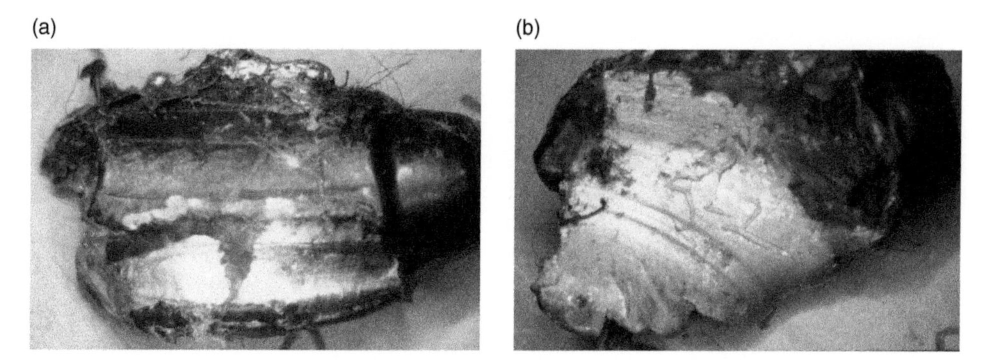

FIGURE 10.35 Two images (a and b) of the Nyclad® (a proprietary nylon coating designed to reduce atmospheric lead pollution in indoor ranges) police bullet that was fired by the officer, which had gone through a railing just below the seventh level. The railing consisted of a painted oak piece of wood hand rail over a wrought iron frame. This bullet first entered through the lower side of the wooden rail, grazed against the wrought iron support, into the side of the defendant's jogging suit, grazed his left flank, and finally impacted a plaster or gypsum wall. Traces retained by the bullet aided in the reconstruction. These included oak chips, paint from the railing, fibers from jogging suit, and plaster or gypsum from the wall. Image B shows a large area of exposed lead with striations and deformities resulting from its forcible contact with the wrought iron support.

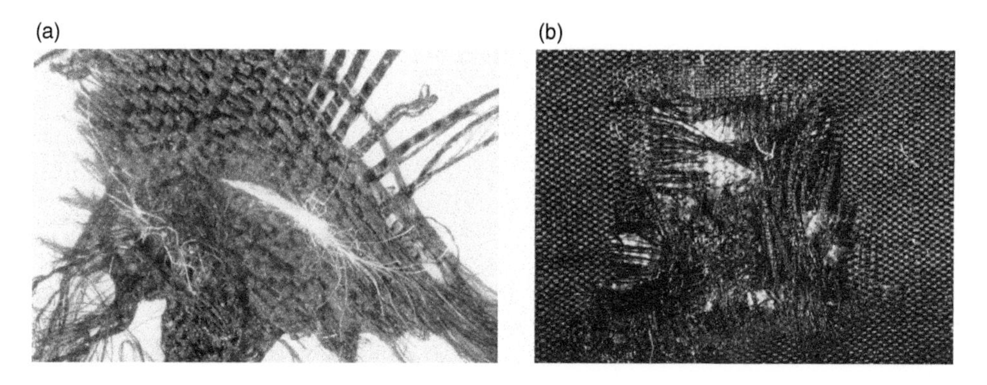

FIGURE 10.36 The bullet from Figure 10.35 had torn and carried a swatch from the defendant's jogging suit.Fig (a) shows a laboratory photograph of the removed nylon swatch and (b) shows an image of the hole in the jogging suit with a rough juxtaposition of the swatch shown in (a).

his ascent the officer fired on the defendant from the stairs between **Levels #6 and #7**. He then continued up the stairs and cornered the wounded defendant who had retreated to a position just below the roof level from where he fired down on the officer in self-defense.

Drawing on the same information from the reconstruction, the prosecution contended that the first shots were fired by the defendant, and that he fired several shots in rapid succession downward at the police officer when he had ascended to a level just below **Level #10**, and that one of these shots struck the officer and caused the fatal wound. The prosecutor argued further that the officer, after being fatally wounded, retreated down the stairwell pursued by the defendant, and that when he reached the area between **Levels #6 and #7**, he fell. The conclusion about the fall was based on the configuration of coughed-up blood just above the stair treads at this location. In this prosecution version, while the officer was down, he looked up and noticed the defendant leaning over the railing at **Level #7** in a menacing fashion. He fired a few poorly aimed shots upward in the direction of the defendant. One of the bullets entered the lower edge of the oak railing, traveled upward through it at an angle, and wounded the defendant. The wounded officer was then was able to get up and continue his descent. At points during his decent, he hesitated long enough to bang on some apartment doors in an attempt to seek help. Some of the occupants heard him exclaim, "Help me I am dying." during these brief interruptions in his descent to the point on **Level #2** where he collapsed.

The reconstruction provided by the consulting criminalists was very valuable, but as is always the case, it had limitations. It was certainly not as richly detailed as the scenarios built on it by the opposing attorneys. The key point of contention between the prosecution and defense versions of the shooting was the sequence of the two sets of shots. The reconstruction, as good as it was, could not resolve this question directly, but it was capable of shedding light on it. During the direct testimony of the consulting criminalist, the prosecutor asked about his opinion concerning the sequence of shots. Although the criminalist was of the opinion that the group of shots from the 25-caliber automatic fired by the defendant preceded those fired from the officer's revolver by a few tens of seconds, he did not feel that this was a scientific opinion. He told the jurors that, if they understood the facts that he had developed from his scene investigation and reconstruction, they were in as good a position as he to draw the conclusion. He felt that common sense was adequate. The pivotal observation shedding light on the sequence was the configuration of the coughed-up aspirated blood on the wall just above the stair treads between **Level #6 and #7**. It showed unequivocally that the officer was down at this point. Were the poorly aimed shots fired from this area the result of them having been fired after he fell during his retreat or were these shots fired from this area during his ascent? The defense scenario requires that the officer's ascent continued for three more floors before he was wounded and began his descent. Further, it is necessary to assume that it was a coincidence that he fell and coughed up blood during his descent, at the same location from which he had fired during his ascent a minute or more earlier. The criminalist felt that he was no more qualified to opine on the likelihood of such a coincidence than the jurors.

10.9.3 Conclusions

Although blood trace configurations were important, it was essential to integrate other traces to arrive at a complete reconstruction of the critical events. For example, the location from where the defendant shot downward at the officer was above the tenth floor landing, as revealed from the bullet impacts on the stair treads and the initiation of the blood trace trail. The injury to the defendant was shown to occur when the officer shot upward from an area of the stairwell between levels six and seven, which was indicated by bullet impact sites and diverse traces on the bullets which provided information about trajectories.

10.9.4 Lessons

A practical lesson demonstrated in this case is the ability to locate events in three-dimensional space from the physical evidence (blood and other traces). This was done in two instances in this case to reconstruct the location of the two significant temporal and spatial shooting events. Although interpretation of the physical traces enabled a spatial reconstruction, the challenges of learning the sequence of the elements of the event could not be scientifically concluded.

The other lesson is a philosophical one, wherein a scientific expert should refrain from giving nonscientific opinions. It is important to remember that the opinion of a scientist is not necessarily a scientific opinion. This should be recognized and reflected on by all scientists when providing testimony. Although we know the positions of the shooter and wounded subject for each of the shooting scenarios, the physical evidence is not adequate to definitively inform us of the sequence of shooting events. This was a critical issue in this case. The defense claimed that the officer had fired at the defendant while he was ascending the stairs and thus the defendant fired in self-defense. For this to be true, the officer would have had to fire from above the sixth floor at the defendant, travel up to between the ninth and tenth floors where he acquired his fatal wound, then descend back to the sixth floor and fall at the same location where the initial shooting occurred. The alternative scenario, favored by the prosecution, had the defendant shooting the officer first from above the tenth level, both descending the staircase to between the sixth and seventh levels, where the officer fell and coughed up blood leaving a horizontal expiration configuration low on the wall near the stair tread level and in the vicinity of where second shooting event occurred which resulted in the wounding of the defendant. Clearly, the defense's version of events is more convoluted, making the prosecution's theory easier to believe. As Occam's razor states, that the simplest solution tends to be the right one. The consulting criminalist felt that the logical sequence of events could be deduced based on common sense; however, it was not a scientific conclusion based on an interpretation of the available traces and thus should not be included in his expert opinion.

10.10 Tissue Spatter from a Large Caliber Gunshot

10.10.1 Case Scenario/Background Information

A confrontation between a father and his daughter's boyfriend resulted in a fatal shooting. This occurred outside of the father's residence, near the garage and boyfriend's vehicle.

10.10.2 The Physical Evidence and Its Interpretation

The father fired two shots at the boyfriend with a semi-automatic 20-gauge shotgun loaded with deer slugs. A 20-gauge projectile is large. It has a diameter of just over 0.6 inches, thus it is equivalent in size to a 60-caliber bullet. Both fired projectiles struck the victim as he stood near the left side of his car. One projectile passed completely through the chest from front to back in a somewhat downward angle. The other struck the victim in the right anterior chest traveling from left to right tangentially. It left a deeply furrowed wound in the pectoral area removing a considerable amount of skin, subcutaneous fat, and muscle tissue in the process. There were several areas on the left side of the vehicle that contained deposits of apparent lipid material and tissue fragments. There were no airborne blood droplet deposits associated with these. Subsequent microscopical examination of histological thin section preparations of the tissue fragments removed from the side of the vehicle revealed red blood cell containing capillaries but no blood outside the capillaries.

10.10.3 Conclusions

The case illustrates that even with massive wounds caused by large caliber projectiles, there can be no absolute expectation of blood spatter. In order for gunshot wounds to produce spattered airborne droplets, it is necessary for there to be coalesced blood along the projectile track during the instant the wound is made. Thus, a bullet can pass through an area of muscle or other tissue containing only very small diameter blood vessels, such as capillaries and arterioles, without encountering coalesced blood. In such a case, no blood spatter in the form of airborne droplets is produced, although there may be fragments of other physiological material, e.g., tissue.

10.10.4 Lessons

The point made concerning the incorrect expectation of gunshot blood spatter in this case is a very important one. In addition to case examples, this point can be demonstrated experimentally using medical dialysis cartridges. These cartridges consist of a large number of hollow fibers arranged in a parallel bundle which is fixed within a clear polymer jacket. The hollow fibers are similar to capillaries and have semi-permeable walls. In normal medical use, a patient's blood is pumped through the hollow fibers while they are bathed in a dialysis buffer passing through the jacket. In gunshot spatter experiments, the jacket can be filled with ballistic gelatin while blood flows through the hollow fibers or capillaries. When a bullet is fired through this experimental assembly, the capillaries are ruptured as would be expected, but no small blood droplets are produced during the passage of the projectile (see also Section 13.1.2). Capillary fragments may be found on the witness panels, but no blood spatter. Of course, immediately after the passage of the projectile blood flows and accumulates in the bullet track. A second shot fired through the previous bullet track will produce a great deal of spatter.

10.11 Shooting of a Driver

10.11.1 Case Scenario/Background Information

In the early morning hours, a taxi cab driver heard a gunshot and saw a man alighting from the right front passenger door of a car at the curb. The taxi driver hailed a nearby

FIGURE 10.37 Photograph of the Colt Cobra six-shot 38-special handgun recovered from the suspect, devoid of any recognizable blood traces.

police patrol car. The police officers promptly confronted and apprehended the man identified by the taxi cab driver. The suspect was a short walking distance away when apprehended, and the officers recovered a warm gun in his pocket. Ultimately, there were no blood deposits detected on the suspect's clothing or on the gun (Figure 10.37), which lead the initial investigators to believe that he was not involved in the shooting. Furthermore, the defendant claimed he had come upon the scene after the shooting had taken place and took the gun from the car. This provided a possible explanation for why the suspect possessed the gun but had no blood deposits on his clothing.

10.11.2 The Physical Evidence and Its Interpretation

The driver was shot five times at close range (four head wounds and one to the right arm) with a compact 38 caliber revolver. Because of the fact that the shots were fired at close range (within a few inches), it was apparent that the shooter was seated in the passenger seat to the right of the driver. There were extensive airborne blood droplet deposits inside the vehicle (Figures 10.38–10.43).

FIGURE 10.38 Photograph of the murdered driver from the vehicle's passenger side showing the blood pool next to his right thigh.

FIGURE 10.39 Photograph of the top of the driver's side seatback and headrest showing bloodspatter.

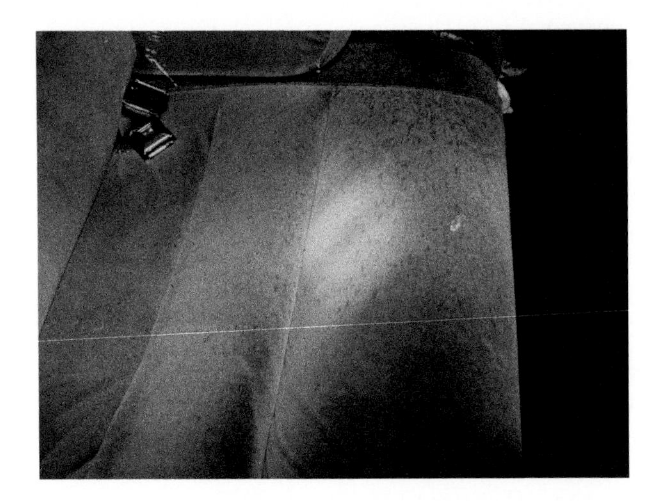

FIGURE 10.40 Photograph of the passenger seat area showing radial blood droplet deposits emanating from the area where blood had pooled on the driver's right hand and seat (top center of the image). It can be noted that there are no blood droplet deposits to the seat on the immediate right of on the arm rest due to the interception of such trajectories by the armrest.

In order to consider a scenario where the suspect was the shooter, there was a need to account for the absence of blood traces on his clothing combined with the extensive airborne blood deposits all over the front interior of the car, especially the passenger seat. It was difficult for the investigators to understand how there could be so much bloodspatter on the passenger seat, if the shooter was sitting there during the event. For this reason, the district attorney's office retained one of the authors as a consulting criminalist.

A later detailed examination by this consulting criminalist of the airborne blood droplet deposits indicated that there were two distinctly different sources of blood trace configurations in space that also occurred at different points in time. One was gunshot

FIGURE 10.41 Photograph of the victim showing vaporized lead deposits on his face, jacket sleeve, and shoulder area. Finely divided blood deposits are visible on the headrest, concentrated on its left side.

spatter created from bullet impacts to the accumulated blood from earlier wounds in the series of four gunshots to the head. This gunshot bloodspatter was confined to areas of the top of the driver's seatback and headrest. The other widespread blood droplet configuration that extended into the passenger side was secondary spatter caused by blood falling directly from the head wound into pooled blood on the driver's seat (Figures 10.44 and 10.45). These airborne droplet deposits were produced after the shooter left the vehicle and closed the door. This interpretation of the blood trace configurations proved that a shooter sitting in the front passenger seat would not necessarily have had any blood droplet deposits on his clothing, thus the suspect could not be eliminated as the perpetrator.

10.11.3 Conclusions

Had the initial investigator's misconception derived from the absence of blood traces on the suspect's clothing or gun not been questioned and thoroughly tested, this suspect would have been wrongly dismissed as the shooter.

10.11.4 Lessons

This case demonstrates how a false and naïve expectation of gunshot bloodspatter could potentially lead to an erroneous reconstruction. Sadly, in this case, there was an

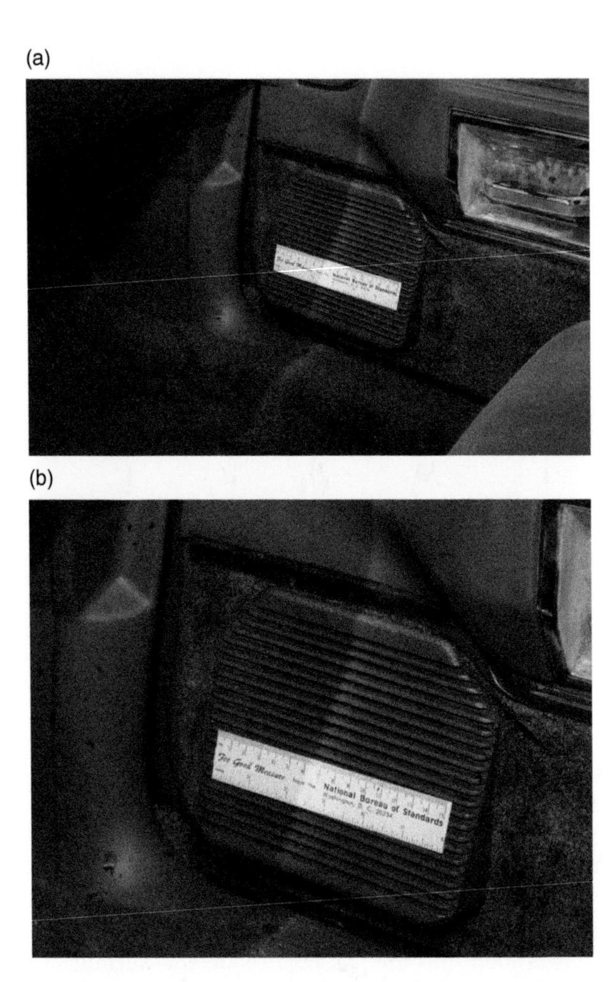

FIGURE 10.42 (a, b) A photograph of the passenger door speaker grill, with tangentially deposited blood droplets, demonstrating that blood droplets were projected onto the horizontal surfaces of the elements of the grill work at a time when the car door was open. This demonstrates that these particular blood configurations were created shortly after the shooter had exited the vehicle and while the door was still open.

FIGURE 10.43 Photograph showing the configuration of blood deposits extended into the passenger area footwell and glove box hatch.

FIGURE 10.44 A simple stringing of selected blood trace deposits in the passenger side of the car, illustrating a back projection to a common area of origin. Of particular note, there is one string representing the blood droplet configurations on the left side of the armrest, showing an upward travel of some of the droplets. This reinforces the idea that the pool of blood was the source of all of the blood deposits on the passenger side of the vehicle, and that the armrest is responsible for the absence of blood on the rear portion of the seat.

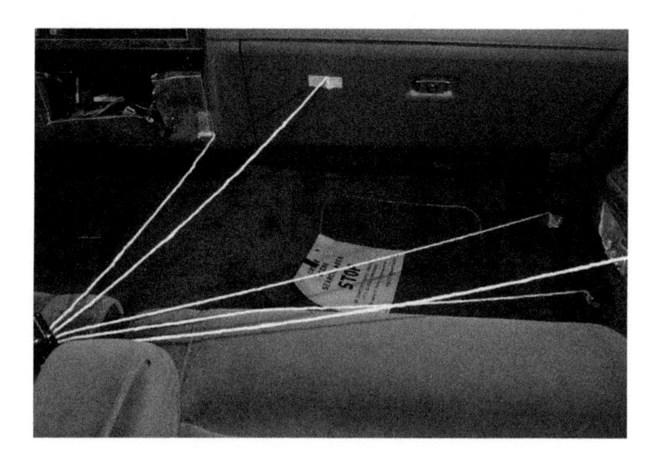

FIGURE 10.45 A simple stringing of selected blood trace deposits in the passenger side of the car, illustrating a back projection to a common area of origin.

unqualified defense expert who testified that "the *perp* must have been covered with blood," thus demonstrating the insidious fallacy perpetuated by the media of large volumes of blood projecting backward toward the shooter from a gunshot wound.

10.12 A Contested Fratricide

10.12.1 Case Scenario/Background Information

A 15-year-old boy was accused and ultimately convicted of murdering his 13-year-old brother. The defendant was home alone, having stayed home from school due to illness. He claimed to have come upon his brother's body in the laundry room shortly after the time he was due to return home from school. The victim had been stabbed over 100 times and had been nearly decapitated from these wounds. The defendant claimed he called the police almost immediately upon finding his brother's body, and the police response was very rapid. The 911 dispatcher and the first police responders described the defendant as being hysterical during the phone call and on arrival. It was determined that he needed medical attention due to hyperventilation. The defendant's bloodstained sweatshirt and socks were removed by an Emergency Medical Technician (EMT) during the ambulance ride, and put into a bag where they were comingled with a towel used by the EMT to clean the defendant's hands.

The suspected murder weapon, a small pocket pen knife, was found by detectives buried in the backyard wrapped in a pair of black-knit athletic gloves. The knife and gloves had significant blood traces, which were eventually typed to the victim. No traces of DNA from the suspect were detected on either the knife or gloves.

10.12.2 The Physical Evidence and Its Interpretation

There was a significant amount of physical evidence in this case that did not include blood traces, such as soil evidence, a cut window screen, fingerprints, and footwear impressions. However, we will be focusing most of this discussion on the blood traces, which were significant in that they possessed meaningful information for understanding some of the relevant details of the event.

One of the present authors was retained shortly after the defendant was convicted in his first trial. He was provided with case documents, photographs, and physical evidence in connection with a post-conviction defense. This was a complex physical evidence case. There was no question that the scene posed challenges. There are some aspects of the scene that raise behavioral issues, which he declined to opine on as they were outside the realm of the science of criminalistics. Examples of these are the choice of weapon, the manner of attack, and the apparent attempt to hide the weapon. According to police investigators, the site where the weapon was buried in the backyard was located by following a trail of luminol-enhanced bloody unshod footprints that extended from inside the house to the evidence burial location and then returned to the house. Subsequent evaluation of the physical aspects of this blood trail was not able to be evaluated because this trail was poorly documented.

There were significant blood trace configurations in the first level of the house. Apparently, the victim's initial attack was in the entry foyer and hallway. Only a scant blood trace trail is documented between the apparent site of the initial attack in the foyer, in the connecting path through the dining room, and the final location in the laundry room where the victim's body was found. The large amount of blood pooled by the victim's head indicated that was the location of the major injuries to the neck including the severing of the right carotid artery. Although these were useful for understanding some details of the attack, they were not helpful for identifying an assailant.

The bloodstained pocket pen knife (Figure 10.46) and knit gloves recovered from the evidence burial site leave little doubt that the evidence burial site was used to hide evidence related to the homicide. However, these items were not definitively connected to the defendant. The knife was similar to others found in a collection maintained by the defendant. It is unknown how common such knives were in the geographic area of the homicide, thus their ultimate significance is beyond the scope of this scientific inquiry.

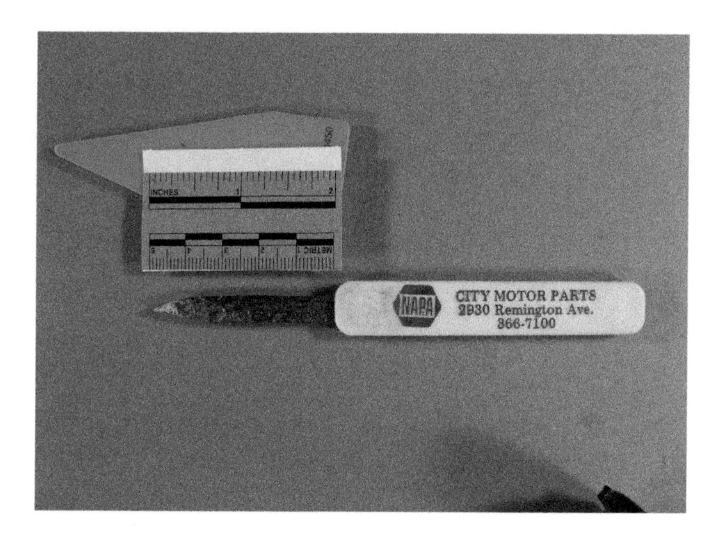

FIGURE 10.46 A photograph of the suspected weapon, a pocket pen knife, recovered from the evidence burial site. This photograph was taken during post-conviction evidence analysis by the author several years after the crime and following police laboratory analysis.

Blood and soil were found on the socks that the defendant was wearing at the time he was first encountered by the scene responders (Figure 10.47). Given the defendant's description of finding his brother's body, the presence of blood would not be unexpected. Sourcing the precise origin of the soil on the socks could have been significant. However, far too little soil remained on these items at the time of the present author's examination to allow analysis. Reports and notes describing soil analyses that were conducted by both the Pennsylvania State Police Laboratory and a private company, R.J. Lee, were provided. It was not apparent from these that any meaningful result was obtained by either laboratory.

The sweatshirt worn by the defendant when he discovered his brother bore evidence of numerous traces made by airborne blood droplets (Figure 10.48). Some of these traces are best explained as resulting from arterial spurting. Others represent impact spatter. The significance of the arterial spurting configuration in the context of the present case depends to an extent on the length of time that sufficient arterial pressure could be maintained following the cessation of the attack. The physiological aspect of this is beyond the expertise of a criminalist, thus a physician was consulted. After being provided with a copy of the autopsy report, the physician opined that systolic pressure sufficient for spurting would only last a few tens of seconds following the transection of the right common carotid artery. The bloodstain pattern expert from the Pennsylvania State Police Laboratory concluded that these configurations (spurting and impact) are consistent with the wearer of the shirt being the attacker of the victim. The consulting criminalist agreed with this conclusion. The state lab expert produced patterns on an exemplar sweatshirt that duplicated aspects of the impact spatter on the evidence shirt. This was done by "stabbing" a blood containing object designed to simulate a body. Such simulations are never exact but can be useful as approximations. The configurations of arterial spurting are best explained by positing that the defendant was in close proximity to the victim at the time of the assault.

(a)

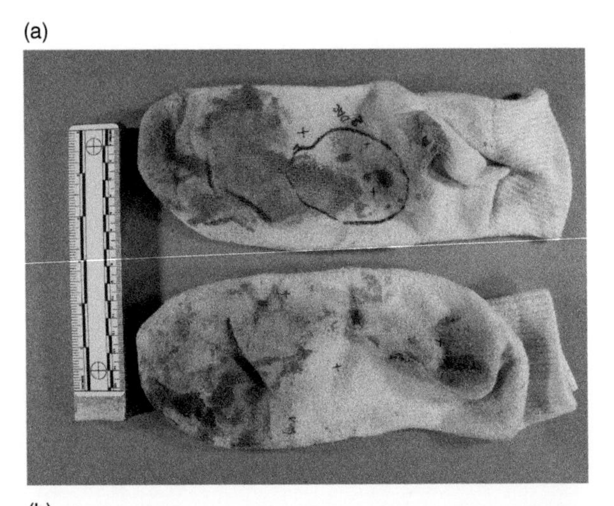

(b)

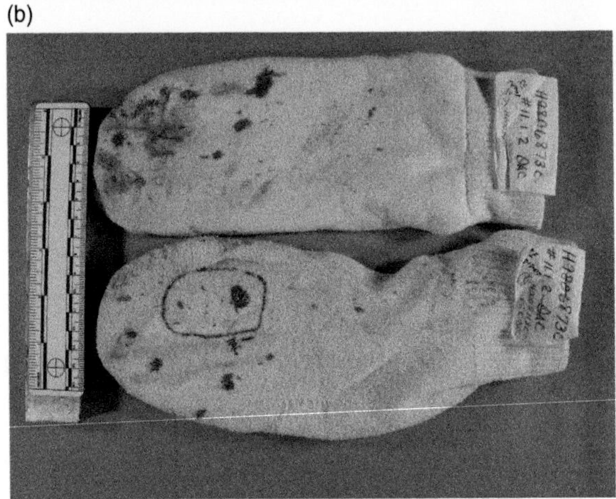

FIGURE 10.47 (a, b) Photographs of the defendant's socks, removed by the EMT during the defendant's ambulance trip. This photograph was taken during post-conviction evidence analysis by the author several years after the crime and following police laboratory analysis.

10.12.3 Conclusions

As stated earlier in this book, the point was made that the totality of the evidence needs to be considered, and it is relatively rare that blood trace configurations alone can resolve all of the relevant issues. However, in this case, the major case question could be addressed by sole reliance on the blood trace configurations. The blood on the defendant's sweatshirt, specifically the apparent arterial spray, indicated that he was in close proximity to the victim during the attack. This refutes the defendant's statements that he came upon the body subsequent to the attack.

10.12.4 Lessons

Unfortunately, this scene investigation suffered from poor photographic documentation. Many of the photographs were either out of focus, exhibited camera-mounted

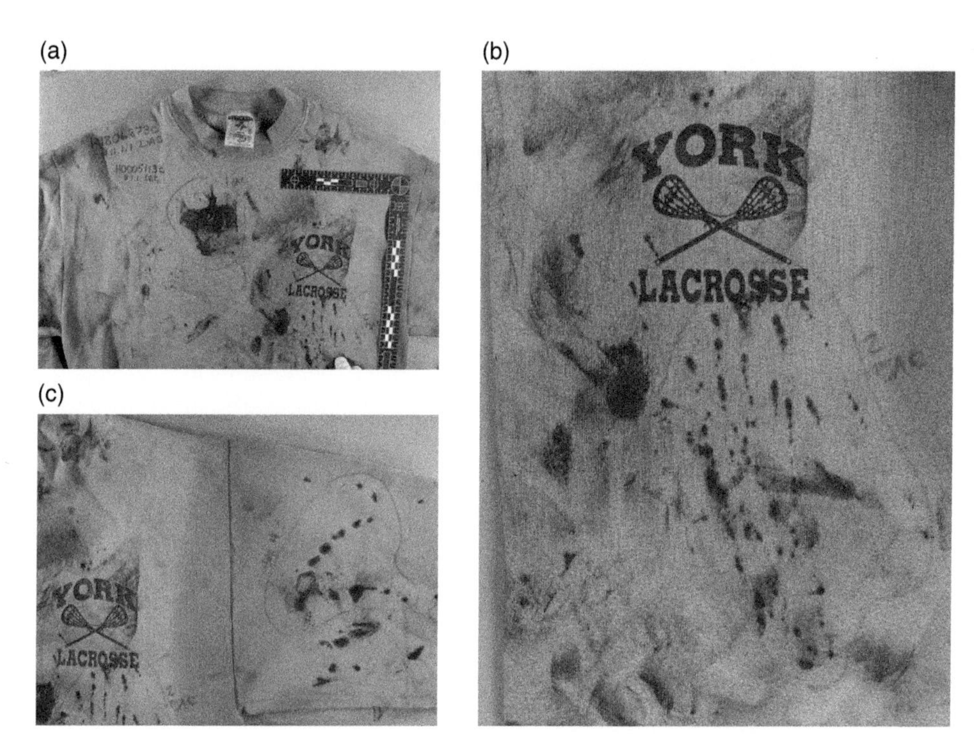

FIGURE 10.48 (a–c) Photographs of the defendant's sweatshirt, removed by the EMT during the defendant's ambulance trip. It has been processed by forensic scientists at the state laboratory with amido black enhancement, apparently to enhance more subtle blood traces, such as the victim's bloody fingerprints (which were not detected). The photographs show (a) the upper portion of the front of the shirt, (b) a close-up picture of the apparent arterial spray on the left front of the shirt, and (b) a close-up picture of the blood spatter deposits on the upper front left sleeve. This photograph was taken during post-conviction evidence analysis by the author several years after the crime and following police laboratory analysis.

flash glare, had lack of flash synchronization with the focal plane shutter, did not contain a scale (ruler), or had a lack of detail in the significant areas of the scene. There was no transition from passive to active documentation of the scene (See Chapter 1) and ultimately no sign of active documentation of the scene in any form. Reportedly, there was a blood trail of poorly defined unshod footprints that was visualized after the treatment with luminol; however, these were not documented with photography. It is unclear what drew the attention of the investigators to this and caused them to apply luminol before visual documentation of an apparent blood trail. It was clear that they had no experience photographing luminol-enhanced deposits, as all provided photographs were grossly underexposed resulting in solid black featureless images. This underscores the need for educated and experienced scientists at the scene with additional expertise in photography.

This case demonstrates the ill-conceived premature use and misuse of luminol and other blood enhancement agents (see Chapter 4). In addition to the undocumented blood trail of footprints, it was reported that luminol was applied to the suspect's light-gray sweatpants. As there are no photographs of these activities, they cannot be reliably evaluated. Additionally, even though patently visible blood trace configurations were observed on the defendant's sweatshirt, a forensic scientist at the state laboratory processed it with amido black, apparently to enhance more subtle blood traces, such as the victim's bloody fingerprints which were ultimately not detected. These were obviously not needed for the patently visible blood traces on the garment. It is critical that all items of evidence be adequately photographed and documented prior to any enhancement because the process could affect the morphology of the blood traces. Unfortunately in this case, this was apparently not done as no examination quality photographs of the blood trail or the defendant's clothing, should they exist, was made available to the consulting criminalist despite requests for all of the physical evidence case materials.

In this case, the consulting criminalist insisted on being able to examine the clothing in person in a laboratory setting, rather than relying on the previous photographs. It is essential to do this. Consultants should not acquiesce to working under the constraints provided by attorneys and instead insist on conditions required for a proper forensic assessment. This is particularly true to airborne droplet deposits on fabric which are difficult to interpret under the best of circumstances. The examination of the evidence directly is necessary to make a scientific conclusion on the mechanism of blood trace deposit.

References

EngagedScholarship @ Cleveland State University n.d. The Sam Sheppard case: 1954–2000. https://engagedscholarship.csuohio.edu/sheppard/#browse (accessed 1 October 2020).

Walraven, J. (1995a). The Simpson trial transcripts: Dr. Fredrich Reiders 7/24 and 8/14. https://www.simpson.walraven.org/ (accessed 1 October 2020).

Walraven, J. (1995b). The Simpson trial transcripts: SA Martz 7/25 and 7/26. https://www.simpson.walraven.org/ (accessed 1 October 2020).

CHAPTER 11

"Bad" Cases – Misleading or Incompetent Interpretations

There is no question that one learns from errors, either one's own or those made by others. For this reason, instructive examples of erroneous or misleading reconstructions are presented and discussed in this chapter. It should be readily appreciated that misleading reconstructions can have serious pernicious effects in the administration of justice. These "bad" case examples are presented in enough detail, so that the reader will hopefully be able to independently identify and understand the problems in the attempted reconstruction.

The examples presented in this chapter were chosen because they represent cases where one of the authors was contacted to attempt to reconstruct the event because of expertise in blood trace configurations.

THE CASES:

1. David Camm
2. Dew Theory
3. Murder of an Off-Duty Police Officer.
4. The imagined "mist" pattern
5. Concealed Blood Traces
6. The Stomping Homicide: Misuse of Enhancement Reagents

Blood Traces: Interpretation of Deposition and Distribution, First Edition. Peter R. De Forest, Peter A. Pizzola, and Brooke W. Kammrath.

11.1 David Camm

11.1.1 Case Scenario/Background Information

On the evening of 28 September 2000, David Camm, a former Indiana State Trooper, discovered the bodies of his wife, Kim, and two children after he returned home from playing basketball with friends. The three victims were found shot to death in the garage of their home. Camm claimed that he found his wife's body on the floor of the garage and his two children in the backseat of the vehicle, a white Bronco. He stated that he thought his son could still be alive and thus pulled him from the left rear seat of the car, over the body of the girl, onto the garage floor in order to perform cardiopulmonary resuscitation (CPR).

David Camm was the primary suspect in the murder of his wife and two children. The first two trials ended in convictions (2004 and 2006, which were subsequently overturned), while the third resulted in an acquittal (2013).

In the second trial, additional evidence was analyzed that supported another individual's presence at the scene. Charles "Backbone" Boney, a convicted felon, was identified by DNA evidence collected from a prison-issued gray sweatshirt found in the garage. In addition, Boney's handprint was found on the right-side C-pillar of the Bronco. Boney admitted to being at the scene in order to deliver a gun to Camm but claimed to not have committed the murders. This was the story that the prosecution proposed in the latter two trials, with there being no apparent wavering in their conviction of Camm's involvement in the murder of his family.

There is much to be learned from this case for anyone involved in criminal justice, from police investigators and crime scene analysts to prosecution and defense attorneys. With respect to the blood trace evidence, this case illustrates a number of points regarding the dangers of an incompetent and/or shoddy investigation and its pernicious repercussions.

11.1.2 The Physical Evidence and Its Interpretation

There was a wealth of physical evidence in this case. This discussion will focus on the most significant blood traces that were seized on by the prosecution in their rush to judgment.

One of the most serious problems with this case was an early focus on unsupported interpretation of the blood droplet trace evidence. The hasty overinterpretation of a limited number of possible blood droplet deposits on Camm's t-shirt was the result of a crime scene photographer, Robert Stites, who was not qualified to render opinions on reconstruction issues. This compounded an unscientific behavioral judgment by investigators, who became convinced of Camm's guilt due to his initial observed behavior that included pacing and failure to make eye contact. Robert Stites was employed by a private consultant, Rod Englert, to document the scene (Camm v. Faith, 937 F.3d 1096 2019). Englert stated that he was unavailable to go to the scene and had sent Stites in his place. Based on Englert's recommendation, police and/or prosecution investigators incorrectly believed Stites to be a bloodstain pattern analyst. During Stites' time at the scene, he told police there was evidence of clean-up activity at the crime scene and "high-velocity impact spatter" on the t-shirt Camm was wearing. This appears to have led the law enforcement investigators to focus prematurely on David Camm as the suspect to the exclusion of other avenues of investigation. As a result of this initial over-interpretation, the prosecution asserted that when Camm shot his daughter, who was seated in the right rear seat area of the vehicle, some of the deposits on his shirt were created from the gunshot backspatter. His son was initially in the left rear seat, before being moved by Camm to the garage floor. By way of contrast to the prosecution's theory, the defense offered the alternative hypothesis that some of the blood traces on his shirt were contact deposits acquired when he pulled his son from the right side of the car over his daughter's body.

There were several areas of blood traces on Camm's t-shirt, with one area on the bottom front being the focus for reconstruction (Figures 11.1–11.7). Of significant importance for interpretation is that the limited number of blood traces on the front of Camm's shirt presented several areas of ambiguity. The source of the blood was shown through DNA to belong to Jill Camm, the defendant's daughter. The prosecution's claims were that these blood droplet traces were due to backspatter from her entry wound; however, there are three significant issues with making this determination.

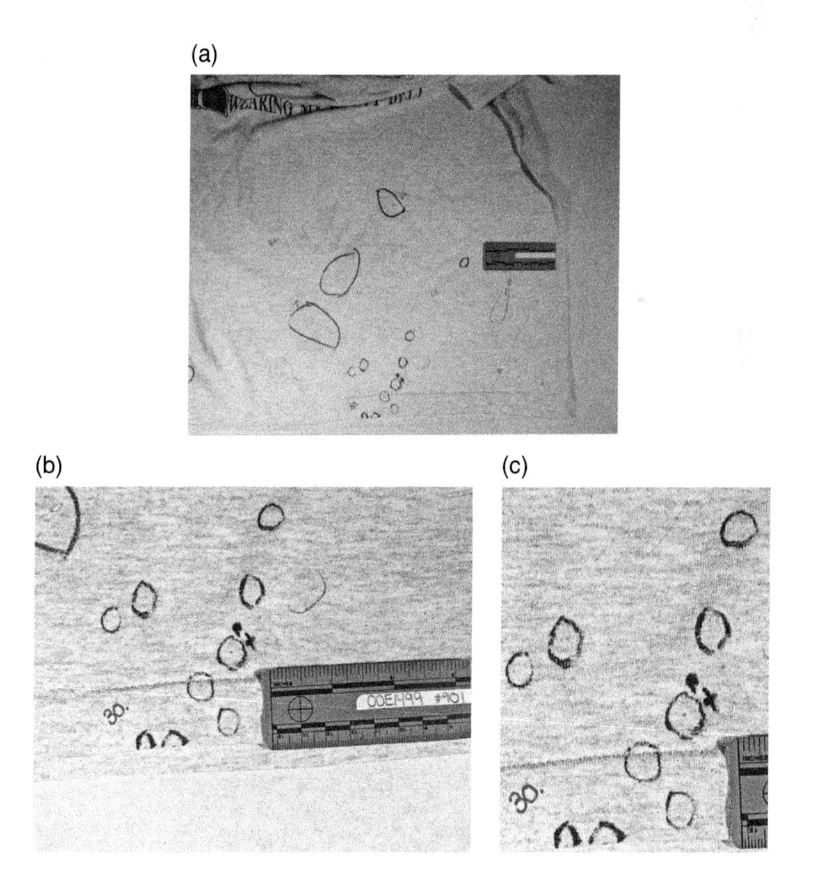

FIGURE 11.1 Mid-range (a) and close-up (b, c) photographs of blood traces on the front of David Camm's t-shirt, with the series of circles on the lower half of the shirt designated as Area 30 having a DNA association to Jill Camm.

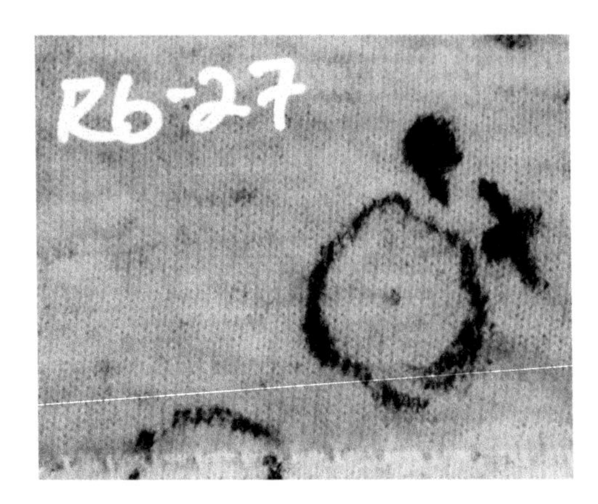

FIGURE 11.2 Stereomicroscope photomicrograph of two blood traces in Area 30 on the front of David Camm's t-shirt.

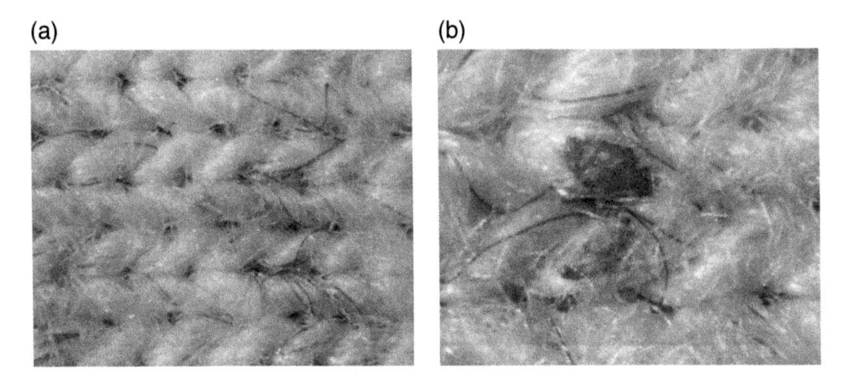

FIGURE 11.3 Stereomicroscope photomicrographs of two deposits (a and b) in Area 30 on the front of David Camm's t-shirt. They appears to be contact deposits due to the traces being very light (in the left image) and on the surface of the fibers. The shape of each of these is dictated by the three-dimensional conformation of the yarn.

FIGURE 11.4 Two blood traces deposited on the bottom of edge of the hem of the front of David Camm's t-shirt, with the right one showing some lateral staining. This argues for contact staining for two reasons: (i) the unlikelihood of two airborne blood droplets striking the lower edge of the t-shirt hem given the limited number of claimed airborne deposits in Area 30, and (ii) the edge of the hem making contact with a surface bearing liquid blood droplets, such as a bloodied head of hair (as seen in the experimental photographs of 11.6–11.7).

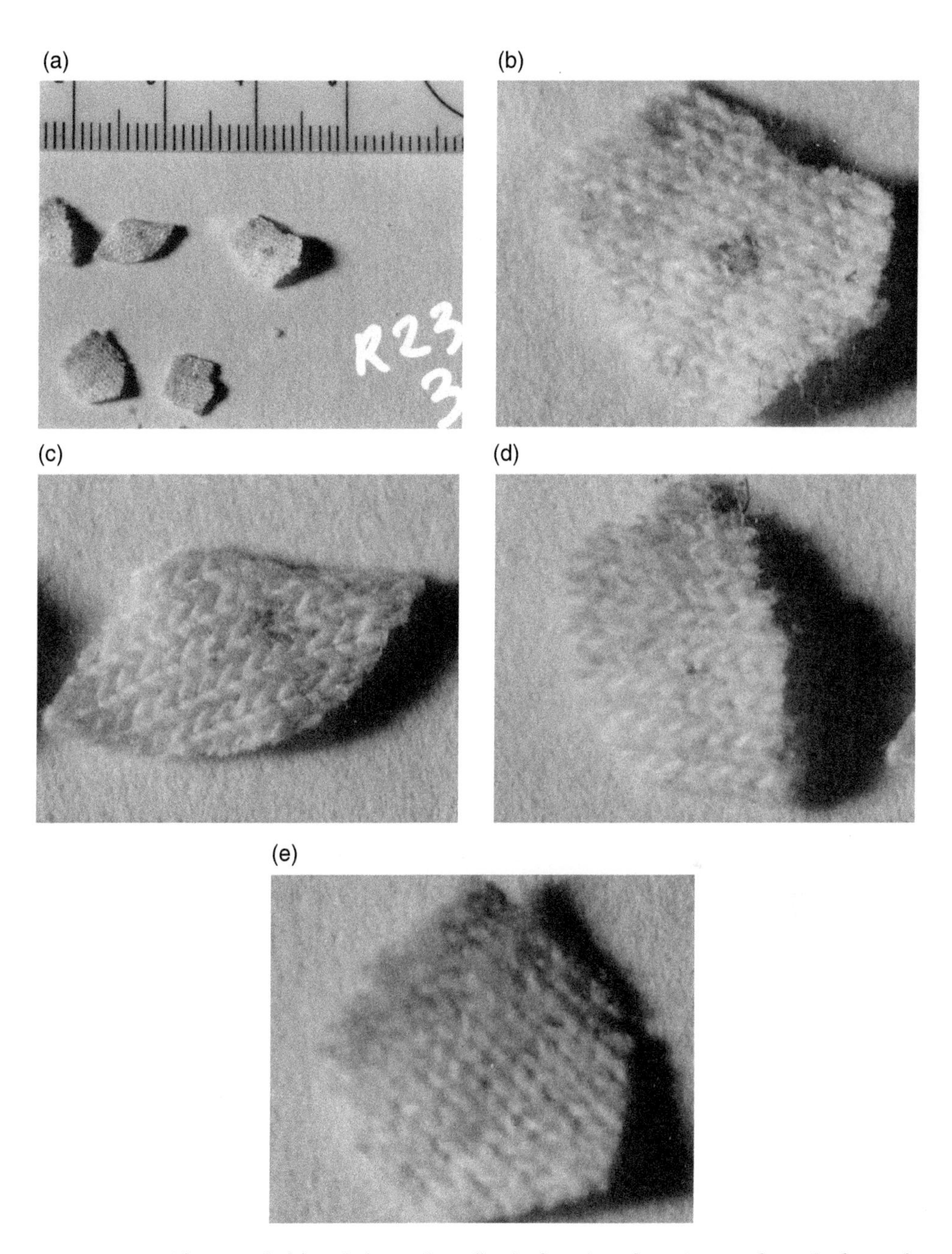

FIGURE 11.5 Photograph (a) and closer views (b–e) of cuttings from Area 30 from the front of David Camm's t-shirt.

The first issue was with the determination of the number of deposits on the shirt, which varied between 3 and 8, depending on the "expert" asked. The second was whether they were in fact even airborne droplet deposits and not contact/transfer traces from brushing against something such as finely divided blood droplets adhering to the daughter's

(a)

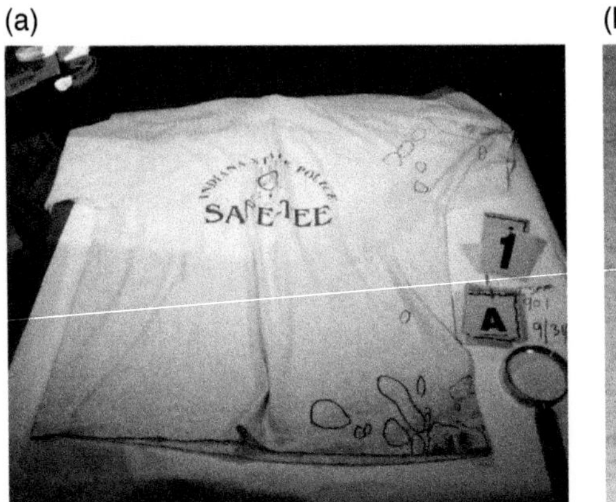

(b)

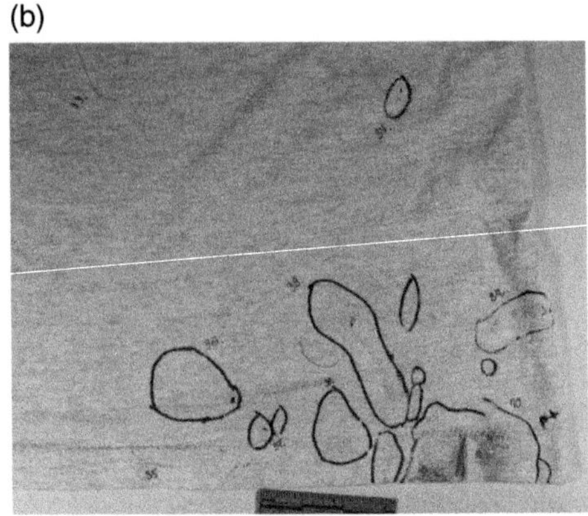

FIGURE 11.6 Mid-range (a) and close up (b) photographs of the back of David Camm's t-shirt, with blood traces in Area 37 and 40 having a DNA association with that of Jill Camm.

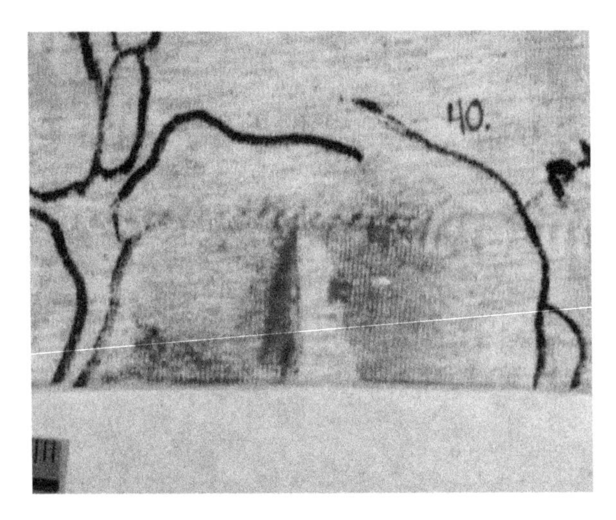

FIGURE 11.7 Close-up photograph of the blood trace deposit in Area 40. It is clear that this is a dynamic contact blood transfer deposit, where the direction of the object making the trace was moving left to right. There is a typical void caused by the folding (or the presence of an existing fold) of material as a bloody object moves across surface.

hair in his efforts to remove his son from the backseat of the vehicle (Figures 11.8 and 11.9). The third, assuming they were airborne droplet deposits, was the determination of the mechanism of production given their limited number. Dr. Robert Shaler, a well-respected forensic scientist and defense witness in the third trial, said "[t]he problem in this case is the number of stains are minimal," he said. "I think you're really on the edge of reliability."

There were additional factors that raised questions about whether there were in fact any airborne blood droplets produced from backspatter from the daughter's head wound. As discussed in Chapter 13, there should never be an expectation of spatter being produced from gunshot wounds. The prosecution experts claimed that there

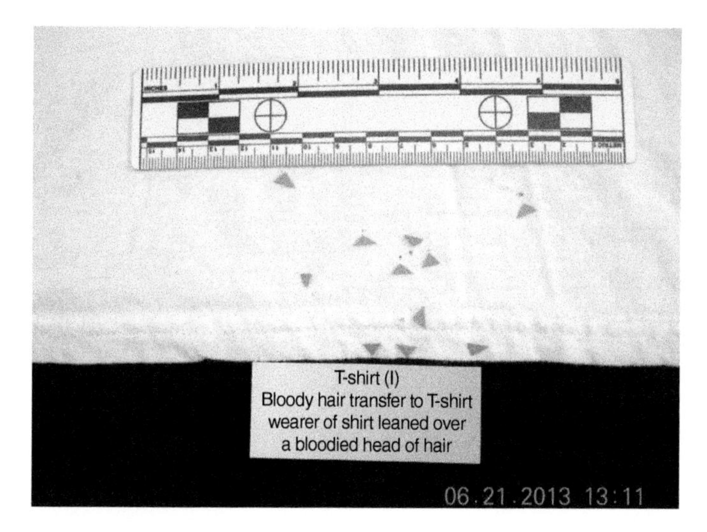

FIGURE 11.8 Photograph of the front of a t-shirt from an experiment by Barie Goetz, which shows the blood trace deposits created when the wearer leaned over a bloodied head of hair. The similarities between the blood traces on this exemplar shirt and the evidence shirt (Figure 11.1) are quite convincing. *Source: Reproduced with permission of Goetz (2017). © Elsevier.*

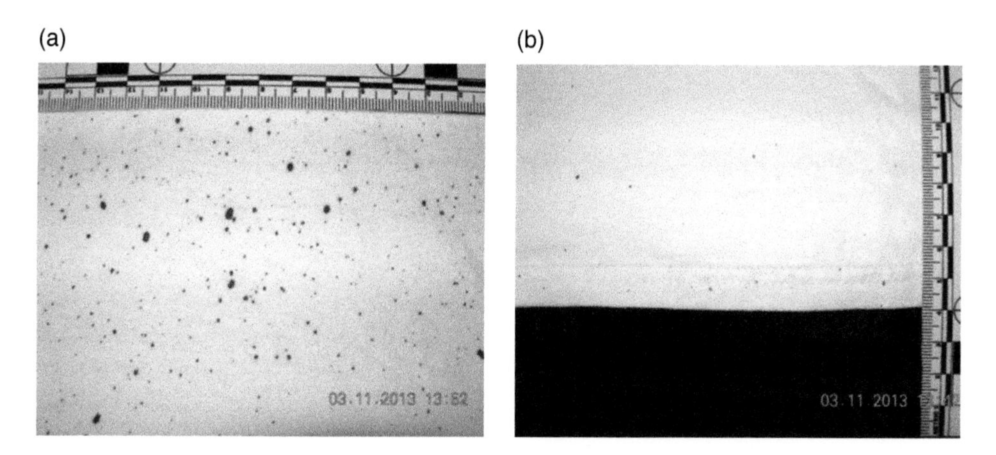

FIGURE 11.9 Photograph (a and b) of the results of an experiment performed and documented by Barie Goetz where fine blood droplets were projected onto a vertical t-shirt from a source above the level of the hem. There would be little ambiguity if the evidence photograph of the front of David Camm's t-shirt was something like this. *Source: Reproduced with permission of Goetz (2017). © Elsevier.*

was an airborne blood droplet pattern on one face of the roll bar shroud in the sport utility vehicle (SUV) interior created by backspatter from the daughter's head wound. However, the lack of adequate documentation does not allow for any meaningful interpretation of these deposits (Figures 11.10 and 11.11). Had there been sufficient documentation of the surrounding area to the roll bar shroud showing the presence or absence of additional airborne deposits, it could have helped to determine whether

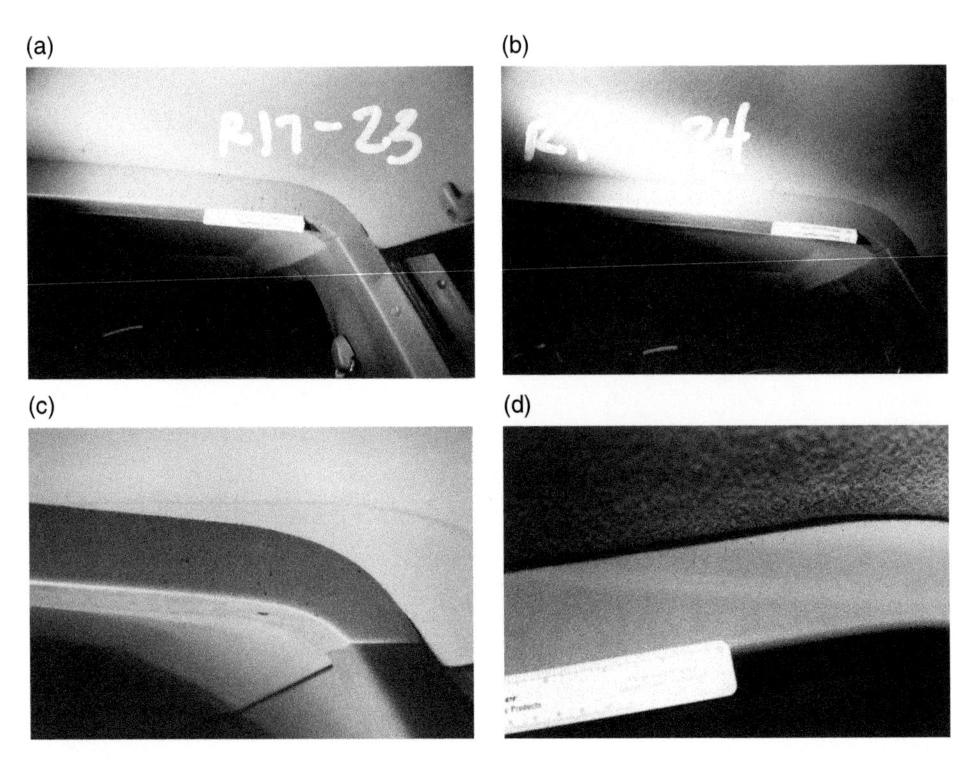

FIGURE 11.10 Mid-range (a–c) and close-up (d) photographs of the inside the Camm vehicle, as viewed from the backseat area, showing blood trace deposits on the rear face of the roll bar shroud. No blood traces are visible in these photographs on the underside of the roll bar or the headliner. This could be due to the absence of such deposits or the poor-quality photography (overexposure and uneven illumination).

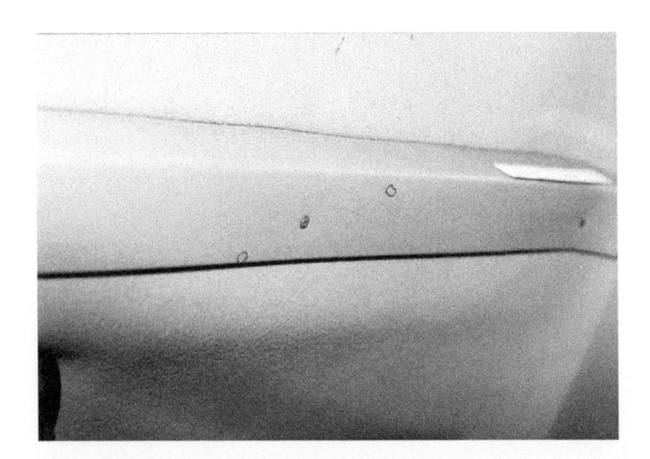

FIGURE 11.11 Photograph of the inside roof area of the Camm vehicle, showing the underside of the roll bar shroud. Although there are two areas circled in black ink, the presence of blood trace deposits inside the circles, if they exist, is not apparent in the photograph.

or not the deposits seen on the posterior face of the roll bar shroud were in fact produced from airborne droplets. If, in fact, these blood traces had been caused by gunshot backspatter, one would have expected there to be additional deposits on the headliner and lower surface of the roll bar shroud (Figures 11.10 and 11.11) as well as critical absences of deposits on the headliner in the immediate area forward of the roll bar protected by the physical presence and blocking action of the roll bar itself. The documentation was inadequate to address whether such contextual deposits were present and demonstrates a lack of active documentation at the scene. Furthermore, there was no documented evidence of any forward spatter from the exit wound in Jill's head on the nearby seatback in the vicinity of the bullet hole. Typically, there is more forward spatter than backspatter associated with shooting events. Thus, the absence of documented forward spatter on the seatback raises the question as to whether or not there would have been any backspatter produced from the entry wound. Mid-range and close-up photographs of the backseat of the vehicle are shown in Figure 11.12, however the blood traces shown did not assist in the event reconstruction.

There was other important blood trace evidence in this case, including but not limited to blood on Camm's sneaker and the door from the garage into the house. There was also significant laser-scanning-based reconstruction analysis completed by Eugene Liscio, who demonstrated the position of the shooter standing outside of the open right rear vehicle door in relation to that of the daughter (Figures 11.13 and 11.14). The results of the laser-scanning-based reconstruction indicated that the shooter's t-shirt would not have received airborne droplets of blood back-projected from the bullet entry wound from the daughter. This was further evidence contradicting the prosecution expert's interpretation of some of the blood deposits on Camm's shirt. This was complemented by the position of the left palm print of Boney on the C-pillar of the vehicle.

FIGURE 11.12 Mid-range (a) and close-up (b) photographs of the back seat of the Camm vehicle.

FIGURE 11.13 Overhead view of possible position of shooter and trajectories for shots to the two children, prepared by Eugene Liscio using laser scanning technology. *Source: Reproduced with permission of Eugene Liscio.*

FIGURE 11.14 A view of the position of the shooter and Jill Camm, prepared by Eugene Liscio using laser scanning technology. *Source: Reproduced with permission of Eugene Liscio.*

11.1.3 Conclusions

The documentation of the blood traces and other physical evidence in this case was exceedingly poor and thus inadequate. The consequences of mediocre photography in this case illustrate the importance of recognizing the variables (lighting, shutter speed, f-stop, International Organization of Standardization [ISO], etc.) involved in acquiring images. In comparing the poor photographs taken of the roll bar shroud (Figures 11.10 and 11.11), it is apparent that different features are obscured or not adequately depicted when exposure is not optimized. When photographs are being taken, one needs to ensure that all features being observed are adequately recorded. The emphasis on active documentation, as previously discussed in Chapter 1, whereby there is a clear understanding of the significance of observations that need to be preserved, is critical to the proper and thorough recording of the essential aspects of the physical evidence in a scene. Too often, if the scene is not adequately documented in active mode through quality photography and other documentation media such as laser scanning technology, the evidence is lost and gone forever.

Another important point that is demonstrated by this case is the need for scientists to do crime scene reconstruction and blood trace configuration interpretation. This point has been previously discussed in this book (Chapters 1, 7–9) as well as by other authors. In a recent book, an author made the following statement:

> *There were some major differences in the backgrounds of the experts who testified for the police and the defense experts. The prosecution experts had their background and experience in police work; they were not scientists. The defense experts were scientists with extensive education who had earned their stripes in the lab, and their expertise was founded in science (Alter 2013). In the end, the jury appeared to be convinced by the defense experts, and after 10 h of deliberation, Camm was found not guilty.*
>
> *(Koen 2017, page 278)*

Had a scientifically based interpretation of the blood trace configurations been appreciated initially, unbiased by the investigators' initial impressions of David Camm's guilt, it is likely that the investigation would have taken a different path. Ultimately, the importance and need for a scientific education for those employed in the field of forensic science cannot be overemphasized.

11.1.4 Lessons

There were many errors made in this case and lessons to be learned about the dangers of overinterpretation from limited data. Again, the comments presented here pertain to one important aspect of the physical evidence and not the totality of the evidence in this case. From the outset, this aspect set the investigation in the wrong direction, thus negatively impacting the documentation of the physical evidence as well as wasting valuable time and resources.

Much of the evidence documentation in this case was passive, and thus this case provides an example of the consequences of the failure to transition to active documentation. An important lesson to be learned is the critical impact of informed selection of areas to be focused on for complete documentation. Additionally, the evidence must be documented in sufficient detail, such as with quality photographs. Current technologies such as laser scanning and total station photographic measurements do not obviate the need for scientific thinking in the selection of areas to be documented in greater detail. When this is not done, aspects of the critically significant physical evidence will fail to be captured and preserved.

Another consequence of the failure to transition to an active documentation mindset at the scene is that without complete information, there is a pernicious temptation to overinterpret from incomplete data. Furthermore, any succeeding experts are similarly handicapped.

This case underscores the important lesson on the value of contributions from experienced scientists at the crime scene.

11.2 Dew Theory

11.2.1 Case Scenario/Background Information

The local police department was called at around 6 a.m. by a husband who said he just found the body of his wife outside of their home. The husband claimed that he was asleep when his wife returned home from her hospital night shift (3–11 p.m.) as a nurse. He further stated that he and his wife had an established routine that involved him putting their two young children to bed while she was at work. Normally, she would come home around midnight, check on the kids, and go to bed. On the night in question, the husband said that he did not wake up until he heard one of his children crying in the morning and realized that his wife was missing. At that time, he searched for her and found her body outside in the driveway, lying beside her car with the driver's door open. He stated that he covered his wife's body with a quilt retrieved from inside the home before calling the police.

11.2.2 The Physical Evidence and Its Interpretation

Although death was due to ligature strangulation with a rope that remained around her neck, the victim also had two bloody forehead wounds. For the purpose of this discussion, only one aspect of the physical evidence will be explored. The blood trace deposits from the two head wounds were the focus of the physical evidence found on the quilt that had been used by the husband to cover the body. A privately retained prosecution expert, Herbert MacDonell, was provided with the bloodstained quilt for his examination of the blood traces.

The blood traces on the quilt were in the form of roughly circular deposits with dark red centers and lighter-colored peripheries, similar in configuration to a "sunny-side up fried egg." MacDonell claimed that the configuration of the two-toned deposits was the result of dew formation on the quilt and resulting diffusion of the blood. This conclusion led to a crime scene hypothesis that the quilt was exposed to dew for a longer period of time during the night, thus contradicting the husband's statement of covering the body at about 6 a.m. To support his hypothesis, MacDonell performed "experiments" where he sprayed misted water above experimentally produced blood traces on a horizontally oriented unstained section previously removed from the evidence quilt. In his flawed attempt to reproduce the formation of dew, MacDonell failed to understand the distinction between condensation (dew formation) and precipitation (depositing misted water). He also failed to understand that for dew to form, the receiving surface must be at a temperature below the dew point (IUPAC 1997).When the temperature of a surface is below the dew point, excess water vapor from water-saturated air will condense on that surface in the form of dew. From an examination of photographs of the experimentally produced traces, they were distinctly different from those in evidence. Careful examination of the experimentally produced traces side by side with the evidence traces should have made it clear that they were different. The evidence deposits had a dark well-defined center and a broad pinkish periphery, with the proportions of a fried egg of the yolk surrounded by the broad band of white albumin. The experiment deposits had a continuous gradient of red color intensity from the center outward, with the steepness of the gradient being greater with the longer drying times (i.e., less diffusion occurred from the dryer deposits).

It is known that after the sun sets, outdoor surfaces that have been warmed by the sun during the day begin to cool by radiating heat into space. On overcast evenings, this is a less efficient process as compared to the much more effective radiant cooling which takes place on a cloudless night. These scientific facts are relevant because the body and quilt were found under a roof overhang, which would have interfered with the radiation of heat into space. The temperature of the quilt would have also been influenced by the body underneath, which would have increased its surface temperature. Consequently, in combination with the shielded surface, the quilt would not have cooled sufficiently to enable the formation of dew.

A defense expert, one of the present authors, proposed that the "fried egg"-like appearance of the deposit configurations was the result of blood that had undergone a cell–serum separation (or prior clot formation) before being transferred to the fabric surface. Observations made by a reporter present at the scene during the investigation indicated that the quilt was repeatedly lifted and replaced on the victim as additional detectives arrived on the scene. Had the traces been formed by diffusion, there would be little difference in the hemoglobin to albumin ratio in the center when compared to the periphery. Laboratory testing of the evidence deposits showed the higher concentration of hemoglobin in the center and serum albumin in the lighter-colored peripheral areas. This analysis strongly supported the defense expert's reconstruction.

Sadly, when MacDonell was confronted with this incontrovertible evidence, and the flaws in his own experimental design, he chose to ignore them rather than revise his opinion. This further illustrates the pseudoscientific nature of these conclusions: first in the misuse of data and second in the refusal to acknowledge problems. Anecdotally, he proudly wrote about it in a chapter in his self-published book on crime scene procedures and his cases.

11.2.3 Conclusions

It is unfortunate that the testimony of the prosecution experts and the defense experts were diametrically opposed. Consequently, this left the decision of the scientific conclusions in the hands of the jury, who are ill equipped for this role. Although the jury has the role of being the triers of fact, this does not mean that they should be left with interpreting complex scientific phenomenon. What is the jury left to decide on if two experts present them with opposing viewpoints, and they cannot understand the science? Some research has shown that the jury uses their preconceived opinions based on the appearance or attire of the expert to draw conclusions about the relative scientific validity of the two opposing assertions (Tanton 1979). Thus, the jurors' conclusions are based not on the scientific veracity but on their biased perceptions. Ideally, there would be an attempt to reconcile the science outside of the court via a discussion of the experts from both sides. Although there are several practicing forensic scientists who disagree with this view and feel that disparities between scientific experts are a logical part of the US criminal justice system, the authors of this text are vehemently opposed to this conclusion. There should be some effort made to resolve the differences in scientific opinions between forensic science experts. The scientist cannot substitute their judgment of legal strategy for that of the attorney, and thus if the attorney objects, a prior meeting of scientists would not be allowed. Still every effort should be made to resolve the difference of scientific opinion, and in general, it is advantageous to the court for this to be accomplished. If this attempt of scientific reconciliation fails, the scientific contribution to justice is thwarted.

11.2.4 Lessons

There is a need for scientists interpreting blood trace configurations to understand additional natural phenomena beyond a simple understanding of specific blood trace configuration or pattern formation. It is not sufficient for an expert to be a scientist. Scientists can engage in promoting pseudoscience. On first blush, the experiment designed by MacDonell has the apparent trappings of science. However, this is what makes it so convincing to a lay person and dangerous as pseudoscience (see Chapter 7).

So-called "experiments" designed to replicate evidence configurations must be designed carefully. In the case just described, the hazard of such "experiments" is that there can be a dangerous tendency to overlook significant differences between the evidence configurations and those produced in the "experiments." If one looks at the experimentally produced configurations in the case described here and compares them with the photographs of the evidence configurations, obvious and significant differences are readily apparent. This lack of replicability demonstrates the pseudoscientific nature of these conclusions, with issue being the lack of critical assessment of the results. It is clear that the prosecution expert was blind to the flaws in his own "experiments." This applied not only to what should have been the obvious disparity in the outcomes, but it also extends to conceptual shortcomings in the design stemming from a failure to understand factors affecting the all important surface temperatures and the fundamental nature of dew formation.

The inclusion of expert testimony that includes undocumented recollections is inappropriate. In this case, one Police Officer testified about the blanket being "damp to touch," but there was no documentation (written or otherwise) of this at the scene. Memories can be fluid, and the tendency to fill in holes is dangerous. This pernicious activity must be guarded against.

11.3 Murder of an Off-Duty Police Officer

11.3.1 Case Scenario/Background Information

A Police Officer was off duty at a social event in a bar. At some point during the evening, some individuals got into a major altercation in the bar. The bouncer in the bar was struck in the back of the head by a broken glass bottle, which angered him. At this point, the bouncer left the bar, went to his car which was parked a few blocks away, and drove to the scene where he double-parked outside of the bar. The bouncer got out of the driver's door of his car and an individual on the sidewalk said, "He's got a gun! He's got a gun!" The bouncer began firing his Glock pistol from a position in the street over the roof of his car. The Officer was a front-seat passenger in the car between the bouncer's car and the curb. The Officer was getting out of the car, and when he got his head above the roof of the car, a bullet caught him in the neck and severed his cervical spine. He fell with his body partially on the sidewalk. Two other victims were wounded and immobilized lying on the sidewalk. The bouncer then ran around the front of both his car and the car at the curb where he attempted to administer *coup de gras* contact shots to the heads of the three immobilized victims. The third victim was able to move his head away from the gun and survived, with the bullet grazing his forehead and producing an impact crater in the concrete pavement.

The bouncer fled the scene and gave his Glock to a friend. The gun was recovered from the friend's custody months after the arrest of the bouncer.

Information about details of the shooting was provided to the police investigators by eyewitnesses. Although this would seem to have been adequate evidence to show that the shootings were deliberate and they were homicide rather than self-defense, a year after the shooting the District Attorney wanted corroborating evidence as he was getting near trial. He was concerned that the witnesses' statements could be compromised or subject to attack because they had been drinking prior to the event. He then hired a consulting criminalist, one of the authors, to complete an independent reconstruction with the expectation that the physical evidence would support the witnesses' accounts.

11.3.2 The Physical Evidence and Its Interpretation

The consulting criminalist found it necessary to go to the scene with some assistants to resolve errors made by the initial crime scene investigators that had been made regarding evidence measurements. The District Attorney arranged for the police to cordon off the area, and a reconstruction was undertaken. The crime scene photographs and additional knowledge and crime scene measurements were used to lay out the positions of the bodies on the sidewalk in chalk (see Figure 11.15).

There were three intact (i.e., unfired) cartridges documented and recovered on the sidewalk in the initial investigation. These had been cycled through the weapon, as evidenced by extractor and magazine lip marks. One of the cartridges (Figure 11.16) also evidenced a "stove pipe" jamb mark. The critical significance of the recovered intact cartridges for the reconstruction was not recognized during the course of the initial investigation and went unrecognized until the involvement of the consulting criminalist over a year after the event. The cycled unfired 9-mm intact cartridges on the ground require an explanation. It is not normal to have to manually cycle a weapon during a shooting unless something has gone wrong with it. This indicates that the weapon, at least in the mind of the shooter, had malfunctioned. This caused him to attempt to clear apparent jambs by racking the slide back which would eject the perceived malfunctioning cartridge (which was in actuality a live round) and then resume shooting as part of the *coup des gras* shots.

The Officer was shot twice. The first shot striking the Officer, that had traveled over the roof of the car from which he was alighting, produced the cervical spine wound that went through his neck. This would have paralyzed him and caused him to fall to the ground. An examination of the photograph (Figure 11.17) reveals important evidence for reconstruction. The avulsion of tissue from the head wound combined with the displacement of a triangular-shaped piece of skull (found in the pool of blood on

FIGURE 11.15 A photograph of the sidewalk showing chalk outlines of the three victims made during a reconstruction effort, as derived from the crime scene photographs. The yellow chalk shows the outlines of the bodies and the pink chalk was used to show the outlines of the blood pools. The body of the Officer was represented by the yellow chalk outline on the left side of the image, next to placard B2, partly on the curb and partly in the street. Placard B2 shows the position of a recovered intact cartridge. The other fatality was represented by the yellow chalk outline at the top of the image. The chalk outline to the right side of the image is where the third victim had fallen, and the shooter had attempted to shoot him in the head, but the victim flinched resulting in a graze wound across the forehead. The bullet impacted the sidewalk by the head area, represented by an "X" in the photograph.

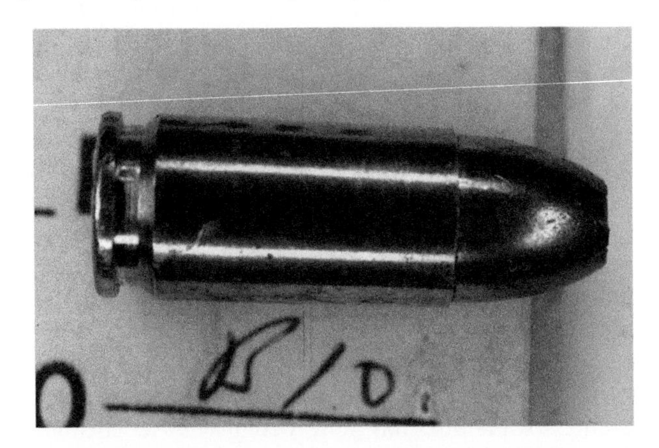

FIGURE 11.16 One of three unfired intact cartridges recovered on the sidewalk in proximity to the shooting scene. The defect on the bottom-left of the image of the casing, near the extraction groove, demonstrates evidence of a "stove pipe" jamb mark. The marks show that this 9-mm hollow-point cartridge had been in the weapon. Additionally, it showed cycling marks from the magazine lips.

the sidewalk) indicates that this was a contact shot. Surprisingly, the pathologist who performed the autopsy did not recognize the significance of the damage done by the handgun in this situation and did not identify it as a contact shot. A 9-mm handgun will not cause this type of extensive wounding unless it results from a near-contact

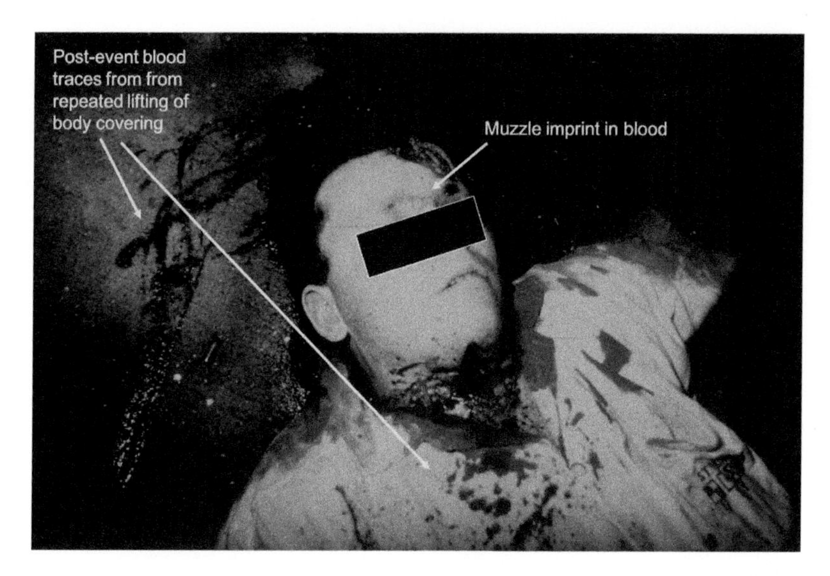

FIGURE 11.17 Photograph of the head of the victim showing an initially unrecognized red ring mark at the root of the nose that was washed off at autopsy. This strongly suggested that the mark was produced by contact with the blood-coated muzzle of the barrel of the shooter's Glock.

or contact shot, where the gases are injected into the wound and cause much more damage than would be caused by the bullet alone. Only a high-powered weapon like a military or hunting rifle could have caused a wound with a displacement of skull material and avulsed tissue, if the shot had been fired from a distance greater than near-contact.

Figure 11.17 images the face and head area of the Officer. The red ring mark at the root of the nose was not recognized until the photographs were examined by the consulting criminalist a year after the event. The shape and approximate size of this mark, combined with the fact that it was washed off during autopsy, suggests that it was produced from contact with a blood-coated muzzle of the shooter's Glock. Since the barrel in the Glock projects beyond the front face of the slide, the muzzle imprint would not be expected to include features of the slide, which is consistent with the observed red ring. The source of the blood on the Glock was not determined because it was washed away during autopsy, prior to it being recognized and sampled for typing. This may have been blood acquired from the gun making contact with the bloody area from the head of the other fatal victim. The blood-like imprint of the muzzle indicates that the gun was pressed to the head of the Officer with sufficient force to take it out of battery. This may have resulted in what appeared to the shooter to be a malfunction, requiring an attempt to clear the weapon, producing one of the ejected intact cartridges.

Other physical evidence included bullet impact marks and damaged bullets recovered at the scene. Figure 11.18 shows an impact mark of a bullet to the siding below the window on the front of a Radio Shack store near the bar. The siding was composed of sandwich of sheet aluminum with a dense polyolefin core. The bullet struck a reinforced area of the siding which lay over a structural wooden member of the framing of the building. It had penetrated only a short distance and then bounced backward where it was recovered on the sidewalk during the initial crime scene investigation (Figure 11.19).

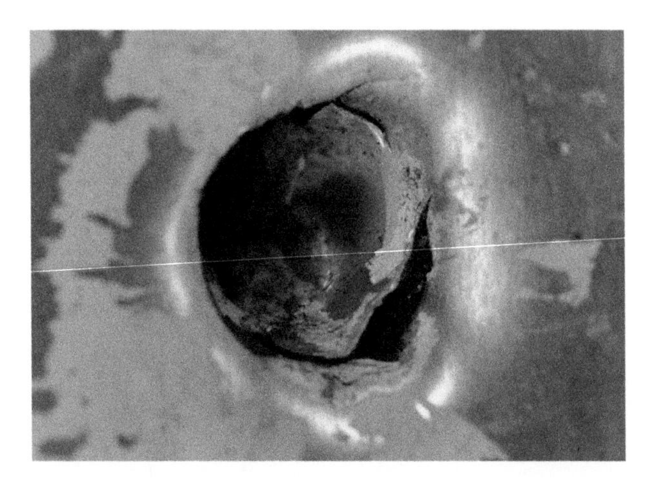

FIGURE 11.18 The bullet impact point on the building's aluminum-clad sheathing.

FIGURE 11.19 Photomicrograph of a bullet that passed through one of the victims, based on clothing fibers, paint and other transferred evidence adhering to the bullet (not in image). The flattened bullet nose indicates that after losing some energy, it impacted a firmly supported surface (e.g., a building sheathing).

This bullet contained both material and deformation-related traces from the building. These included the damage to the copper jacket and the lead core, and the acquisition of aluminum, paint, and polyolefin material from the siding of the building. Polyolefin material was wedged between the bullet jacket and the core. There were also fibers that were consistent with passage through the other fatal victim's upper clothing. This indicated that this bullet traveled through the victim's upper torso (which was consistent with the autopsy observations) before hitting the side of the building and then ricocheting onto the sidewalk.

There was an impact crater in the pavement by the third surviving victim's head. This is shown by a yellow "X" next to head of the chalk outline in Figure 11.15. The bullet hitting the pavement ejected a number of particles of concrete, which then impacted the building with the aluminum-clad polyolefin siding, near to the previously discussed bullet impact mark. This shot did not produce a fatal head wound but left a graze mark across his forehead.

As previously mentioned, the Glock was recovered months after the event. The exterior surfaces of the gun had no visible blood traces suggesting it had been cleaned subsequent to the murders. Criminalists in the city's DNA laboratory swabbed the external surface of the gun for DNA analysis by reverse dot-blot typing. This yielded a confusing mixture of DNA types that were difficult to interpret with the available technology of the day. But this was sufficient to associate the weapon with the homicides.

The consulting criminalist was the first to disassemble the weapon to gain access to its interior surfaces. The interior contained additional traces, including hairs in the recoil spring and tissue fragments in the recess of the forward portion of the frame (Figure 11.20). Histological slides were made of some of the tissue fragments after paraffin embedding, thin sectioning, and staining (Figure 11.21). These stained tissue sections exhibited numerous nuclei, thus indicating that they would have been good candidates for nuclear DNA typing. Unfortunately, this was not done because the consultant was only retained a few weeks prior to trial, and consequently there was not sufficient time to complete any DNA typing of the recovered tissue. Had those fragments been DNA-typed individually, more information would have been obtained from them than had been achieved from the rote swabbing. This eliminates the need for mixture interpretation and makes a very strong argument for not simply getting DNA types on things where the actual somatic origin or tissue origin of the DNA is not known.

The reconstruction of this event requires an integration of all of the physical evidence. The shooter initially shot the three victims from the vicinity of the left side of his double-parked car. This was supported by bullet paths through the victims, bullet impact marks, and the recovered bullet's traces and deformation. The shooter then went onto the sidewalk where he approached each of the fallen victims, placed the muzzle in contact with the victim's heads, and attempted to shoot them.

FIGURE 11.20 Glock's recoil spring chamber containing particulate traces of tissue. These were isolated and subjected to histological examination.

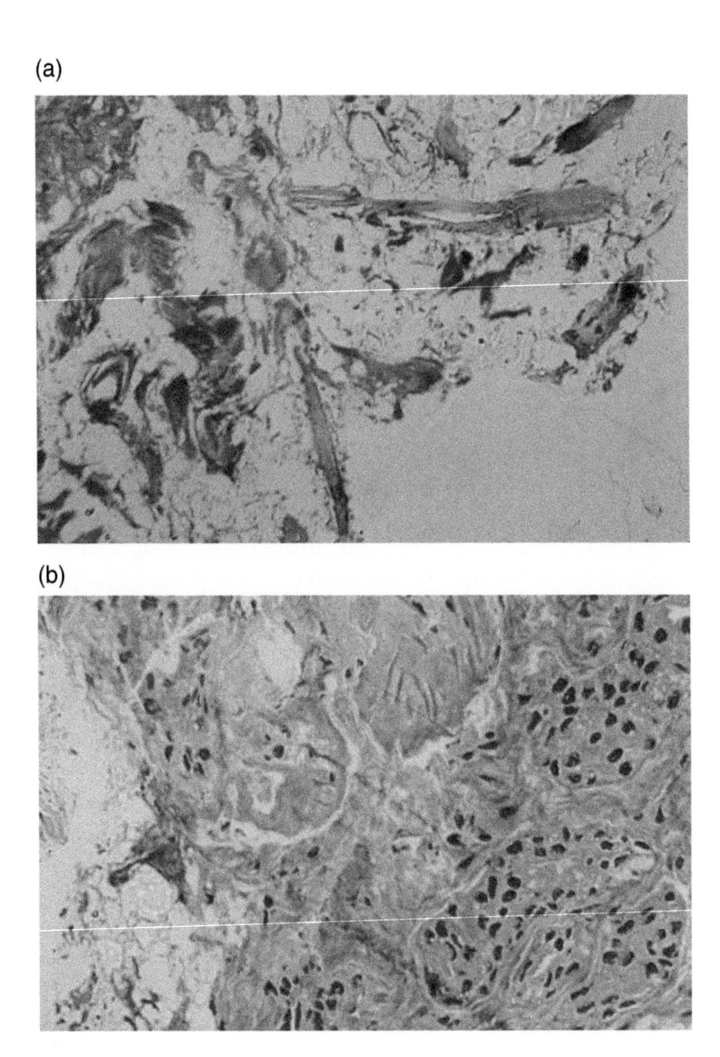

FIGURE 11.21 Photomicrographs of thin sections prepared from tissue particulates obtained from the recoil spring chamber of the Glock 9-mm handgun. Image (a) includes images of hair follicles and (b) shows numerous nucleated cells.

As previously mentioned, the presence of the three unfired cartridges which were manually cycled through the weapon provides significant information for reconstruction. First, they indicate that the shooter thought that there was a malfunction of the weapon which required manual opening of the action and ejection of the unfired cartridges. There are two things to consider here when considering the cause of this "malfunction." One is that the weapon was placed against the heads of at least two of the victims. The red ring on Officer's face is evidence of this. Further, this Browning-type action in an autoloading handgun is designed to unlock during recoil, and they also unlock when the barrel and the slide are pushed back a little bit. This disconnects the trigger mechanism from the striker or hammer which prevents the firing of the weapon. This would create an apparent jamb for the shooter, accompanied by only a "click" when the trigger is pulled. The unfired cycled cartridges indicate that the jambs were cleared prior to the shooter leaving the scene. Clearly, the action of the weapon was manually cycled at least three times during the event. The only hypothesis which takes into account the entirety of the physical evidence is that when the muzzle of the shooter's gun was pressed against one or more of the victim's heads, an apparent malfunction of the gun occurred which required the manual ejections of the three unfired cartridges.

11.3.3 Conclusions

The totality of information including the bullet flight paths and impact marks, combined with analysis of the victims' wounds, show that the initial shots were fired from the left side of the double-parked car. Three of the victims were immobilized and were lying on the sidewalk when the shooter approached them. Following the initial shots, the evidence of the wounds to the victim's heads, the red ring between Officer's eyes, the crater in the concrete, and the three unfired but cycled cartridges show that the shooter pressed the weapon against the head of at least one of the victims, the gun did not fire, and therefore manually operated the action to clear the apparent jamb, before he was able to successfully fire the gun. To account for all three of the cycled but unfired cartridges, this may have been repeated with other victims.

A reconstruction of events was only possible after previously unrecognized physical evidence (e.g., the significance of the ejected intact 9-mm cartridges and the red ring imprint on the root of the Officer's nose) was appreciated by the consulting criminalist. This was achievable because he was a generalist-scientist with extensive crime scene experience. Although each of the described items of physical evidence may support several hypotheses, when taken together, there is only one logical hypothesis that fits with all of the observations to provide a reconstruction of the events. An effective integration of the diverse types of physical evidence (ballistics, DNA, tissue, fibers, and building materials) necessitated that reconstruction be completed by a generalist-scientist in order to get a complete understanding of the case events.

The assailant was convicted of one count of second-degree murder and one count of second-degree manslaughter and sentenced to 30 years to life.

11.3.4 Lessons

This was clearly a poorly handled case from a physical evidence point of view. In addition to the failure to recognize the significance of certain physical evidences (e.g., the red ring muzzle imprint on the forehead and the cycled unfired cartridges), there were numerous errors made at the crime scene. This included poor evidence handling, specifically with the use of body coverings for the victims. The use of body coverings created additional complications in this case, as it created drip marks and possibly obliterated other traces on the sidewalk besides the victims and on their bodies. The repeated covering and lifting of body coverings, although common, is a bad practice. Other ways of protecting the victim's privacy, if deemed necessary, should be explored.

It needs to be appreciated in this and other case examples described here that the reconstruction of the event was not understood until the physical traces were integrated and synthesized by a generalist-scientist. Neglect of the importance of the integration of the physical evidence has the potential to lead to both unsolved cases and wrongful convictions. It is vital to consider the interrelationships of the physical traces when attempting a reconstruction. A reconstruction is made more robust when it is supported by multiple lines of scientific reasoning based on the totality of the traces. Again, this requires the knowledge and expertise of a generalist-scientist. Sadly, in the last several decades, specialists have been replacing the generalist-scientist in forensic laboratories without any realization as to how this loss has negatively impacted the effectiveness of science in physical evidence investigations (San Pietro et al. 2019).

11.4 The Imagined Mist Pattern

11.4.1 Case Scenario/Background Information

In some cases, scientific investigation is used for dealing with subtleties. In many cases, the question may revolve not around the issue of "who" but instead details of "how" the event took place. The case to be discussed here is one of these. It involved the fatal shooting of a man by his wife with a 25 ACP (auto Colt pistol) auto-loading pistol. It was clear that five shots had been fired in rapid succession from a fairly close range, and that all of the bullets fired had struck and wounded the victim.

There was never an issue about the fact of the wife shooting her husband. It was undisputed that she had shot him. The woman claimed that she was subjected to both mental and physical abuse by her husband, and that she had become fearful of him. On the day of the shooting, the husband had just returned from an extended trip with his mistress in California. The wife was not at home when he returned and only became aware of his return when she herself returned from running an errand. She became afraid of entering the house and drove to another location where she called the local police department. She requested a police escort to enter the house. The dispatcher put her call on hold and made a telephone call to the husband who asked "oh, has she been drinking again" and assured him that there would be no problem. The dispatcher then told the woman that he had spoken with her husband and been assured that he would not harm her. He also told her that they would be unable to give her a police escort. Ultimately, the wife claimed that when she entered the home her husband swore at her and verbally threatened her. She said that she shot him out of fear for her safety. She asserted that when he made a threatening move, she aimed the gun at him, closed her eyes, and pulled the trigger five times. The five small caliber bullets entered the victim's body, but none of them exited. The prosecution claimed she shot her husband unprovoked, and she was charged with aggravated manslaughter.

11.4.2 The Physical Evidence and Its Interpretation

Two privately retained experts became involved. One was retained by the prosecutor and the other by the attorney for the defendant. Neither of these experts was involved in the initial investigation. It was necessary for them to rely mainly on photographs and a crime scene video. In the opinion of the defense expert, there were no blood trace deposits resulting directly from the five shots. There were extensive blood trace deposits at the scene, but he concluded that all of these were the result of post-shooting bleeding and dripping. Contrary to this interpretation, the prosecution expert pointed to an apparent blood trace present on the left palm of the victim as recorded in the crime scene video tape. There was no detailed still photo of this. The low-resolution still image was a "frame grab" from a video home system (VHS) video that incorporated a view of the victim's left palm of the victim lying on the floor after he had fallen from his desk chair. There was no indication that the videographer had attributed any particular significance to the blood on the left palm at the time the body on the floor was being videotaped. The blood trace deposit on the palm occupied a relatively small area of the video frame as a result of a large subject-to-camera distance and consequently lacked detail. Unfortunately, this was the best image of the left palm recorded. The prosecution expert seized on this image to offer the pseudoscientific opinion that the blood trace on the palm was a "mist" pattern resulting from gunshot spatter, despite the fact no individual droplet deposits were documented in the image. The defense expert challenged this interpretation in a private conversation with the prosecution expert prior to trial testimony and asserted that it was equally likely to have been a contact transfer deposit. During this conversation, the prosecution expert acknowledged that the image was ambiguous and stated that it was an unimportant issue in the case. However, during subsequent testimony, the prosecution expert stated that the "mist" pattern allowed him to draw the conclusion that the victim, rather than presenting a threat to the defendant, was shot "unawares" while he sat at his desk. The defense expert concluded that there was nothing in this poor-quality image to suggest a fine "mist" pattern from gunshot backspatter. Additionally, the defense expert questioned the possibility of the small caliber, low-energy projectile producing backspatter through the entry hole in the overlying clothing.

11.4.3 Conclusions

This was a highly publicized trial that was featured as a "battered woman" case on Court TV. Ultimately, the defendant was convicted of aggravated manslaughter. The jury was apparently persuaded that the defendant's version of events was not true, based on the prosecution's story and expert's reconstruction.

11.4.4 Lessons

The prosecution expert's unsupported and unscientific opinion could be attributed to "wishful thinking" driven by expectation bias and a desire to please the "team" that retained him. This resulted in an unsubstantiated opinion that was not based on the demonstrable physical evidence: there was no documented "mist pattern" on the victim's hand in the low-resolution video frame. Furthermore, this claim is not falsifiable because there is no conceivable test for proving or disproving the existence of this "mist pattern" in the poor-resolution images and thus classifies as pseudoscience. Fundamentally, there was a lack of appreciation for the scientific method and the robust requirements for making a quality scientific reconstruction. This must be guarded against.

Additionally, there was a lack of knowledge by the prosecution expert of ancillary areas such as wound ballistics. There were no exit wounds, thus there could not have been any forward spatter. Any claimed blood deposits produced as a direct result of the gunshots would have had to be the result of backspatter. However, as per principle #4 (*The initial wounding may not produce any immediate or useful blood trace configurations*), there should be no expectation of backspatter. The 25 ACP projectiles are small caliber with low velocity and the victim was wearing clothing, thus the creation of backspatter would be unlikely. The totality of blood deposits present at the scene were best explained as being derived from post-event bleeding and dripping from the wounds. This case further underscores the inherent complexity of deriving reconstruction information from blood trace deposits.

11.5 Concealed Blood Traces

11.5.1 Case Scenario/Background Information

The body of the young woman was found beside a turnout on a mountain road in the 1960s. She had been stabbed and her body was dumped down an embankment. Immediate suspicion fell on the young man she had been with the previous evening. His car was seized as evidence and examined by police investigators. No blood traces were found in the vehicle to suggest that the stabbing had taken place in the vehicle or that the body had been transported in the vehicle. Ultimately, the car was returned to its owner. It was then sold and taken to another state by the new owner. After several months, the investigation came to a dead end. At this point, the investigators considered that they had released the vehicle prematurely and wanted it reexamined. They obtained a court order to obtain the car and bring it back to the original jurisdiction for a more detailed examination.

11.5.2 The Physical Evidence and Its Interpretation

Upon examination of the returned vehicle, the investigators speculated that the original suspect had gone to an automotive supply store and had aftermarket seat covers installed the morning following the homicide. The local criminalist was brought in to examine the vehicle. When the new seat covers were removed, suspicious looking traces were observed on the front passenger seat. It appeared that an attempt has been made to obscure the deposits on the bluish colored seat upholstery using a blue-colored pigment or dye. Laboratory personnel cut samples from this area for presumptive blood testing. This testing was negative for blood, and although suspicion remained, attention turned to other areas of the vehicle interior.

In the meantime, the suspect’s attorney had retained a private forensic science laboratory to conduct an independent examination. The seat was removed from the vehicle and submitted to the private laboratory. As part of the examination of the seat, a “mapping” technique was employed. A large sheet of moistened filter paper was placed over the area containing the deposits and covered with a plastic film. Uniform pressure was applied to ensure sufficient contact with the upholstery. The contact was maintained for over an hour. When the filter paper was removed from the seat, brown stains were observed over an extended area. The filter paper was then sprayed with benzidine, a catalytic color test then in use as a presumptive reagent for blood. Benzidine has since been identified as a carcinogen and is no longer used as a presumptive blood test. A positive result was obtained with the benzidine corresponding to the brownish-colored areas. Cuttings taken from the areas that yielded the strongest presumptive result were tested with additional reagents. These included the Takayama microcrystal test and the precipitin test for human blood. The positive results of these tests combined with visual examination of the stained cuttings confirmed that large quantities of human blood had been deposited.

In addition, a blood encrusted knife was recovered by the defense criminalists from the juncture of the seatback and the bench. This posed a dilemma for the criminalists. Their instinct was to secure the knife, document it, and package it separately for return to the prosecution separate from the seat itself. However, on learning of this plan, the defense counsel threatened legal action against the criminalists for disclosure of privileged information developed by the defense. The rules of discovery enable the defense to not share its discoveries due to Fifth Amendment claims, thereby protecting the defendant from self-incrimination. Informing the prosecution of the discovery of the presumed murder weapon by the criminalists retained by the defense attorney certainly qualifies as self-incrimination. As a result, the criminalists initialed and dated the blood-encrusted knife and returned it to its original location.

11.5.3 Conclusions

When the consulting criminalists informed the defense of the results, the attorney did not want any further laboratory work carried out. Because of legal constraints involving asymmetrical discovery, the prosecution did not learn of these findings during the trial preparation.

Interestingly, the knife was discovered in the bench seat assembly when it was being brought into the courtroom by a curious bailiff during the trial. This occurred during a recess, so the jury had no knowledge of this event. It apparently did not influence the outcome of the case.

The defendant was ultimately found not guilty of the homicide.

11.5.4 Lessons

In certain circumstances, a filter paper transfer and mapping technique can be very valuable. It preserves geometric relationships and allows destructive testing on a portion of a specimen while leaving the original sample essentially unaltered. It is not applicable to all circumstances. For example, it would not be appropriate for a thin or sheer cloth specimen where the risk of a nearly quantitative extraction and transfer to the filter paper would be great. Informed scientific judgment is necessary.

This case also demonstrates the negative consequences of prosecutorial interference with scientific assessment and decision-making. The local criminalist was inundated with lists of daily tasks from the District Attorney and never given the opportunity to look at the evidence holistically. The laboratory director did not insulate the newly hired local laboratory criminalist from this meddling. It is the responsibility of laboratory leadership to enable scientific inquiry to proceed unhindered by outside non-scientific interference. It is interesting to speculate that given the opportunity for an unhindered scientific investigation, the local criminalist could have found the blood evidence indicating the victim was stabbed in the vehicle.

11.6 A Stomping Homicide – Misuse of Enhancement Reagents

11.6.1 Case Scenario/Background Information

A violent homicide took place in a run-down apartment being shared by three young men. Two of them were suspected of stomping the third man to death. A reconstruction was possible due to a very large amount of blood spatter evidence along with contact transfer blood evidence in the form of several whole handprints (palm and fingers).

11.6.2 The Physical Evidence and Its Interpretation

Most of the violence inflicted on the victim took place adjacent to a living room wall. There was a pool of blood soaked into the carpet at the base of the wall. The lower portion of the wall itself was covered with a large number of airborne droplet deposits. This area also contained numerous bloody hair swipes. Well above this area on the same wall were several blood-containing contact transfers in the form of easily visible (patent) whole handprints. It was readily apparent from their observed positions above the floor that one set of handprints was left by a relatively tall individual, whereas the other was left by someone of significantly shorter stature.

Although it was not known where the initial assault took place, by examining the blood traces, it was apparent that the major stomping of the victim took place while the body was on the floor at the base of the wall. The two assailants stood over the victim and kicked him while partially steadying themselves with their hands against the wall for support. The fact that there were whole hand contact configurations in blood on the wall made it clear that their hands were in contact with the victim's bloodied body before the stomping was completed.

At the outset of the investigation, the victim's body was not present at the base of the wall. Strangely, the blood traces made it clear that the victim had been dragged from that area to the bathroom in an apparent effort to clean him up. There were remnants of diluted blood in the bathtub and deposits of excess gel-type shampoo on the floor.

A private consultant was contacted by the state prosecutor's office and asked to come to the scene which was three hours away by car. He was told that the scene would remain secured and undisturbed until his arrival. Unfortunately, when the consultant arrived with his team, there were two law enforcement investigators already present inside the scene conducting their own examination. They had recently been introduced to two enhancement reagents at a lecture given in their area. The two reagents were leucocrystal violet (LCV) and amido black. As discussed elsewhere in this book, LCV is a catalytic presumptive blood detection reagent and amido black is a protein stain. They are valuable enhancement reagents where latent or poorly visualized blood contact transfers are suspected of being present. However, they must be used knowledgeably and judgment is necessary and deciding when they should be used, if at all. Misuse can result in the destruction of evidence. This was the situation encountered here. It became clear that these two investigators had no experience using these reagents. With the patent bloody hand contact transfer configurations, there was no need for any enhancement. They were easily documentable photographically. When these investigators sprayed the reagents onto the configurations on the wall, the excess liquid with the soluble blood components ran down the wall destroying the friction ridge details in the evidence deposits (Figure 11.22). Fortunately, there were ample unsprayed areas on the wall which allowed for facile association of the minutiae of the friction ridge detail the patent bloody handprints with the responsible individuals. Despite this misuse of enhancement reagents, the overall configuration of the blood traces at the scene allowed for a detailed reconstruction supporting the stomping and attempts at cleaning up the body.

FIGURE 11.22 Photograph of the wall, showing the misuse of enhancement reagents which destroyed many of the blood trace deposit configurations.

11.6.3 Conclusions

The overall conclusion of this case was that the inappropriate use of enhancement reagents by nonscientists destroyed potentially valuable physical evidence. These are useful tools for the scientific investigator when properly applied; however, they are not appropriate in all blood trace cases. Although it should be obvious that patent prints do not require enhancement, unfortunately that was not appreciated in this case. Luckily, not all of the blood trace deposits were destroyed, thus a scientifically based reconstruction was possible for a timely understanding of this event.

11.6.4 Lessons

Enhancement reagents, although readily available commercially, should only be applied to blood trace deposits by well-educated and trained scientists who can evaluate the appropriateness of their application in the context of the specific case. Furthermore, an experienced scientist would know that there are versions of these reagent formulations that contain protein-denaturing components to fix the protein components to prevent or limit them from diffusing or flowing and thus altering the original blood configuration. Additionally, these enhancement reagents can be carefully sprayed in a succession of light mistings in order to prevent any runoff. This controlled application is advisable in the use of all blood trace enhancement reagents in order to avoid the flow of excess liquid. Lastly, clearly patent blood trace deposits do not require any chemical enhancement.

References

Alter, M. (2013). Camm trial in review. http://www.wlky.com/news/local-news/ david-camm/camm-trial-in-review/22429352 (accessed 1 October 2020).

Camm v. Faith (2019). 937 F.3d 1096.

Goetz, B. (2017). Bloodstain pattern analysis. In: *Forensic Science Reform* (pp. 279–298) (eds. W.J. Koen and C.M. Bowers). Academic Press https://doi.org/10.1016/B978-0-12-802719-6.00009-1.

IUPAC (1997). *IUPAC Compendium of Chemical Terminology (the "Gold Book")*, 2nde (eds. A.D. McNaught and A. Wilkinson). Oxford: Blackwell Scientific Publications. Online version (2019-) created by S. J. Chalk. ISBN 0-9678550-9-8. doi: https://doi.org/10.1351/goldbook.

Koen, W.J. (2017). Bloodstain pattern analysis: case study: David Camm. In: *Forensic Science Reform* (eds. W.J. Koen and C.M. Bowers), 272–278. Academic Press https://doi.org/10.1016/B978-0-12-802719-6.00009-1.

San Pietro, D., Kammrath, B.W., and De Forest, P.R. (2019). Is forensic science in danger of extinction? *Science & Justice* 59: 199–202.

Tanton, R.L. (1979). Jury preconceptions and their effect on expert scientific testimony. *Journal of Forensic Sciences* 24 (3): 681–691.

CHAPTER 12

More Broadly Assessed Cases: Going Beyond the Request

This chapter will present case examples where the initial contact and consultation with the expert sought by investigators or attorneys was directed toward the interpretation of blood trace configurations, but where the expert recognized other evidence that proved to be much more valuable. These cases underscore the point that one cannot separate blood trace configuration expertise from that necessary for crime scene reconstruction. To be maximally useful an expert analyzing blood trace configurations in a given case cannot operate in a silo and focus solely on blood traces. In some cases such a narrow focus may result in crucial evidence not being recognized. This is the situation that applies to the case examples selected for presentation and discussion here.

In any case investigation, it is important to remember the most relevant evidence may not be the most obvious and may not be emphasized by the police investigator or attorney. An open-ended holistic approach to crime scene investigation is critical for all complex case analyses.

THE CASES:

1. Gunshot to the Forehead and the Runaway Car
2. The Obscured Bloody Imprint
3. The Murder of a Deputy: Shooting in a Hospital Room

Blood Traces: Interpretation of Deposition and Distribution, First Edition. Peter R. De Forest, Peter A. Pizzola, and Brooke W. Kammrath.

12.1 Gunshot to the Forehead and the Runaway Car

12.1.1 Case Scenario/Background Information

A private consultant with recognized expertise in interpreting the configuration blood traces was retained by a local prosecutor's office to reconstruct the fatal shooting of the driver of a passenger vehicle who was the sole occupant of the vehicle. It was envisioned by the prosecutor that the reconstruction would be based on the configuration of hundreds of airborne blood droplet deposits in the vehicle interior. The event took place in a depressed portion of a medium-sized city. There was a single gunshot wound to the driver's forehead. There was no visible gunshot residue or propellant surrounding the wound indicating that the muzzle of the gun was approximately a foot or more away from the victim's forehead. The shooter was readily identified and arrested nearly immediately. The police and the prosecutor suspected that there had been a confrontation between the driver and the shooter who would have been standing outside the vehicle beside the driver's open window. The driver's window was in fact rolled down. The suspect disputed this account. He claimed that he had fired an un-aimed shot from a location an appreciable distance from the right side of the vehicle and that he had not intended to shoot anybody. Following the shooting the victim's foot depressed the accelerator pedal, and the car sped away for several blocks in more or less a straight line interacting with parking meters and sign posts on the right side of the street before coming to rest in a parking lot after having flattened a chain-link fence.

12.1.2 The Physical Evidence and Its Interpretation

The expert assessed the blood traces in the vehicle, and determined that they were uninformative. Although there were significant droplet deposits in various areas of the interior, these may have been due to body movement during the multiple collisions of the vehicle and could not be attributable to the gunshot to the head. The gunshot to the head was a penetrating wound with no exit. As discussed in Chapter 5 and in relationship to Principal #4, one should not necessarily expect blood backspatter from a gunshot wound.

There was no physical evidence to refute the prosecution's supposition that the shot had been fired through the open driver's window and struck the driver in the forehead, assuming that his head was turned to the left. Dealing with the defense's assertion that the fatal bullet was fired from the right side was not straightforward. Significant physical evidence from the scene had not been recognized and retained during the course of the investigation. For the bullet to have struck the victim in the forehead from a firing position at a location on the right side of the vehicle, it would have been necessary for the bullet to have passed through the passenger door window opening, and the driver's head would have had to have been turned to the right.

Examination of the vehicle revealed the presence of tempered glass "dice" in the window channel at the top of the right door. This showed that the window had been fully raised at the time it was shattered. If the window had been shattered by the bullet, this would have supported the defendant's version of the event. No shattered glass was recognized or collected from the vehicle interior or from the scene. Had such evidence been present and recognized, and subsequently documented and preserved, a relatively straightforward resolution of the question of how the window was shattered might have been possible. For example, if a pile of broken tempered glass "dice" had been found at the right curb along the path of the runway car, it would have been clear that the right side window would have been intact at the time of the shooting and was broken later, refuting the defendant's account of the event. However, because no glass was recovered, the question was left open. If possible, another means of addressing this important question was needed.

The front of the vehicle was more or less uniformly damaged from striking the chain-link fence nearly straight on and bringing it to a stop at the end of its travels, whereas the right side of the vehicle bore superficial damage from interacting with or "sideswiping" parking meters and signposts. The direction of all the damage was from front-to-back. It was not clear how the passenger side window could be broken under these circumstances. This would have required a force at right angles to the direction of travel of the car.

12.1.3 Conclusions

Additional reflection by the consulting expert allowed the development of a hypothesis with a possible explanation for the window being shattered during the runaway event rather than resulting from the bullet passing through it. This was that the normally non-pivoting mounting of the right side rear-view cast aluminum mirror housing could have been forcibly pivoted and folded inward pressing against the glass of the window as a result of striking parking meters and sign posts. However, there was no evidence of this in the photographs provided to the expert by the prosecutor's office. At this point, the consulting expert suggested that the vehicle be made available for him to examine. Arrangements were made and he traveled to an impound garage where the vehicle was being stored. This examination revealed bending and compression damage to the lower inner corner of the cast aluminum mirror housing. This damage to the aluminum housing would not have been present if there had been no glass pane for the mirror housing to fold in and press against (Figures 12.1 and 12.2). With these observations, it was now clear that the right side window was raised and intact at the time of the shooting. It was evident that the window had been broken later in the course of the runaway event by the action of the mirror housing being folded inward. Thus, it was concluded that the shot had been fired through the open driver's window with the driver's head turned to the left.

12.1.4 Lessons

It is important for the expert to take a holistic or global view of the totality of the physical evidence record and not restrict the scope of one's inquiry to the request. To do so would assume that the request was a scientifically informed and knowledgeable one. This would imply that the requester had scientific expertise and experience superior to, or at least equal to, that of the expert performing the examination. This rarely, if ever, is the case. Despite this, many experts do not step back and consider questions or issues that are not part of the request.

FIGURE 12.1 Photograph depicting one of the areas inside of the vehicle containing uninterpretable blood traces.

If experts were to limit their examinations to the scientifically naïve requests of an attorney or investigator, the resulting analyses would risk being incomplete. It should be clear that if the scope of the inquiry is limited to that of the request, critical evidence may go unrecognized. It would then be "lost and gone forever," thus resulting in an irredeemable fatal flaw in the investigation.

Clearly, an "expert" who's education, training, and experience are limited to interpreting the configuration of blood trace deposits is not properly equipped for this type of endeavor. Stated somewhat differently "bloodstain pattern analysis" is not a stand-alone specialty of crime scene reconstruction. It must be fully integrated with the reconstruction activity.

This case demonstrates the importance of thorough scene investigation and the recognition of valuable physical evidence. Had the investigators recognized the value of glass evidence, and found a pile of broken glass along the route of the vehicle, the reconstruction of this case would have been easier to accomplish.

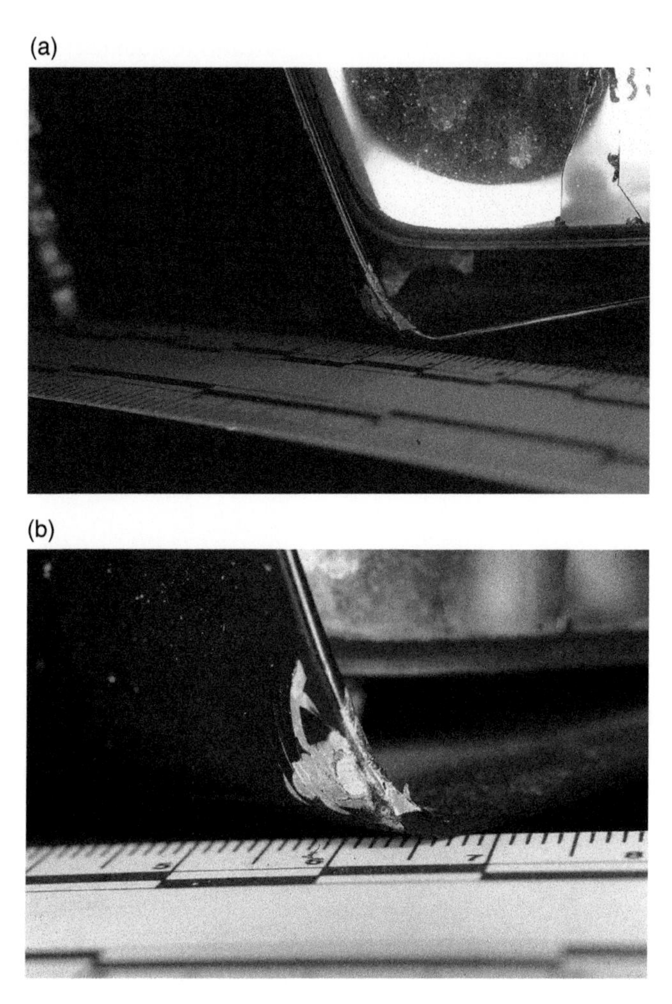

FIGURE 12.2 Two photographs (a and b) of the passenger side rear-view mirror, depicting the damage to the cast aluminum housing. As the vehicle was moving and striking a succession of undocumented structures along the right side of the vehicle's path, the mirror housing was bent back and forced against the passenger side window.

12.2 The Obscured Bloody Imprint

12.2.1 Case Scenario/Background Information

A reportedly reclusive woman was found stabbed to death in her home. The homicide was discovered by her common-law husband when he returned home from work. Based on the woman's reclusiveness, it had been thought that she would be extremely reluctant to let anyone into her home under any circumstances. Initially, investigators suspected the husband had killed her, but quickly eliminated him as a suspect based on investigative details that do not need to be discussed here. A suspect whom the investigators felt very strongly had committed the homicide was developed within a couple of days. He was interrogated at length and steadfastly denied any involvement. He even denied ever having been present in the victim's home. Several hours into the interrogation the investigators lied and told the suspect that his fingerprints had been found at the homicide scene. He countered that he now remembered that he may have been at the scene a few weeks earlier. He claimed to have been invited in for a cup of tea. At that point the investigators then told him that the fingerprints were in blood. Ultimately, the suspect gave a detailed confession. When the case was forwarded to the prosecutor, he became concerned that the confession would be inadmissible because of the deception involved in the interrogation. There was little other evidence pointing to the guilt of the suspect, so the prosecutor felt he needed to utilize the confession. He sought ways of getting corroboration for it. The scene photographs illustrated that it was a bloody scene. He thought that perhaps a reconstruction of the event based on an interpretation of the blood deposits present at the scene could serve to confirm important aspects of the suspect's confession. An expert was consulted and retained.

12.2.2 The Physical Evidence and Its Interpretation

A meeting was arranged with the prosecutor and his chief crime scene investigator at the expert's office. To familiarize the expert with certain details of the case they brought case related documents and numerous crime scene photographs with them. There were actually two different sets of crime scene photographs taken by different photographers. One set was produced by photographers from the local police department during their crime scene investigation. The other set was produced by photographers with the county prosecutor's crime scene investigation unit. There were many dozens of photographs in each set. During the meeting the consulting expert made a preliminary examination of the photographs in both sets. From the airborne droplet deposits and other configurations, it was apparent that the attack had taken place initially on the right side of the bed (as viewed from the foot of the bed) and then moved to an area on the left side of the bed. The presence of extensive pooling of blood on the left side of the bed made it clear that the victim had bled out in this area. Little in the way of evidence of airborne droplet deposits was apparent on the left side of the bed or in the immediate area. This suggested that the actual stabbing had ceased by this point, or that airborne droplet deposits that may have been present were obscured by the subsequent flowing and pooling blood. No preliminary opinion was offered at this point, as it was evident that a great deal of study would be necessary in order to understand the details sufficiently to know whether or not it would even be possible to develop a full reconstruction.

Other aspects of the scene were considered such as disturbances of the carpet pile and displacement of the bed frame and box spring. As found, the victim was face down on the floor wedged between the bed and a small bookcase (Figures 12.3 and 12.4). Configurations produced by bloody hair were observed on some of the bookcase shelves (Figure 12.3). During the continued examination of the photographs, the consulting expert asked the investigator and the prosecutor what they had done, if anything, with a faint pattern that appeared to be a partial footwear outsole imprint on the victim's back (Figures 12.3–12.6). They responded that they were not aware of the existence of such a pattern. The expert pointed out the faint pattern which was within a larger area of a thinly spread blood distribution. It was detectable in multiple photographs in each of the two sets. Subsequent autopsy photographs (Figure 12.7) revealed that it had been

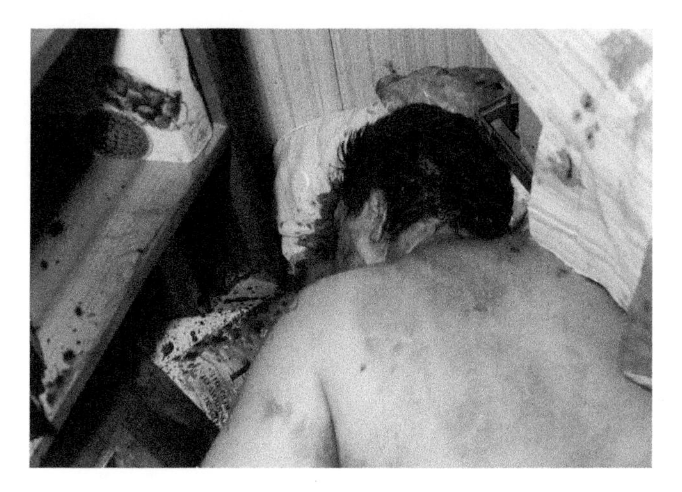

FIGURE 12.3 A photograph showing the back and left shoulder area of the victim. Also in view is the bookshelf with bloody hair imprints or contact traces.

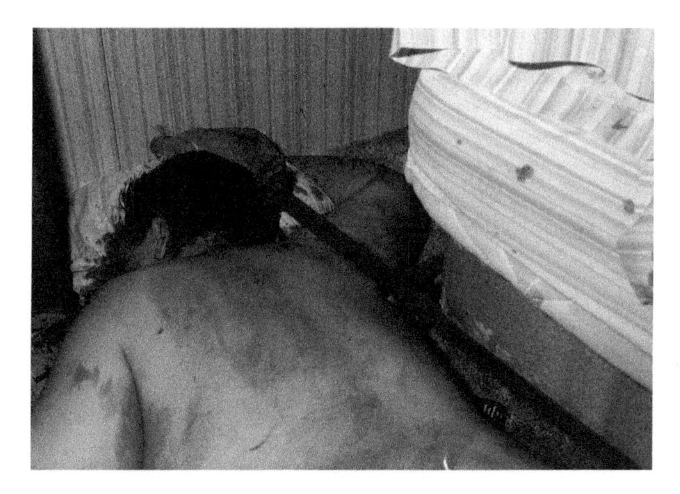

FIGURE 12.4 A photograph showing the back and left shoulder area of the victim. The partial footwear outsole pattern is faintly visible.

washed away with the other blood on the victim's back during the autopsy. The value of a bloody footwear outsole imprint on the back of the victim has obvious relevance and probative value.

It was suggested by the consultant that the faint image in some of the pre-autopsy photographic frames be enlarged and enhanced to make the apparent pattern more visible (Figure 12.5). The consultant also asked that he be provided with whatever footwear the defendant was wearing at the time of his arrest. Due to the prior lack of awareness of its existence, it was clear that no investigator present at the scene during the investigation or responsible for photography had recognized any pattern at that time or even during any subsequent follow-on review of the photographs. Since the faint pattern had not been recognized or had any significance ascribed it to it during the course of the investigation, no effort had been made to employ special photographic techniques to

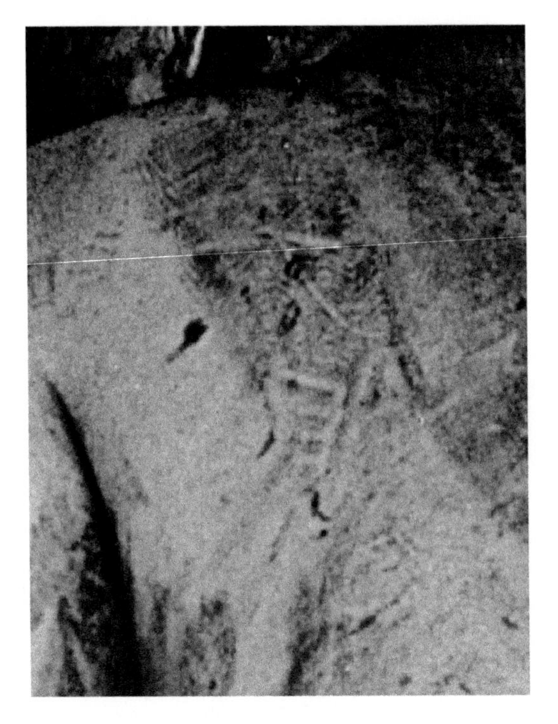

FIGURE 12.5 An enhanced version of the previous photograph, showing one partial impression of a footwear outsole pattern.

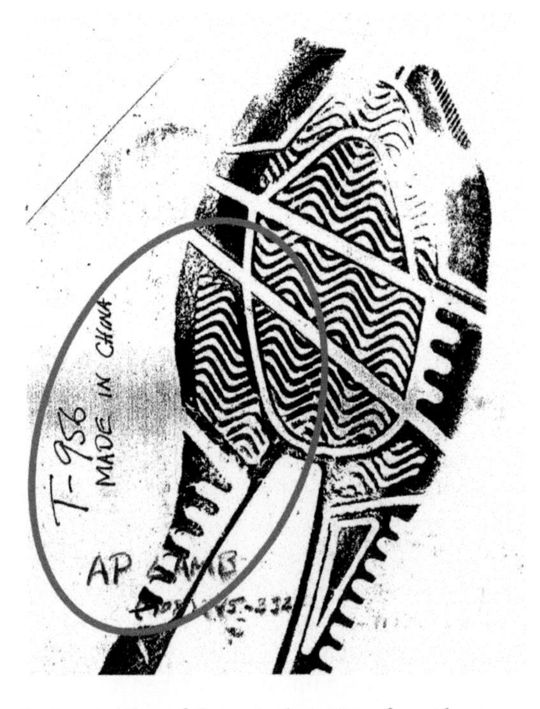

FIGURE 12.6 Exemplar impression of the outsole pattern from the suspect's right shoe collected by police investigators when he was initially detained. The red circle shows the area that corresponded with features seen in the previous figure.

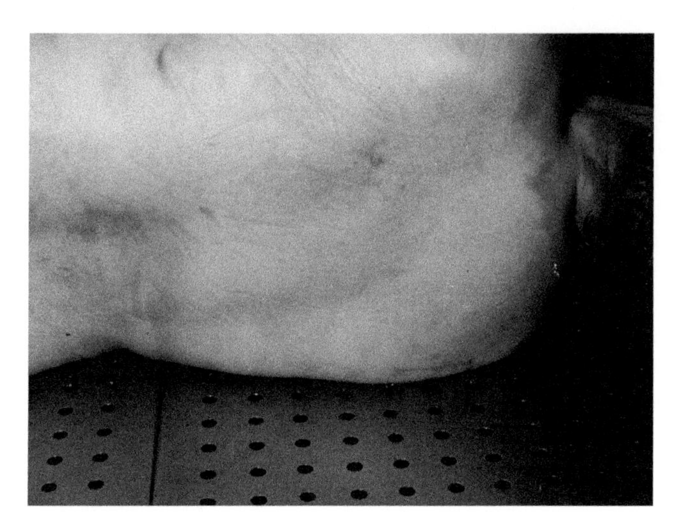

FIGURE 12.7 A photograph of the victim's back at autopsy, after the body had been cleaned. The evidence of the bloody footwear outsole imprint and surrounding widely dispersed blood deposits were washed away.

improve the contrast while the photography was underway, and no reference scale had been placed next to it in any of the many photographs that had been taken. This limitation made it impossible to determine the size of the footwear responsible for the partial pattern detected. There were further limitations that will be discussed below.

When the athletic shoes being worn by the defendant at the time of his arrest were provided to the expert, exemplar prints were made (Figure 12.6). In conducting a comparison of the exemplar prints with portions of the enhanced and enlarged photographs of the particular pattern area from the victim's back, it was clear that the pattern "could have" been created by the outsole of the right side evidence shoe or by a shoe of the same design. Stated somewhat differently, the class characteristics of the shoe outsole were consistent with the partial pattern and therefore the defendant's shoe could not be eliminated as a source of the partial class pattern. Despite the inherent limitations, the consulting expert felt that the faint pattern constituted potentially valuable evidence that should be evaluated further. The evidence footwear submitted was not a name brand. The labeling indicated that it was manufactured in China and bore the designation "T–956" as its brand. On the face of it, the fact that the evidence footwear was not a name brand would tend to enhance its value as associative evidence. An effort was made by the consulting expert to get information to assess the frequency of occurrence of this brand of footwear in the overall population. This proved to be very difficult. Neither the outsole pattern nor the T-956 brand designation could be found in any of the forensic footwear databases. Efforts made by the prosecutor's office and the attorneys for the defendant to purchase an exemplar shoe and learn more about the brand's distribution and sales were to no avail. All of this suggested that it would be a rare event to encounter one of these at random. This served to enhance the value of the evidence and offset several limitations.

At the time the case went to trial the defendant's confession was admitted into evidence by the trial judge and the consultant testified for the prosecution concerning the partial footwear outsole pattern observed in blood on the victim's back. The testimony about the partial pattern had to be presented carefully to point out its limitations, strengths and make sure that these were understood by the jury.

There were five significant limitations of the footwear outsole pattern. First, no size determination of the outsole responsible for producing the faint evidence imprint was possible. Second, there were no individual features or "accidental characteristics" which could be used to associate a specific footwear with the imprint was present on either the evidence shoe outsole or in the questioned or evidence imprint. The association was limited to class characteristics. Third, the total number of shoes manufactured with this outsole design was unknown. Fourth, the number of shoes with this outsole design sold in the United States was unknown. And finally, no distribution information within the United States for such shoes could be obtained.

There were three meaningful strengths of the footwear outsole pattern. First, the faint imprint was made in apparent blood contemporaneously with the homicide and it was washed away with the blood at autopsy. It was clearly not an artifact resulting from some post event activity at the scene. It was unquestionably relevant. Second, the a priori likelihood that the shoes from the defendant would have the same outsole pattern as a candidate source of the faint imprint, provided that he was not present at the crime scene contemporaneously with the homicide, would be expected to be vanishingly small but cannot be known. Third, the apparent rarity of the outsole pattern, evidenced by the failure by both the prosecution and defense investigators to locate a matching pattern in a database or an actual shoe of the same make and model, made it a very valuable association.

12.2.3 Conclusions

The trial jury found the defendant guilty and he was sentenced to prison. About two years later the conviction was overturned by the state appellate division citing the deceitful interrogation that led to the confession. At a retrial in which the confession was not introduced, the consulting expert testified again carefully explaining the limitations and strengths of the comparison to the jury. The defendant was found guilty for the second time.

12.2.4 Lessons

The overarching lesson demonstrated by all cases in this chapter, but certainly here, is that the expert should not confine the inquiry to the scope of the request. The consultant's generalist knowledge, keen observation skills, and evidence awareness enabled him to study and scrutinize the crime scene photographs for all relevant traces and recognize a bloody footwear outsole pattern. The initial focus by investigators and attorneys was on the blood spatter. This resulted in the footwear outsole pattern going unrecognized by investigators through several phases of the investigation. The consulting criminalist, considering the configurations beyond the commonly considered airborne droplet deposits, recognized significance of the blood on her back. Consistent with the scientific method, he logically questioned himself as to how it got there (*e.g.*, was it spattered or through a contact transfer?). It was only through the process of scientific inquiry that the partial bloody footwear outsole pattern was recognized.

There is a lack of respect for and recognition of the body as a source of valuable physical evidence. The traces on a body, in this case the bloody footwear outsole pattern, was not documented at autopsy and instead was washed away. Prior to the autopsy would have been a useful time to take evidence quality photographs of the imprint with a scale. In the authors experience, too often potentially valuable physical traces go unrecognized and are lost by the handling and washing of bodies in preparation for autopsy. In most cases, we would recommend a generalist criminalist be present at the scene or autopsy to recognize, assess and secure any of these physical traces.

The case also illustrates the value of associative evidence in certain case circumstances even when statistical data for expressing the strength of the comparison cannot be adduced. It needs to be understood that there are many situations with compelling associative evidence where numerical expressions are not possible. In this case, the apparent rarity and contextual relevance of the bloody footwear outsole pattern (limited to class characteristics) combined to leave little doubt that the defendant's footwear produced the mark on the victim's back.

12.3 The Murder of a Deputy: Shooting in a Hospital Room

12.3.1 Case Scenario/Background Information

The crime scene in this case was a hospital room. The patient in the room had been shot and wounded in a shootout with the police a few days earlier. He was being held as a prisoner under a 24 hour watch as he was recovering from his wound. He was being restrained with leather cuffs around one wrist and one ankle. These restraints were typically used for disturbed psychiatric patients, rather than criminal offenders or prisoners. Various deputies from the sheriff's department were assigned to guard the prisoner around the clock in a succession of shifts. One night the nurses heard a series of gunshots. They fled to a more distant area. Looking down the hall back toward the area, which was the source of the gunshot sounds, they observed the prisoner fleeing the building via a stairwell wearing only a hospital gown open in the back. Because it was winter time, and there was snow on the ground, the prisoner did not succeed in going very far. He left a distinct trail in the snow. He was arrested very shortly hiding in an outbuilding on the hospital grounds with the deputy's service revolver in his possession. Officers checking the hospital room found that the deputy assigned to guard the prisoner had received fatal gunshot wounds. There were extensive blood traces in several locations around the room and on the bedding. The room itself showed signs of a struggle. The upper half of an intravenous (IV) stand was separated from its base, and was found lying on the floor. Damage to the top portion of it was evident in the form of bending one of the arms that would hold the IV bag. Items of potential evidential value, including the bedding and the upper half of the IV stand were taken as evidence and sent to the state laboratory.

The prisoner became a homicide defendant when new charges were filed concerning the shooting death of the police officer. Ultimately, in his defense he claimed that the officer became abusive and a struggle had ensued as a result of an argument stemming from the defendant's frequent demands for use of the bed pan. During the claimed struggle, the defendant asserted that the officer threatened to shoot him with his service revolver. He claimed further that he gained control of the weapon and shot the officer in self-defense during the struggle. This statement seemed contrived and not credible to the prosecutor and his investigators. However, it could not be easily dismissed or ignored.

12.3.2 The Physical Evidence and Its Interpretation

State forensic scientists observed and documented blood traces and gunshot residue on the bedding from the hospital room, but nothing was found that would clearly refute the defendant's claim. Ultimately, the examination of the evidence in the state laboratory did not result in an understanding of significant details of the event. The evidence items were subsequently returned to the prosecutor's office. The prosecutor then contacted and retained the services of an outside expert on blood traces.

As a preliminary step the consulting expert was asked to examine the evidence items in the prosecutor's office. The consultant did not wish to confine his inquiry to blood traces only and asked to see the shooting scene photographs as well as the physical evidence removed from the hospital room and examined in the laboratory. All of this was provided for review. The bedding included three bedsheets: a bottom sheet, a draw sheet (which is commonly used to assist in moving or positioning patients), and a top sheet. A light grey colored irregular shaped trace was observed on the draw sheet. No cause or explanation for this was immediately apparent. All of the sheets bore airborne droplets blood deposits. The consultant suggested that it would be very useful to return to the hospital room and place and position the evidence bedding on the bed using the shooting scene photographs to duplicate the original arrangement. This was done, and it became clear that the grayish trace was on the upper side of the folded draw sheet in the original bedding configuration. Nothing particularly useful for reconstruction purposes was learned from seeing the blood droplet configurations in their original configuration.

In studying the grayish trace further in conjunction with the scene photographs, a testable hypothesis as to its cause was developed by the consulting expert. In the scene photographs the upper portion of an intravenous (IV) stand which had been separated from its wheeled base was observed lying on the floor. It was hypothesized that a grease-like lubricant observed on the lower end of this portion of the IV stand could account for the grey-colored trace. Several means of evaluating this hypothesis were explored. The first of these was the production of an exemplar deposit using another IV stand component and a fresh bed sheet placed over a hospital mattress. The test IV stand portion was laid horizontally on the clean sheet and pressed against it forcibly. A configuration essentially duplicating that observed on the evidence sheet was produced. Several additional tests were proposed by the consulting expert, and the cooperation of the laboratory was enlisted to carry these out. In the first of these small cuttings were taken from the exemplar and evidence traces and analyzed using scanning electron microscopy (SEM) combined with energy dispersive x-ray spectroscopy (EDS). The SEM imaging revealed the presence of an amorphous, loosely adhering, material on the cotton yarns of the specimen cuttings recovered from both exemplar and evidence sheets. In subsequent laundering tests this material was not removed by machine laundering as assessed visually. However, in examining these specimens with SEM, a change in appearance was noted relative to the specimens from the pre-laundering examination. There was less of the adherent material which now appeared more tightly bound to the cotton yarns. EDS data revealed the presence of aluminum. Material extracted from small portions of the deposit for Fourier transform infrared spectroscopy (FTIR) analysis using an aliphatic solvent was consistent with a grease-like lubricant as the amorphous matrix observed with SEM.

Based on the visual observations and the data obtained by the laboratory analyses, the following reconstruction was offered by the consulting expert. During the night of the shooting the prisoner quietly separated the top or upper part of the IV stand from its base without being observed by the guarding officer. This would have taken considerable time as the threaded sleeve and collet connecting the two portions of the IV stand required 13 turns to affect a separation. Following success in separating the two portions of the IV stand, the prisoner hid the upper portion in the bed by lying on top of it and pressing it into the draw sheet which had been stretched across the midportion of the bed. He kept it hidden until an opportunity to use it as a bludgeon against the officer presented itself. It should be noted that the top cross member of the IV stand component was bent and that there were two wounds on the officer's forehead separated by a distance consistent with the separation of the two suspected features at the top of the IV stand component.

12.3.3 Conclusions

The recognition of the significance of the gray-colored trace on the sheet was a major key to the solution of this case. The full development of this trace deposit made it clear that the guarding officer had been deliberately attacked in a premeditated fashion totally contradicting the defense claim. When confronted with the consulting criminalist's reconstruction, the defendant plead guilty.

12.3.4 Lessons

As with the other examples presented in this chapter, the key to the case solution was not letting the scope of the physical evidence inquiry be circumscribed by others. Had the consulting criminalist only examined the blood traces, the key information for a reconstruction of the event would not have been possible. Furthermore, this case demonstrates the value of a generalist forensic scientist when examining the physical evidence. Ultimately, overspecialization and a narrow focus can result in critical evidence going unrecognized.

CHAPTER 13

Widely Held Misconceptions

In this chapter, several topics will be focused on that the authors feel require considerable discussion as they are and continue to be examples of inadequate scientific or ethical thought. Several of these concepts have contributed to poorly designed experiments carried out for research and/or ad hoc purposes relative to casework. It is not a quantum leap to infer that some of these ideas resulted in wrongful convictions or wrongful exonerations. It is disconcerting to think that much of what is iterated here will continue to negatively influence research and casework.

13.1 Blood Traces Produced by Gunshot Wounds

13.1.1 Introduction to Firearms and Wounding

In order to understand gunshot-produced blood traces, it is necessary to have at least a rudimentary knowledge of the types of firearms and the wounds they produce. There are three basic types. These are shotguns, high-powered rifles, and handguns. Basic shotguns are designed with a smoothbore to fire cartridges which contain an array of metallic pellets of a range of sizes. Shotgun wounds can vary from being extraordinary devastating at close range to being relatively minor due to pellet impacts from shots fired from great distances. The distant wounds may be from only a few pellets that have spread laterally and lost energy with distance from the muzzle. Extreme amounts of tissue destruction results from wounds where the muzzle distance is quite short. For contact or very close-range shots, massive tissue destruction is caused by both muzzle gases and the relatively dense pellet charge. With such shots, extensive amounts of blood and tissue can be deposited on nearby surfaces. The distinction between entrance and exit may be blurred.

Blood Traces: Interpretation of Deposition and Distribution, First Edition. Peter R. De Forest, Peter A. Pizzola, and Brooke W. Kammrath.

The ammunition for rifles and handguns is designed to fire a single projectile. The term "handgun" in our usage will be the classic one that would include all handheld guns including single-shot dueling pistols, revolvers, and "automatics" (semi-automatics or auto loaders). The discharge of a single projectile is also possible with "rifled slug" shotgun ammunition, although shotgun cartridges loaded with pellets are much more common for sporting and hunting shooting.

Some of the terms and nomenclature used in firearms have evolved over several hundred years and can be inconsistent and confusing. For example, most handguns have a rifled barrel but are not normally referred to as rifles. The interior of a rifled barrel contains helical grooves that are designed to impart an axial stabilizing spin to the bullets. For our purposes in considering the blood traces from firearms wounds, a distinction between handguns and "high-powered" rifles is important. This distinction relates to the velocity of the projectile. So-called "high-powered" rifles would include military rifles and hunting rifles. The projectile velocities of these rifles are vastly greater than the velocities of bullets issued from handguns. Velocities for projectiles from handguns are commonly subsonic or transonic. Those from high-powered rifles are hypersonic.

Wounds from hypersonic bullets fired from high-powered rifles can create extensive tissue damage from shockwaves in addition to the injury created by the physical passage of the bullet. Handguns can also create extensive damage to the cranium, typically only from close-range or contact shots. This is due to not only from the impact of the projectile but also the forces from the muzzle gases. Hypersonic bullets from high-powered rifles can create skull fractures and bone displacement even from shots fired from a distance.

Much confusion exists concerning bloodstain patterns produced by gunshots. This confusion can lead to misinterpretations in attempts to reconstruct past shooting events transpiring at crime scenes. Confusion on the part of laypersons, such as lawyers, judges, or the members of a jury, maybe at least be partially attributable to graphic media portrayals of shootings in film and television, where large volumes of blood immediately spurt out of the wounds. Confusion among experts charged with the responsibility of reconstructing shooting also exists. The latter may be attributable to poor training, poorly designed experiments, and inappropriately published photographs in training materials. For example, published photographs of a bullet traversing a blood-saturated sponge are very misleading (Pizzola et al. 1988; De Forest et al. 2015).

The blood spatter from bullets can be divided into three categories. These are forward spatter, backspatter, and blowback. There is confusion in published information on these that needs clarification. Forward spatter refers to droplet traces produced by material traveling down range from the exit wound. The term backspatter has been applied to two different phenomena resulting in the retrograde projection of blood from an entrance wound. The term backspatter can be applied to describing blood traces projected backward as a result of the bullet impacting coalesced liquid blood near the entrance wound. Stephens and Allen (1983) recognized blowback as a component of backspatter. This is the delayed retrograde emission of blood traces from the entrance wound caused by the exiting of gases that were initially projected into the wound by the gunshot discharge for contact or near-contact wounds. Following the disruption caused by the projectile, blood flows into the wound channel and is then available to be projected backward by the escaping gases. It does not take place for shots where the muzzle-to-target distance is more than a few inches, because at distances much beyond contact, no significant gunshot discharge gases may be forced into the wound.

13.1.2 Microvascularization and Experimental Laboratory Models

Experiments incorporating tissue simulants that more closely resembled the microvascular network (Wiedeman 1963) without the larger blood vessels would provide a more valuable and realistic insight into the production of gunshot spatter. Two different ways of achieving this have been utilized by two of the present authors, and others, as described below.

The first tissue simulant focused on the construction of simulated capillaries in gelatin. The basic idea was to use gelatin as a casting material surrounding fibers; the fibers would then be removed and then the vacated channels would be filled with blood to simulate blood in capillaries. To dissolve, the gelatin physiological saline was utilized to limit the amount of hemolysis of the red blood cells. The gelatin had another advantage since it had been used to simulate human tissue for wound ballistic studies. The use of several different size fibers was considered under different conditions and one can see the problem in trying to simulate actual capillary tissue channels when you consider that if you are going to produce capillary voids on the order of red blood cell diameters for a 1-l block of gelatin, you would need approx. 10 million fibers. Obviously, that is not practical. Thus, several different fiber diameters were considered as well as the percentage that was needed to approximate on the order of say 5–7%. It was determined that something practical would be about 0.2 mm fibers and selected polyester fibers of that diameter for this study. Such fibers are used commercially in the manufacture of paint brushes; Dupont Corporation provided the fibers in bulk. To arrange the fibers in a regular array, electroformed perforated nickel plates were used – two sheets of it. Specific plates were selected so that the percentage of fibers desired could be obtained in a cube of gelatin. The perforated plates used for this preliminary study possessed 60 k holes. The size of the holes was 0.3 mm in diameter and so not more than one fiber could fit in any one hole. The tips of about 14 K fibers (6 inches long) were fused, creating a bulbous head that prevented the fiber from falling out of the two plates, and dropped through the holes of the two superimposed nickel plates. One of the plates was used to hold down the gel, while the other was used to raise the fibers up out of the gelatin after solidification. A section of gelatin was prepared as indicated and dipped into a tray containing a shallow pool of blood (Figure 13.1a). By withdrawing the fibers while the bottom of the gelatin was submerged, blood flowed into the channels by a combination of vacuum (negative pressure creating by withdrawing the fibers) and capillary action (Figure 13.1b). Preliminary studies conducted with firearm discharges at muzzle distances of 61 cm utilizing small sections of gelatin infused with blood, like shown here in Figure 13.2, did not yield backspatter. A practical problem that obstructed further efforts was that when the number of fibers was increased, insufficient force could be exerted to remove the fibers from the gelatin block (Pizzola et al. 1988).

The second capillary tissue simulant was the narrow inner diameter hollow fibers used for dialysis purposes. Dialysis tubing cartridges constructed of a cylindrical jacket in which a large parallel bundle of hollow fibers (synthetic capillaries) was contained were used for experimentation (Diaczuk et al. 2002). In the medical application, two ports in the jacket allow a buffer solution to enter one port, bathe the exterior of the hollow fibers, and then flow out the other port while blood from the patient flows

FIGURE 13.1 For experiments involving blood trace deposits, a more realistic model of human tissue consists of capillary tubes embedded in ordnance gelatin. These pictures show both empty "capillaries" or channels in ordnance gelatin (a) and those that are blood-filled (b).

FIGURE 13.2 Blood-soaked polyurethane foam is a very poor model for gunshot experiments.

through the hollow fibers. The fibers had an approximate internal diameter of 225 μm and with a wall thickness of 15 μm; outside diameter was approximately 255 μm. Instead, the buffer solution in the circumferential area, in the dialysate compartment, was first rinsed with saline and then filled with liquid ballistic gelatin which congealed with refrigeration. A small window was then created, back and front, which allowed the passage of a bullet without engaging the rigid plastic outer covering eliminating unnecessary shock waves and debris (Figure 13.3). All firearm discharges were made at a distance of 6 feet. The dialysis cartridge was then exposed to a series of gunshots while being perfused with blood (Figure 13.4). When the capillaries of one of these dialysis cartridges were filled with blood and then used as a target in a shooting experiment, no droplets of blood impact the witness papers. Only bloodstained fragments of the capillaries were found by careful examination of the witness papers after the first discharge (Figure 13.5). In Figure 13.6, significant bloodspatter can be readily observed that was generated from subsequent firearm discharge to an area where blood has coalesced due to a previous discharge.

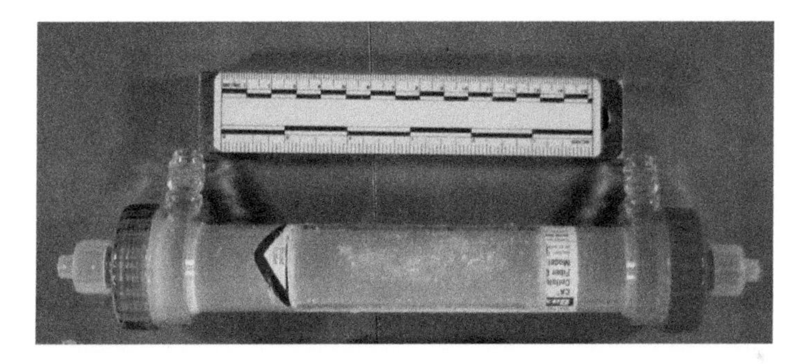

FIGURE 13.3 Dialysis cartridge with a section of the plastic cover removed from the central area exposing the "capillaries." Concept by Zvi Hershman.

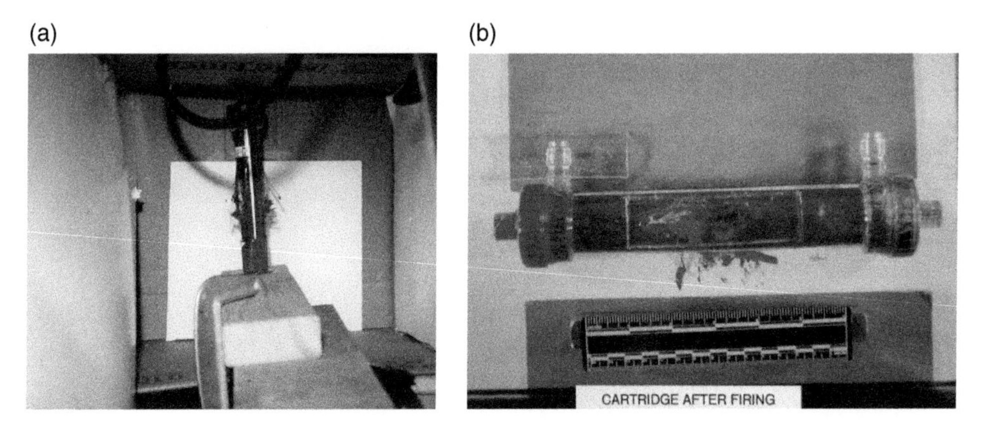

FIGURE 13.4 Blood-infused dialysis tubing post-firearm discharge, in the experimental setup (a), and a closeup view showing one of the windows cut out from each side to prevent secondary projectiles or interference from any fragments projected from the fractured polycarbonate cylinder (b).

FIGURE 13.5 Witness paper on the muzzle side of the firearm discharge to detect retrograde spatter. There were no droplet deposits of blood projected backward; however, there were some fragments of blood-contaminated hollow fibers in some of the experiments.

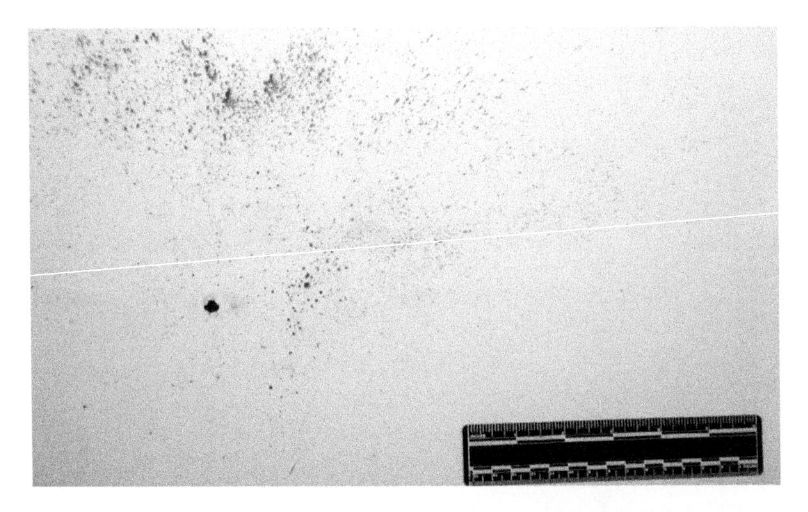

FIGURE 13.6 Witness paper on the muzzle side of the firearm discharge to detect retrograde spatter in a companion experiment to that shown in Figure 13.4. Subsequent discharge to the same area of the dialysis tubing where coalesced blood has accumulated following the earlier damage. Significant spatter has now been produced because of the impact to coalesced pool of blood.

13.1.3 Proposed Models and Their Failure to Consider Microvascular Structures

There has been relatively little discussion of the serious limitations of models used to simulate the human body. One of the overarching problems is the inability to demonstrate the equivalence of any of the models by direct comparison to human tissue, i.e., bone and/or soft tissue. Some researchers (Karger and Brinkman 1997; Karger et al.

1996, 1997a, b, 2002) have recognized the weaknesses of various models and attempted to circumvent those shortcomings by conducting experiments with livestock. While animal models do not provide the equivalence of human organs and tissues, they are vastly superior to crude inanimate models such as blood-soaked urethane foams (sponge) or styrofoam heads filled with blood. It would seem obvious to those who have thought long and hard about these models that experimentation with mammalian livestock would yield the most valuable and scientifically reliable information in comparison to humans albeit there are significant anatomical differences. Of course, the bones of other mammals have a different configuration, but bone is nevertheless bone; the major blood vessels are spatially different but perform the same function; livestock also have capillary and brain tissue that are similar to that of humans. We do not encourage the killing of animals for experiments. However, from a scientific perspective, there needs to be some fundamental understanding of the distribution of bloodspatter and tissue spatter (including brain, muscle and adipose) that can be reasonably relied upon as a standard to which results from experiments involving inanimate models can be compared. Otherwise, the results of experiments from inanimate objects, such as urethane foams, are naïve and inappropriate. Once a reasonable number of detailed studies have been done, comparing results from suitably designed models to results obtained with animals, then reasonable inferences or conclusions can be drawn about reliability. At a minimum, scientists must understand the limitations of the inanimate models and try to mimic the impact and passage of a bullet through models developed to simulate capillary tissue.

Despite the claims of some that urethane foam is an acceptable substitute for animal models, this assertion remains unsubstantiated. Some have even criticized Karger et al. for conducting research involving livestock (Augenstein 2015). Apparently, the critics either did not carefully read the Karger et al. studies or did not accept the results. The weaknesses of the urethane foam substitute can be explained from a theoretical perspective and its impact from a pragmatic real perspective. It is surprising to the extent that blood-soaked urethane foam have been and continue to be used to conduct experiments, in the study of gunshot phenomena (forward and backspatter), particularly with respect to discharges to the cranium, for both research and case studies in crime scene reconstruction. The utilization of urethane foam that is blood-soaked is not an adequate model for simulating human tissue (Pizzola et al. 1988). From elementary considerations of individual cell dimension and the mass of blood per unit volume of foam, it is clear that blood-soaked urethane foam is an exceedingly poor model for regions of tissue devoid of any blood vessels with dimensions significantly larger than capillaries and other fine vessels. Perhaps, because urethane foam is so inexpensive and convenient to use, it helps explain why its shortcomings are overlooked.

In many areas of the human body, all of the blood contained in that region is within dimensionally small arterioles, veins, venules, and capillaries. Approximately 50% of the blood volume in the body resides in these narrow tubular structures, with diameters as small as 5 μm and not in a large volume of coalesced blood. The dimension of a typical airborne blood droplet is substantially larger than the vessels described above, and droplets this large cannot be formed from blood contained in vessels with smaller diameters than the droplets themselves. Blood released from small vessels ruptured by the passage of the bullet cannot coalesce to form larger volumes during the short interval taken by the transit of the bullet. The blood released would be available to coalesce to form droplets if it is impacted by a second shot to the same area. Something in the vicinity of 90–95% of the weight of a blood-saturated foam is due to the blood contained therein. This is not a good experimental model for shots traversing many regions of the human body. Only about 5–7% of the weight of a human body is blood. In addition, the blood is not evenly distributed throughout the body. In most regions and tissues in the body, the blood is confined to fine vessels and capillary beds. The function of the fine diameter capillaries is to provide more intimate contact between the oxygen carrying red cells with the vessel walls and the somatic tissues dependent upon oxygen. Life-sustaining oxygen transport would be inefficient without hemoglobin and small diameter capillaries. The capillaries have inside diameters of approximately 5 μm. Some publications discussing blood spatter produced by gunshot wounds refer to droplet trace diameters of 1 mm as being diagnostic for gunshot-derived blood spatter. This dimension is roughly 200 times that of the capillary diameter. Thus, for a bullet traversing capillary supplied tissue, no coalesced volumes of liquid blood are encountered. The bullet may carry tissue and fragments of capillaries with it and deposit them on a surface near the exit wound (or even on a surface near the entrance wound). In this circumstance, there will be no droplets of blood deposited. In such cases, histological examination of the deposited tissue fragments has revealed fragments of capillaries with red cells lined up single file within them. We have encountered multiple cases in which little to no blood droplet traces were generated while concomitantly observing skin cells, hair follicles, muscle tissue, and even red blood cell filled capillaries, for example, that has been projected onto nearby surfaces (see Figure 13.7 below).

On the other hand, if the bullet encounters a volume of coalesced blood along its path, such as that contained in larger blood vessels or even the heart, fine droplets of blood can contribute to the spatter pattern. To reiterate, after a bullet has traversed capillary supplied tissue, blood will flow into the wound channel, coalesce to form larger volumes, and may be available for spattering during a follow-on shot to the same area. Clothing which becomes saturated with blood flowing from a wound may behave like the blood-soaked foam discussed above, if it is struck by another bullet. The point is that special conditions are required in order for gunshot to produce airborne droplet

FIGURE 13.7 Several gunshot wounds from a large bore shotgun from close range produced no bloodspatter, but in areas remote from the victim, various fragments of physiological tissue and pieces of the sweater were projected onto nearby objects.

patterns to be produced and observed. First, there must be a volume of coalesced blood along the bullet path to produce droplets. There must also be suitable nearby surfaces to record the configuration of any blood droplets that are projected from the wound for this to be detected.

Many authors have focused their attention on the phenomenon of backspatter on a deceased victim's hands, since it is commonly believed, incorrectly, that an absence of it demonstrates that the firearm discharge was made at a distance obviating the likelihood of a suicide. The term "backspatter" itself is ambiguous. The word "spatter" is essentially a term utilized to describe blood traces resulting from airborne droplets. When one considers all the physiological materials that may be generated from gunshot, these would include blood and tissue in various forms – including but not limited to muscle, brain, adipose, and bone. With respect to firearm discharges to the cranium, fragments of teeth may be included as well as those from hair, cartilage, vitreous and aqueous humor, spinal fluid and trace materials extraneous to the body. Obviously, firearm discharges to other aspects of the anatomy can produce other physiological tissue and fluids generated from the various organs. Albeit blood traces may not be present, the locations of physiological tissue particles may be quite valuable in reconstructions. Potentially, for example, the approximate position of a victim when wounded could be determined.

Unclad foam models, used by some, eject unrealistically large amounts of spatter when struck by bullets. Much of this is ejected from areas somewhat removed from the bullet entry hole. Many authors have and continue to advocate the use of sponge (MacDonell and Brooks 1977; Stephens and Allen 1983; Das et al. (2015)). It should be realized as well that merely using an inanimate model that possesses a morphology (such as hollowed out styrofoam heads) similar to a human cranium is an exceedingly naïve model. Other ill-conceived models have been designed to approximate the configuration of a human head; the shape of the object is largely irrelevant. These models ignore the microanatomy of the human head, such as its dimensions of vascularization, which is a critical issue in understanding phenomena such as backspatter or forward spatter. In 2015, Kunz et al. incorporated a sponge in a plastic bag as a simulant for studying backspatter and concluded the following:

> *Backspatter can be seen in contact or near contact gunshot discharge. Based on the presence or absence of blood on the shooting hand and gun, it is possible to distinguish between a close-range and a distant gunshot. Furthermore, the position and orientation of the gun and the shooter can be reconstructed by analyzing the intensity and direction of the backspatter. In consequence for the forensic investigator at a crime scene or during an autopsy, a careful bloodstain pattern analysis of backspatter findings on the shooting hand and weapon (or at least a careful documentation) is an essential tool to reconstruct the position of the weapon and to possibly distinguish between a homicide, suicide, and accident.*

It seems clear that Kunz et al. (2015) are assuming that if backspatter or blowback is absent, it can be safely concluded that the gunshot was not close range. Unfortunately, it is rarely safe to make these assumptions and make such interpretations. As previously discussed in Chapters 5 and 7 (Principle #4), an initial injury may not necessarily produce any immediate blood flow or the production of airborne droplets.

Das et al. (2015) evaluated a variety of models for the study of backspatter but also incorporated "sponge" to contain a blood substitute. The blood substitute was 100 ml of water to which three drops of food coloring were added. There was no apparent concern about the difference in viscosity compared to that of blood. One set of experiments, consisting of two sample runs, incorporated what the authors refer to as a "full cranium model" using silicone as simulant for skin, the sponge containing the blood substitute as described above, a skull/bone simulant in the form of polyurethane resin, and gelatin acting as simulant for brain tissue. They observed initial backspatter followed by secondary ballooning which generated additional backspatter. Das et al. also evaluated computer-driven simulations utilizing smoothed particle hydrodynamics – known as "SPH" and proposed that it is an "effective numerical method" for studying high-speed impacts. They evaluated medium-density fiberboard as a simulant for bone and found that although it fractures at the site of the bullet impact, it does not generate radial fractures (in contrast to the human cranium); but it is similar in behavior to bone, because of its brittleness, in the creation of "material backspatter." Another material they experimented with is polycarbonate and found that it does not generate any "material backspatter" nor "material fracture" and its ductile property is an inadequate substitute for bone. They also experimented with two skin simulants: natural rubber and Lorica leather (synthetic). In summary, they determined that the rubber's elasticity and resistance to tearing were too great making it unacceptable as a skin simulant. On the other hand, they found that the synthetic leather did tear like human skin and provided an entrance hole through which backspatter can be generated.

Das et al. (2015) stated that key findings of Karger et al. experiments (1996, 1997a, b) were that "... backspatter results varied with each shot despite a controlled environment, but the pattern was a consistent fine mist with every shot immediately after bullet impact". However, in a subsequent report authored by Karger et al. (2002), no backspatter was found on the hands, firearms, or right sleeve in two of nine experiments with livestock; in two other experiments one droplet was found on either sleeves but not on the firearm or gloves. The absence of spatter might best be explained by passage of the bullet through the cranium consisting of only fine vascular tissue which could have been confirmed by dissection. In the latter paper, Karger et al. concluded that "The presence of backspatter stains establishes a clear link between a person or object and a clearly defined gunshot, especially if the stains are individualized by DNA analysis... However, *no conclusions can be drawn from the absence of backspatter stains on the firearm and hands, which was the case in two of nine gunshots* [italics added for emphasis]." Regarding instances where backspatter is not found on the firearm or hands, the authors advised the careful examination of trousers and footwear of the suspect. The authors examined the firearms with a magnifier at the site where the discharges were made and used stereromicroscopy in the laboratory to study the gloves and sleeves of the shooter. The utilization of microscopy to detect blood traces is a highly recommended practice. However, we are critical of the degree of certainty associated with the attribution "backspatter stains."

The variation that Karger et al. observed is extremely important – it should not be cavalierly ignored as it may be a manifestation of the non-homogenity of the cranial media (large arteries vs. capillaries) through which the projectile has penetrated or perforated. Similar results might be obtained from humans depending on a variety of factors. The researchers' results would have been more meaningful, if the crania had been examined by a forensic pathologist postmortem to determine exactly what kind of tissue was disrupted by the projectile.

Schyma et al. developed what they refer to as the "Triple Contrast" method for studying backspatter and wound ballistics (2015). In summary, the authors mixed freshly drawn blood without anti-clotting agent, acrylic paint, and barium sulfate. The liquid mixture was placed into a thin foil bag and attached to the outside of a plastic bottle which had been filled with ordnance gelatin. The entire object was then covered with "a layer of silicone" (2–3 mm thick). In the discussion section of the paper, the authors admit that they are aware of the experiments conducted by Karger with livestock as well as Radford; the latter author experimented with livestock as well as severed pig heads. However, Schyma et al. claim that both Karger et al. and Radford's works, in addition to being complex, because of anatomical differences with humans, are of "...only limited informative value." It is perplexing that Schyma et al. apparently made no effort to measure any physical properties of their experimental mixture such as surface tension, viscosity, or density. Moreover, the proportions of the three components are not provided. The authors recognized that they might be criticized regarding the question of physical properties and preemptively noted "...however, the viscosity of the triple mixture is astonishingly low when compared to the pasty acrylic paint. This might explain the successful emulation of backspatter in- and outside of the weapon." Schyma et al. also refer to prior publications with ballistic models using this foil bag model, of which Schyma is a co-author on both. One of the articles was published in the same journal as the paper being discussed here and the other in *Forensic Science International* (2014). The assertion by Schyma et al. that Radford's research is of "limited informative value" must be evaluated. Although the former authors are correct that there are significant anatomical differences between humans and pigs, it would seem that a pig head is an exceedingly better model for experimentation in comparison to most of the synthetic models contrived to date.

Radford et al. (2015) specifically chose to experiment with pig models because of the claimed similarities of the properties of their bone and skin to that of humans, as well as availability for study. They conducted experiments with both live and slaughtered pigs. The live pig experiments utilized five adult specimens after appropriate sedation, application of anesthetic gas, and administration of IV fluid. To attenuate the influence of muzzle gas for experiments conducted with three of the live pigs, the muzzle-to-target distance was 1.2 m.

One of the notable findings of the Radford et al. (2015) experiments was the production of what the authors deemed "secondary bloodspatter." In essence, they fired three rounds at the live pig's cranium. The first discharge was contact; the second and third discharges were made from 1.2 m. All experiments were documented during and after discharge via high-speed video. They reported that there was a substantial blood flow from the first entrance wound a few seconds post-discharge. The subsequent shots were made while blood was still flowing from the entrance wound (up to 15 min later). The second discharge did not impact the area where the blood was flowing, since it struck an area above the first wound site. However, the third discharge impacted an area of surface blood below the first site and generated substantial "backspatter" of blood in a conical distribution. Also, when the third discharge struck the cranium, blood issued from the second site but in a much smaller amount. The production of substantial backspatter from a projectile impacting a flow of blood is not at all surprising and should be anticipated. Since the slaughtered pig heads were not reinfused with blood, those experiments were confined to the study of the retrograde ejection of both soft tissue and bone. According to the authors, in summary, "A small amount of backspattered material was produced with all targets, and blood backspatter was seen in a few cases." As we have previously mentioned regarding these kinds of experiments with livestock, it would have been of additional value to have examined the cranium to determine what kind of tissue the bullet perforated, i.e., capillary-rich tissue, larger diameter vascular tissue, or a combination of both.

Firearm discharges to the sponge model were compared to a "Complex Model" by Rubio et al. (2014). Not surprisingly, they determined that the sponge model was inadequate for studying backspatter. The complex model, that Rubio et al. utilized incorporated the resin calvarias, and was based on the work of Thali et al. (2002a, b), Jussila (2004) and Jussila et al. (2005) which incorporated resin calvarias. Reportedly, this reference material is very similar to the human cranium with respect to both morphological and mechanical properties. Rubio et al. also used cowhide and ordnance gelatin as substitutes for skin and brain tissue, respectively. Their model also included sponges of different thicknesses to hold the blood (pig's blood with heparin as an anticoagulant). The muzzle-to-target distance was 2 m to obviate gas effects on the production of backspatter. Their experiments were videotaped at very high speed (420 k/s) permitting them to identify three different time frames when backspatter is formed. Based on their observations, it was their opinion that the results more closely mimicked casework compared to the data obtained when conducting experiments with common sponge. Finally, they agreed with Stephen and Allen that common sponge is inadequate model for backspatter studies and that they intend to experiment with biological models for an improved comprehension of the phenomenon of backspatter.

Comiskey et al. (2016) proposed a theoretical model for the interpretation and prediction of blood traces resulting from gunshot wounds – backspatter; the numerical solution of the theoretical equations predicted the Weber number for trace formation, angle of impact, stain morphology, and distribution, etc. The theoretical information obtained was compared by the authors to experimental data obtained from sponge soaked with blood. The authors assumed "...that there is no significant difference whether blood is being emitted from a free surface or from a highly porous sponge ..." We would agree with this assumption if blood is residing in a preexisting pool on the surface of the skin, or clothing, that has been created from a previous injury. It should be obvious that successive discharges to the same region of the cranium will encounter coalesced blood that has pooled and generate what is in essence impact spatter. However, if the projectile does not impact a preexisting pool and only passes through capillary tissue, little to no backspatter may be produced and we cannot agree with this assumption under that condition.

A novel approach to evaluating blood-soaked sponge as model for the human cranium, in the context of backspatter, was introduced by Rossi et al. (2018) as a technical note by reinfusing the crania of two human cadavers. Unfortunately, this study was severely limited because of a number of factors. Having only two cadavers available for experimentation, in addition to the inability to adequately reinfuse one cranium, is a serious obstacle. Furthermore, the authors did not examine the brain tissue post-discharge. This was necessary for two reasons. First, in order to determine what type of vascularization the bullet passed through, i.e., large vessels (veins or arteries) or merely capillary-rich tissue? Secondly, the vascular tissue should have been examined post-discharge to ensure that hemolysis had not occurred; if hemolysis had occurred, it is likely that ruptured red blood cells entered tissue external to the capillaries.

Moreover, some vascular tissues may have ruptured prior to reinfusion because of decomposition allowing blood to reside outside the vessels. If this actually happened, then the bullet is impacting essentially a preexisting pool of blood in stark contrast to interacting with finely separated blood cells distributed in capillary tissue. According to the authors "The fact that Donor #2 produced very little backspatter is likely the result of the low amount of blood reinfused" – the little backspatter was attributed to the volume of blood infused rather than another possibility of the fine subdivision of the capillaries and other fine blood vessels. Based on the aforementioned weaknesses of these limited experiments, little can be inferred from their results. However, the concept represents a more realistic way to evaluate various models that have been proposed as substitutes for the human cranium – assuming of course if its inherent issues can be eliminated.

In summation, the scientist should not expect forward or backspatter unless dealing with a contact or near-contact discharge involving blowback. For future studies, experiments that capture the idea of a firearm discharge into simulated capillary tissue infused with blood that have obviated the mechanical obstacles that were experienced as described above regarding experiments with "capillaries" in ordnance gelatin would be highly advantageous. One conceivable way to do this would be to create narrow channels in gelatin with lasers, 3D printing, or by vasculogenesis in a collagen–fibrin gel (Lee et al. 2014; Peterson et al. 2014).

13.2 The "Normal Drop" Claim

Based on limited and questionably designed experiments, MacDonell and Bialousz, in 1971, proffered the concept of a "normal drop" in their popular LEAA pamphlet. It was not merely an academic issue that this concept was presented, since its existence permitted the further incorrect extrapolation that one could then use the deposit diameter to determine the height of fall of the drop that produced the trace. This notion was further propagated by Bevel (1983) in the FBI Bulletin and still later by others. Since scientists did not question the validity of this normal drop claim, it is hard to fault non-scientists for being misled.

It should be clear that it is not enough to conduct a series of "experiments" that point the worker in a particular direction. There must be a well thought out experimental design that poses relevant questions to be addressed. The MacDonell experiment clearly did not achieve this goal. It did not address the relevant variables to be considered such as a range of pipet tips and inevitably provided an answer to the wrong question. To understand what happened and how not to design a series of experiments, we need to examine what MacDonell actually did. MacDonell and Bialousz stated that most physical and physiological properties of blood have been reported, but the volume of a single drop of blood had not been found in the literature and thus was determined by them experimentally (p.3, 1971). They continued

> *Volume of a single drop of blood was determined as 0.05 ml using a simple volumetric technique. The 95% confidence limits for this value are 0.0495–0.0516. High speed photographs of single drops were taken as an independent method and confirmed this range.*

It can be seen that their initial reasoning was faulty to begin with. The existence of a "normal" drop was unquestiongly assumed. What made them think that there was such a thing as the volume of a drop of blood? This kind of reasoning presumes the existence of a certain drop volume prior to conducting any experiment. Then if we critically assess how they conducted their experiments, we readily see a major flaw in the design. Recall that they used a simple volumetric technique. That they failed to mention how they did this is another major issue with this claim. It was not until two years later that they provided some sketchy details of how they carried out these experiments in a laboratory manual (MacDonell and Bialousz 1973). Essentially, what they did was to produce single droplets gently with an unspecified glass volumetric pipette, collect them together in a glass beaker, and weigh them in toto. It is then a simple matter to estimate the average drop volume. They did not consider the formation of drops produced by pipettes with a range of tip configurations or orifice sizes nor did they consider pipettes composed of plastic. By designing a series of experiments using the same type of the glass pipette, they locked themselves into obtaining the same result consistently within the same range. The high-speed photography would add no new information, since the drops they photographed were formed by the same exact process. The sources of error and uncertainty with this method are many. It is intriguing to note that MacDonell years later claimed that he had trouble seeing blood drops photographically and had to use milk (2011).

13.3 MacDonell Priority Claims Relative to the Seminal 1939 Balthazard et al. Paper

The work of Balthazard et al., published in 1939, predates the work of MacDonell by three decades. MacDonell claimed on a number of occasions that he "discovered" or "found" the Balthazard paper (1992, 1995, 2011). Such claims are unsubstantiated by the facts. In 1971, MacDonell co-authored his first LEAA report but did not cite Balthazard et al. Eleven years later, he published a revision of the LEAA report entitled "Bloodstain Pattern Interpretation," and again did not cite the Balthazard paper. It was not until 1992 that he cited it in the IABPA's newsletter but claimed that he was aware of the paper in 1982, if not as early as 1980. In the United States, John Thornton cited the Balthazard paper in 1975. A forensic scientist must be careful to not make exaggerated claims as it lessens one's credibility and casts doubt on one's integrity. De Forest and Pizzola (1995) provides additional information on the history of this topic.

13.4 The Claimed Equivalence of Deposits Diameters and Drop Diameters

It should never be assumed that a deposit diameter is equivalent to the drop diameter as has been claimed by some and noted below. To make such an assumption, it may cause an examiner or researcher to form an unsubstantiated incorrect hypothesis.

In most instances, it is likely that the trace diameter will be substantially larger than the drop diameter from which it originated. In other cases, as with hydrophobic substrates, the resulting trace may be much smaller than its maximum spread at impact. In the latter situation, during the drop collapse, the periphery will spread out substantially but then retract since there is little affinity between its surface and the substrate. It may even bead up into a sphere-like shape ultimately being considerably smaller than its original diameter upon the initial impact. The elliptical trace size is clearly related to the diameter of the drop from which it originated. This relationship is not a direct linear one. With respect to oblique droplet impacts involving non-hydrophobic substrates, spreading occurs dramatically in the direction of travel of the collapsing droplet as well as, to a lesser extent laterally, along its minor diameter (width). MacDonell's model for the production of elliptical stains made this inference: "Width of the bloodstain ellipse A', B' is identical to the diameter of the blood drop sphere A, B" (1982, p. 52). Subsequently, Carter stated that "The projection of the spherical drop onto the inclined plane is an ellipse of width W, equal to the diameter of the drop" and related that "It is assumed for the purpose of this discussion, that the blood droplet, upon striking the inclined plane with the impact angle α degrees, will produce a stain that resembles the projected ellipse whose width is equal to the diameter of the spherical drop" (in James (ed), 1999, pp. 20–21). These reports, and others, coupled with other concepts, apparently led Buck et al. (2011) to decide that they could use trace diameter to infer drop diameter as an integral aspect of their proposal for ballistic reconstruction of blood drop trajectories for the purpose of "...determining the centres of origin of the bloodstains." Pizzola et al. (2012), in a commentary, challenged the validity of the assumption of equivalence between trace diameter and drop diameter.

In the case of oblique impacts, the degree of spreading is not symmetrical (see vector discussion in Chapter 6). Simple experiments, such as conducted by Marin et al. (2011), amply demonstrate why this is of concern. For example, at a seven-foot height of fall, a 25-μl drop (3.60 mm D) onto paper at an oblique impact angle of 50° produced on average a trace diameter (width) of over 14 mm. This represents an increase of approximately 300% of the trace diameter vs. the drop diameter. The fact that when droplets collapse on impact that there is dramatic spreading is nothing new in either the milieu of fluid dynamics or criminalistics. Thus, to use the trace's width for back projection into three-dimensional space for the purpose of reconstructing the area of origin is a concept with which we need to be concerned.

13.5 Ambiguous Trace Configurations

13.5.1 Configuration Issues

Single droplet traces are, in general, readily recognized and documented in terms of travel (left vs. right) and the approximate angle of incidence of the droplets that produced them. In many instances, too, the appearance of hundreds of these traces as a group, comprising overall distribution configurations, are easily and accurately recognized. These may include trail, arc, arterial, and shooting configurations (backspatter and forward spatter). However, in many cases, the origin of a large group of traces is uncertain. One of the most common errors is for the examiner to assume that a group of traces present on a vertical or horizontal surface is the direct and nearly instantaneous result of some force or act exerted by an assailant on the victim. The great potential for misinterpretation to occur in cases involving many wounds can be learned from those cases in which victims were only shot or struck once; yet, in many of these cases, we see what is ostensibly several different locations from which large groups of traces came that are clearly many inches or even feet apart. If the victim was only struck once and has only a single entrance wound (and no exit wound), how can there be more than one site where groups of deposits originated from? In these specific instances, it is clear that one pattern is more clearly associated with the act itself, and the others occurred subsequently – at times many seconds or even minutes later. These ambiguous patterns may be the result of blood being projected, falling or dripping directly from the wound site for a considerable time after the wound was acquired. Also, we may see evidence of secondary spatter that may be overlooked.

If there is more than one wound site, which is often the case, then it would seem that our confidence in the origin of various patterns should be less secure than in the single wound case.

The following serves to illustrate how misinterpretations can occur with multiple wound sites albeit this case involves only one wound – an entrance wound from a single gunshot. The victim was shot in the back of the head while apparently standing as clearly demonstrated by the presence of bloodspatter deposits at that height on the left door jamb and wire rack. However, on the same door jamb, at about 3 feet from the floor are blood traces that originated from that much lower height along with numerous blood traces along the adjacent wall that clearly originated from that same approximate position (36″ from the floor). But if there is only one wound, how can blood droplets be projected from a position at about 69″ from the floor and from another position at approx. 36″ from the floor? Both sets of traces could not have been created at the same exact time. After careful analysis of the blood traces and a discussion with the forensic pathologist who conducted the autopsy, it became clear that the victim was shot while standing, producing the traces at approximately 69 inches, and then fell with his head in close proximity to the door jamb; as this occurred, additional blood droplets were projected from the wound site producing a set of traces on the lower portion of the door jamb and on the adjacent wall. This conclusion was not very difficult to reach because of the distribution of deposits at the much higher position and the fact that there was only one wound site. Interpretation of the distribution patterns could have been much more complex, if there had been additional sites for blood and/tissue to be projected from, such as penetrating wounds.

13.5.2 Fabric Issues

For many years, blood traces on various fabrics have presented forensic scientists with a number of issues. One of the most prominent problems, often encountered in casework, is the difficulty in distinguishing between contact transfers and airborne droplet deposits. Several cases that we have discussed in this book are examples of this issue such as the David Camm case (Chapter 11) and the Passive Documentation Case (Chapter 10). A stereomicroscope is an excellent tool for making this distinction because airborne droplet traces will appear on both the high and low points of a fabric's weave, while blood from a contact transfer are more likely to appear on the high points only. As a caveat, this is dependent on the contact pressure, which can compress the high points and make transfer to the low parts possible. When taking photomicrographs, the illumination angle and the use of polarizing filters (discussed in Chapter 4) can minimize glare. The scientist should also look for signs of the disruption of the blood droplet upon impact with the surface of the textile, such as satellite droplets.

It is also apparent from Figures 13.8–13.11. that the original deposition configuration of blood traces can become dramatically different as the diffusion within a fabric takes place (Reynolds and Silenieks 2016). Diffusion is dictated by the structure and composition of the fabric. For example, in fabrics where the warp and weft are composed of different fiber types, the diffusion of blood may be preferentially directed. The same holds true when the tightness of weave varies with warp and weft. The lateral spreading of blood within a fabric, in the direction of the warp and weft in addition to within the depth of the textile, should be carefully examined and interpreted. This can be used to inform conclusions pertaining to the side from which the deposit is made. It has been reported that 100% polyester fabrics generate the greatest directional asymmetry and diffusion and thus the most striking changes in blood traces (Holbrook 2010, p. 11). Another pressing problem with traces on fabric involves ambiguities (or even pareidolia, Chapter 7) of presumed knife traces and folded fabric.

In many cases, blood traces are observed on bedding or garments that resemble the shape of the planar surface of a knife blade. Unfortunately, what these configurations appear to be may not represent reality. Very sharp longitudinal edges to these configurations often arise from dynamic contact transfer of some object, such as a bloody hand, across the surface of the fabric while it is folded and then unfolded as seen in Figure 13.12. It should be noted too that there is no indication of an image of a hand in the resulting configuration.

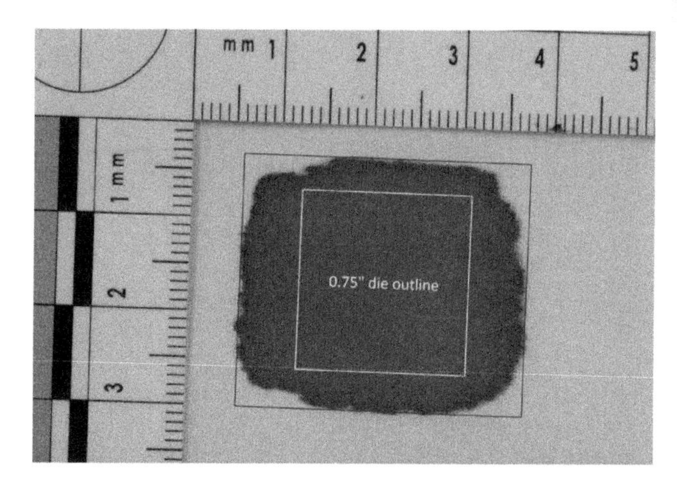

FIGURE 13.8 Diffusion of a contact transfer on Broadcloth (65/35 cotton–polyester blend). The original area of the transfer, made with a 0.75 inch metal die, is depicted by the square outlined in yellow superimposed over the ultimate trace image. *Source: Photograph courtesy of Swathi Murali, Graduate Program, Pennsylvania State University.*

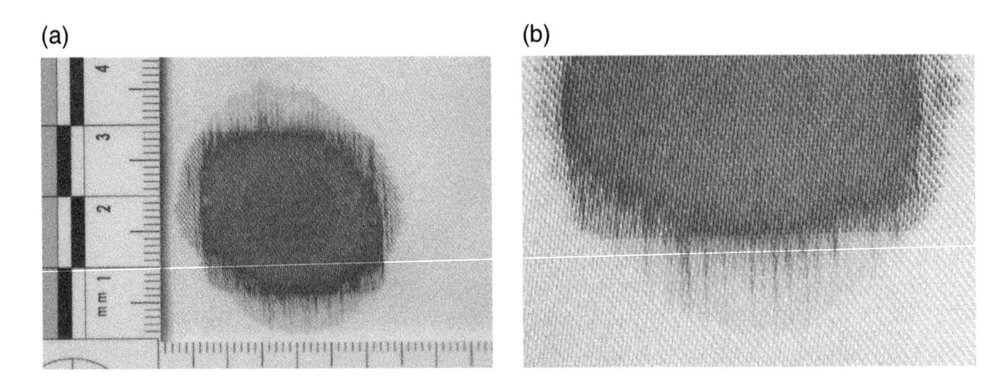

FIGURE 13.9 Diffusion of a 45-μl droplet that fell 50 cm onto satin (polyester composition) (a); closeup view of lower periphery of same trace (b). *Source: Photographs courtesy of Swathi Murali, Graduate Program, Pennsylvania State University.*

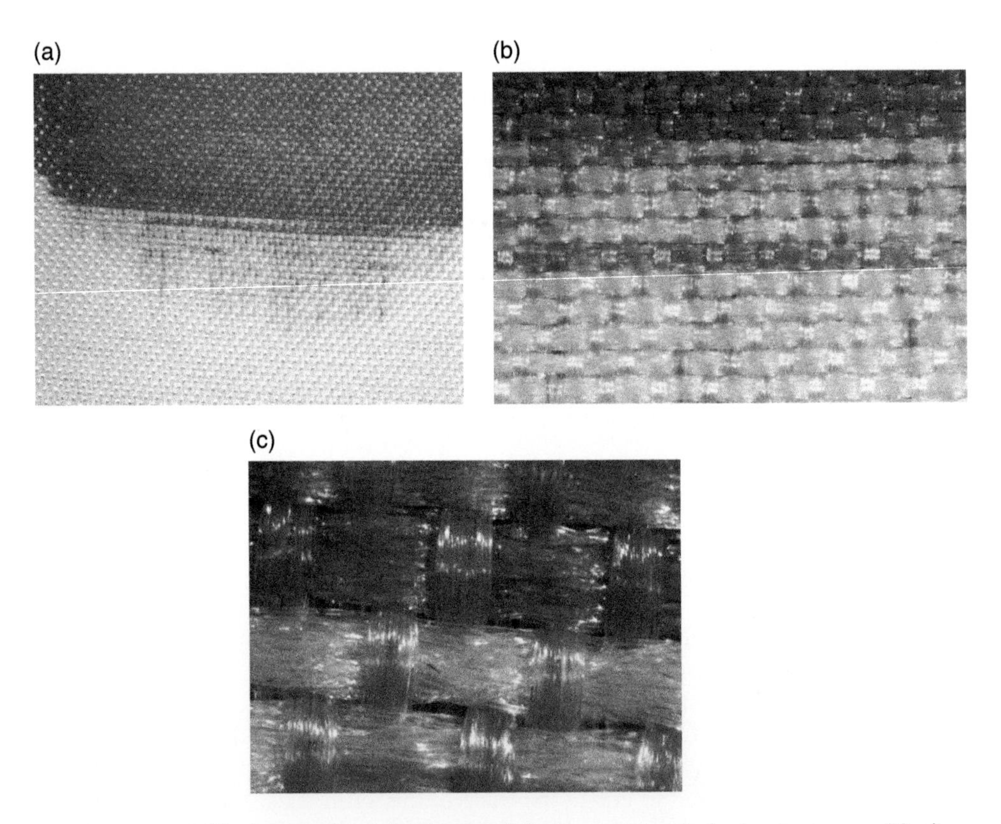

FIGURE 13.10 Diffusion of a 45-μl droplet that fell 10 cm onto satin (polyester composition) as viewed with a stereomicroscope at 8 times magnification (a), the lower periphery at 30 times magnification (b), and at 80 times magnification (c). *Source: Photomicrographs courtesy of Swathi Murali, Graduate Program, Pennsylvania State University.*

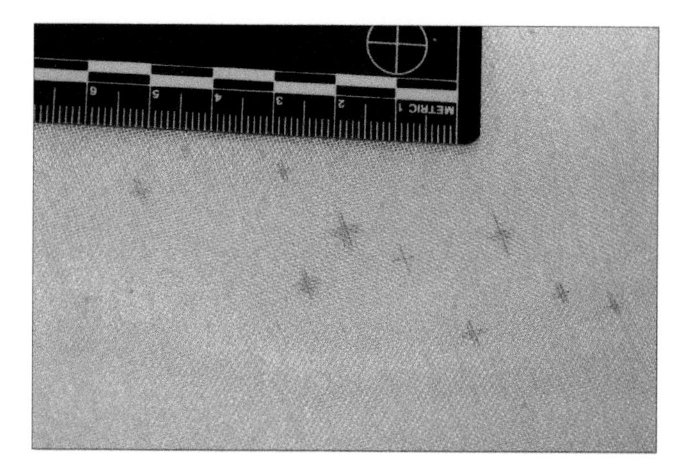

FIGURE 13.11 Differential diffusion in polyester–cotton blend. *Source: Photograph courtesy of Norman Marin & Jeffrey Buszka.*

FIGURE 13.12 In the above illustration, it might be tempting to conclude that the above blood trace deposit was made by the knife, but that would be incorrect. The above blood trace was produced by moving a bloody gloved hand across the folded fabric which was subsequently unfolded.

Blood traces shown in both images in Figure 13.13 originated from contact between the same bloody knife and the fabric. The left is from static contact transfer to flat fabric, while the right image is from dynamic contact transfer, i.e., movement across the fabric while it was folded and then unfolded. Without being told, it would be somewhat difficult to conclude that the traces in the right image originated from a knife – in fact the same one that was used to create the trace in the left image. As previously noted and shown in Figure 13.12, transfer traces from other objects coupled with folds in fabric may produce traces that resemble a knife blade. From a case perspective, even when there is an evidence deposit showing a clear pattern, such as that in Figure 13.13, extreme care must be taken when drawing a conclusion as to its source.

(a) (b)

FIGURE 13.13 Are these traces from two different objects? No. Both of these blood traces came from the same knife, where the first is a static contact transfer (a) and the second is a dynamic contact transfer that includes traces of the serrations of the knife blade (b).

13.6 Issues with Interpretation of Asymmetrical Blood Projections from Impacts

It should never be assumed that the projection of blood from a wound impact site is always symmetrical, as discussed in Chapter 5. Because of this, traces are not always deposited where one might expect. The cause of a lack of blood traces may be ambiguous. Some of this may be due to the presence of intervening objects creating voids. Alternatively, assymetry may also result from the manner in which the assault is occurring, the configuration and the nature of the wound among other factors.

References

Augenstein, S. (2015). *Studying gunshot backspatter by shooting live pigs. Forensic Magazine.* http://www.forensicmag.com/articles/2015/09/studying-gunshot-backspatter-shooting-live-pigs (accessed 21 October 2015).

Balthazard, V., Piédelièvre, R., Desoille, H., and Dérobert, L. (1939). Étude des Gouttes de Sang Projeté. Annales de Médecine Légale de Criminologie, Police Scientifique, Médecine Sociale, et Toxicologie 19: 265–323.

Bevel, T. (1983). Geometric bloodstain interpretation. F.B.I. Law Enforcement Bulletin 52 (5): 7–10.

Buck, U., Kneubuehl, B., Näther, S. et al. (2011). 3D bloodstain pattern analysis: ballistic reconstruction of the trajectories of blood drops and determination of the centres of origin of the bloodstains. Forensic Science International 206 (1–3): 22–28.

Carter, A. (1999). Bloodstain pattern analysis with a computer. In: Scientific and Legal Applications of Bloodstain Pattern Interpretation (ed. S.H. James), 20–21. Boca Raton, FL: CRC Press.

Comiskey, P.M., Yarin, A.L., Kim, S., and Attinger, D. (2016). Prediction of blood back spatter from a gunshot in bloodstain pattern analysis. Physical Review Fluids 1 (4): 043201.

Das, R., Collins, A., Verma, A. et al. (2015). Evaluating simulant materials for understanding cranial backspatter from a ballistic projectile. Journal of Forensic Sciences 60 (3): 627–637.

De Forest, P.R. and Pizzola, P.A. (1995). Letter to the editor. Journal of Forensic Science, Response to Letter to the Editor by Herbert MacDonell, November 1995 40 (6): 928–931.

De Forest, P.R., Pizzola, P.A., Ristenbatt III, R.R., and Diaczuk, P.R. (2015). *Bloodstain patterns associated with gunshot wounds – misconceptions. Proceedings of the 67th Annual Meeting of the American Academy of Forensic Sciences*, Chicago, IL (February 2003).

Diaczuk, P.J., Herschman, Z., Pizzola, P.A., and De Forest, P.R. (2002). *A new experimental model for evaluating mechanisms of gunshot spatter. Proceedings of the 99th Semiannual Seminar of the California Association of Criminalists*, San Francisco, CA (7–11 May 2002).

Holbrook, M. (2010). Evaluation of blood deposition on fabric: distinguishing spatter and transfer stains. IABPA Newsletter 26: 3–10.

Jussila, J. (2004). Preparing ballistic gelatine–review and proposal for a standard method. Forensic Science International 141 (2–3): 91–98.

Jussila, J., Leppäniemi, A., Paronen, M., and Kulomäki, E. (2005). Ballistic skin simulant. Forensic Science International 150 (1): 63–71.

Karger, B. and Brinkmann, B. (1997). Multiple gunshot suicides: potential for physical activity and medico-legal aspects. International Journal of Legal Medicine 110 (4): 188–192.

Karger, B., Nüsse, R., Schroeder, G. et al. (1996). Backspatter from experimental close-range shots to the head I. Macrobackspatter. International Journal of Legal Medicine 109: 66–74.

Karger, B., Nüsse, R., Tröger, H.D., and Brinkmann, B. (1997a). Backspatter from experimental close-range shots to the head II. Microbackspatter and the norphology of bloodstains. International Journal of Legal Medicine 110: 27–30.

Karger, B., Rand, S.P., and Brinkmann, B. (1997b). Experimental bloodstains on fabric from contact and from droplets. International Journal of Legal Medicine 111 (1): 17–21.

Karger, B., Nüsse, R., and Bajanowski, T. (2002). Backspatter on the firearm and hand in experimental close-range gunshots to the head. The American Journal of Forensic Medicine and Pathology 23 (3): 211–213.

Kunz, S.N., Brandtner, H., and Meyer, H.J. (2015). Characteristics of backspatter on the firearm and shooting hand—an experimental analysis of close-range gunshots. Journal of Forensic Sciences 60 (1): 166–170.

Lee, V.K., Lanzi, A.M., Ngo, H. et al. (2014). Generation of multi-scale vascular network system within 3D hydrogel using 3D bio-printing technology. Cellular and Molecular Bioengineering 7 (3): 460–472.

MacDonell, H.L. (2011). Credit where it's due. Journal of Forensic Identification 61: 210–221.

MacDonell, H.L. and Bialousz, L.F. (1973). Laboratory Manual on the Geometric Interpretation of Human Bloodstain Evidence. Painted Post, NY: Laboratory of Forensic Science.

MacDonell, H.L. and Brooks, B.A. (1977). Detection and significance of blood in firearms. In: Legal Medicine Annual 1977 (ed. C.H. Wecht), 185–199. New York: Appleton-Century-Crofts.

Marin, N., Buszka, J., and Pizzola, P.A. (2011). *Measurement of the spread of blood droplets on impact.* Unpublished experiments conducted at NY OCME, Special Investigations Unit.

Peterson, A.W., Caldwell, D.J., Rioja, A.Y. et al. (2014). Vasculogenesis and angiogenesis in modular collagen–fibrin microtissues. Biomaterials Science 2 (10): 1497–1508.

Pizzola, P.A., Perkins, J.C., Kodet-Sherwin, L., and De Forest, P.R. (1988). *A critical assessment of the phenomenon of gunshot backspatter. Proceedings of the 40th Annual Meeting of the American Academy of Forensic Sciences*, Philadelphia, PA (February 1988).

Pizzola, P.A., Buszka, J.M., Marin, N. et al. (2012). Commentary on "3D bloodstain pattern analysis: ballistic reconstruction of the trajectories of blood drops and determination of the centres of origin of the bloodstains" by Buck et al.[Forensic Sci. Int. 206 (2011) 22–28]. Forensic Science International 220 (1–3): e39–e41.

Radford, G.E., Taylor, M.C., Kieser, J.A. et al. (2015). Simulating backspatter of blood from cranial gunshot wounds using pig models. International Journal of Legal Medicine 130 (4): 985–994. https://doi.org/10.1007/s00414-015-1219-s.

Reynolds, M. and Silenieks, E. (2016). Considerations for the assessment of bloodstains on fabrics. Journal of Bloodstain Pattern Analysis 32: 15–20.

Rossi, C., Herold, L.D., Bevel, T. et al. (2018). Cranial backspatter pattern production utilizing human cadavers. Journal of Forensic Sciences 63 (5): 1526–1532.

Rubio, A., Esperança, P., and Martrille, L. (2014). Backspatter simulation: comparison of a basic sponge and a complex model. Journal of Forensic Identification 64 (3): 285.

Schyma, C., Lux, C., Madea, B., and Courts, C. (2015). The 'triple contrast' method in experimental wound ballistics and backspatter analysis. International Journal of Legal Medicine 129 (5): 1027–1033.

Stephens, B.G. and Allen, T.B. (1983). Back spatter of blood from gunshot wounds—observations and experimental simulation. Journal of Forensic Science 28 (2): 437–439.

Thali, M.J., Kneubuehl, B.P., Zollinger, U., and Dirnhofer, R. (2002a). The "skin–skull–brain model": a new instrument for the study of gunshot effects. Forensic Science International 125 (2–3): 178–189.

Thali, M.J., Kneubuehl, B.P., Zollinger, U., and Dirnhofer, R. (2002b). A study of the morphology of gunshot entrance wounds, in connection with their dynamic creation, utilizing the "skin–skull–brain model". Forensic Science International 125 (2–3): 190–194.

Wiedeman, M.P. (1963). Dimensions of blood vessels from distributing artery to collecting vein. Circulation Research 12 (4): 375–378.

CHAPTER 14

Resources

14.1 Bloodstain Pattern Analysis Groups

The forensic science community recognized the need for consensus and standards dating from the late 1960s and thus initiated the formation of groups of interested parties (practitioners and educators). The earliest of these was American Society for Testing and Materials (ASTM) E-30 on Forensic Science that was created in 1970 under the pre-existing ASTM which was founded in 1898 with the goal of establishing standards and analytical methods for the manufacture and testing of industrial materials. Subsequently, the National Institute of Justice formed Technical Working Groups (TWGs) which broadened the scope of standard building to include areas of forensic science outside those of strictly analytical methods, such as education and training. In addition, the purpose of TWGs was "the establishment of professional forums in which local, state, and federal government experts, together with academic and commercial scientists, can operate to address operational issues arising within specific forensic disciplines" (SWGSTAIN, 2010). The bloodstain pattern TWG consisted of 20–30 practitioners of bloodstain pattern analysis from a variety of local, state, and federal agencies in addition to those from other countries and academia. In the early 1990s, the Federal Bureau of Investigation (FBI) took over the function of the TWGs and transitioned them into Scientific Working Groups (SWGs), with the continued goal of improving forensic science practices and creating agreed upon standards for practitioners. The first government-funded bloodstain pattern analysis group was a SWG called SWGSTAIN that was formed in 2002. Inexplicably, the title of the bloodstain pattern SWG was a misnomer due to the inclusion of the word "scientific" because the membership consisted of many non-scientists.

In addition to government-sponsored bloodstain pattern analysis groups, there are several associations of practitioners. Some of these groups, which will be discussed in more detail later in this chapter, have a general focus on criminal investigation and physical evidence. The first professional membership organization devoted to the field of forensic identification is the International Association for Investigation (IAI) and was founded in 1915 in Oakland California (http://www.theiai.org). In addition, there is one association, the International Association of Bloodstain Pattern Analysts (IABPA), which is specific to practitioners of bloodstain pattern analysis that was formed in 1983.

In 2009, the National Research Council of the National Academy of Sciences (NAS) issued a report entitled "Strengthening Forensic Science in the United States: A Path

Blood Traces: Interpretation of Deposition and Distribution, First Edition. Peter R. De Forest, Peter A. Pizzola, and Brooke W. Kammrath.

Forward" was highly critical of some aspects of forensic science, with specific comments on the area of bloodstain pattern analysis. The NAS noted that "...many sources of variability arise with the production of bloodstain patterns and their interpretation is not nearly as straightforward as the process implies." Some of the key items they delineated being required, *at a minimum,* for the interpretation of bloodstains and for the integration into a reconstruction are the following (2009, p.177):

1. "An appropriate scientific education"
2. An understanding of:
   - **a.** limitations of the measurement tools
   - **b.** use of significant digits and applied math
   - **c.** fluid transfer physics
   - **d.** wound pathology

The report made the following pertinent recommendations (National Research Council of the National Academies 2009, page 2):

> (1) assess the present and future resource needs of the forensic science community, to include State and local crime laboratories, medical examiners, and coroners;
> (2) make recommendations for maximizing the use of forensic technologies and techniques to solve crimes, investigate deaths, and protect the public;
> (3) identify potential scientific advances that may assist law enforcement in using forensic technologies and techniques to protect the public;
> (4) make recommendations for programs that will increase the number of qualified forensic scientists and medical examiners available to work in public crime laboratories;
> (5) disseminate best practices and guidelines concerning the collection and analysis of forensic evidence to help ensure quality and consistency in the use of forensic technologies and techniques to solve crimes, investigate deaths, and protect the public;
> (6) examine the role of the forensic community in the homeland security mission;
> [examine] interoperability of Automated Fingerprint Information Systems [AFIS]; and
> (7) examine additional issues pertaining to forensic science as determined by the Committee.

As a result of criticisms, including those articulated in the NAS report, a government commission was formed. The National Institute for Science and Technology (NIST) established the Forensic Science Standards Board which oversees the Organization of Scientific Area Committees (OSAC) which were created in 2014. SWGSTAIN was disbanded with the creation of the Bloodstain Pattern Analysis subcommittee of the Physics/Pattern Evidence OSAC (Figure 14.1).

14.1.1 SWGSTAIN

As previously detailed, the first government-funded bloodstain pattern analysis group was SWGSTAIN which was supported by the FBI from 2002 through early 2015. As stated on their website, "[t]he mission of SWGSTAIN is to promote and enhance the development of quality forensic bloodstain pattern analysis (BPA) practices through the collaborative efforts of government forensic laboratories, law enforcement, private industry, and academia." (SWGSTAIN 2015).

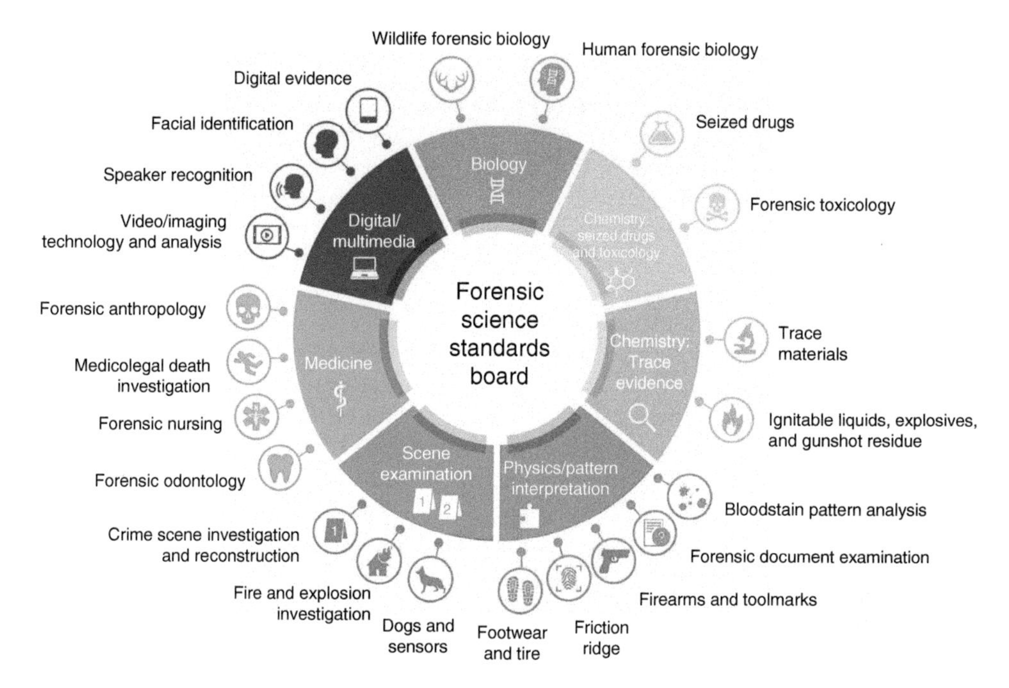

FIGURE 14.1 A graphic displaying the organization of NIST's OSACs (as of June 2021). It is disappointing that the Bloodstain Pattern Analysis subcommittee is placed in the "Physics/Pattern Evidence" committee rather than the "Scene Examination" one. This misplacement implies that the analysis of blood traces can be separated from the crime scene examination, which is patently absurd. *Source: https://www.nist.gov/topics/organization-scientific-area-committees-forensic-science/osac-organizational-structure. Reprinted with permission from the Organization of Scientific Area Committees for Forensic Science.*

"Through its membership, SWGSTAIN (2015) seeks to:

1. Discuss, share and compare stain pattern analysis methods, protocols, and research for the enhancement of forensic bloodstain pattern analysis (BPA) techniques.
2. Design and encourage the implementation by practitioners of a quality assurance program in bloodstain pattern analysis and to advise the forensic bloodstain pattern analysis community of emerging quality assurance issues.
3. Discuss and share strategies for presenting bloodstain pattern information to meet Frye, Daubert, or other jurisdictional admissibility challenges.
4. Address the development and/or validation of forensic bloodstain pattern analysis methods.
5. Design and encourage the adoption of guidelines to insure the quality of specialized training in the field of bloodstain pattern analysis."

The most significant benefit from SWGSTAIN was the efforts at developing standards and collected resources for bloodstain pattern analysts. As of the writing of this book, the resources from SWGSTAIN are still available but they will likely be subsumed by the NIST OSAC Bloodstain Pattern Analysis Subcommittee. A significant criticism of SWGSTAIN has been the lack of appreciation of the need for formalized scientific education for those offering opinions on bloodstain pattern reconstructions, and consequently some of the standardized documents that were produced.

14.1.2 NIST OSAC Bloodstain Pattern Analysis Subcommittee

As of the writing of this book, the Bloodstain Pattern Analysis Subcommittee of the NIST OSAC is in its sixth year, after having its first face-to-face meeting in January 2015 in Norman, Oklahoma. At its commencement, it was hoped that the bloodstain pattern analysis OSAC would make a positive impact on the field, specifically by addressing and rectifying the shortcomings of current practices and its antecedent organizations. At this point, the progress has been disappointing and misdirected exhibiting an unwillingness to consider the need for formal scientific education – the same major issue with the SWGSTAIN. Too much time has been squandered hashing out pattern classification and a corresponding process map at the expense of the heart of the matter – interpretation. Since many of the members of the OSAC lack a scientific education, it should not be surprising that they would be of the opinion that a formal scientific education is unnecessary. Figure 14.2 shows a list of thought-provoking questions about casework approaches that one of the members of OSAC sent to other members. Not a single response was obtained.

The failure to respond to any of the above questions by any members of the OSAC strongly suggests an unwillingness to interact with those posing probing questions that are designed to evoke thought. Perhaps they felt "insulted" that anyone would have the audacity to ask such questions. It appears that the OSAC was "impenetrable."

Consider your past casework in terms of each of the following five categories of bloodstain pattern analysis situations or problems as initially presented to you as an analyst. How many of your past cases fit into each of the five categories?

1. Cases where you conducted your pattern analysis with no contextual knowledge concerning the details of the case.

2. Cases where you were presented with the pattern evidence and with contextual knowledge, but no questions were posed or scenarios proposed. You were simply asked to reconstruct the event from the bloodstain pattern evidence.

3. Cases where you were presented with the pattern evidence and contextual knowledge along with specific questions to address. In other words, as you began your analysis you were asked to see if your work with the bloodstain patterns could support or refute a particular scenario.

4. At the outset of your work on a particular case you were asked to review and perhaps challenge the work done previously by an opposing expert.

5. In how many cases was your bloodstain pattern analysis or reconstruction reviewed by a subsequent expert? In these cases, did you revisit, reconsider or re-examine your work?

Reflect on how each of the above situations could influence the casework approach taken.

FIGURE 14.2 Questions posed to members of the OSAC on bloodstain pattern analysis - casework approaches.

In 2019 and 2020, the ASB-ANSI published its first set of standards involving quality assurance (#030), report writing (#031), training program for bloodstain pattern analysts (#032) and procedure validation (#072) (AAFS Standards Board 2019a, b, 2020a, b). Many of the members of the ASB also serve on the OSAC concurrently. In our opinion, the "educational requirement" standard (#032) in bloodstain pattern analysis is defective. Not now, or in the future, does it require a bachelor's degree in a natural science. It does instead require a bachelor's degree (does not specify in what major) in the year 2025 with coursework in trigonometry, and "...science related coursework and laboratory work in biology, physics, and chemistry from an accredited institution." So, in essence, a bloodstain trainee could obtain a bachelor's degree in sociology, take the aforementioned coursework and meet the educational requirement as written in Section 4.1 of the standard. Allowing 5 years for the educational requirement to take effect should be sufficient time for anyone interested in becoming a bloodstain examiner to obtain a bachelor's degree in a natural science – nevertheless, the ASB decided a science degree was unnecessary. In our opinion, this is hardly a sufficient step forward. Pursuant to a footnote included with the standard it will be applied prospectively (not retroactively). As succinctly stated by one of our colleagues under the new standard "It is possible that someone with a B.A. degree in Acting or Hospitality Management could easily meet these requirements and undertake the training to become an analyst; the absurdity should be obvious." (R. R. Ristenbatt, personal communication, July 2020).

Morrison et al. (2020) are highly critical of two of the aforementioned standards, i.e., pertaining to quality assurance (#030) and validation of procedures (#072) referring to both standards as "vacuous." It is certainly reasonable that the primarily non-scientist composition of the board may be a contributing factor to the root cause of the establishment of the inadequate validation and quality assurance standards.

14.1.3 Organizations

Organizations such as the IAI and the Association for Crime Scene Reconstruction have memberships composed of those concerned with physical evidence at crime scenes, while the IABPA is a specialized group for those focused solely on bloodstain pattern analysis. The authors maintain that the implicit overspecialization that the focus of the IABPA membership has on bloodstain patterns to the exclusion of other aspects of physical evidence at a crime scene is objectionable and short-sighted. This is a considerable criticism of the IABPA as opposed to other organizations. Blood trace configurations are not context-free and may not be properly interpreted without considering other physical evidence (De Forest 2018). This is a fundamental concept discussed throughout this book.

The value of these organizations are that they provide a forum for discussion and an exchange of knowledge among colleagues. In addition, a significant benefit of these organizations is that they have codes of ethics for governing their members as well as providing education and scholarship opportunities, with information about these available on their websites (http://www.theiai.org, http://www.acsr.org/, http://www.iabpa.org/about).

There are issues with some current organizations, specifically with respect to their vetting of applicants for membership. Historically, there has been a tendency to accept individuals as members in certain organizations with minimal qualifications and no scientific education; thus, these organizations have been guilty of enabling dangerous self-delusion of pseudoscientists and bloodstain pattern "enthusiasts." Ideally, the membership of bloodstain pattern, crime scene reconstruction, and forensic science organizations would consist of true professionals with appropriate scientific education, training, and experience.

A well-publicized extreme example of a problematic and misleadingly legitimate organization is the American College of Forensic Examiners Institute (ACFEI). This group claims to offer numerous advantages for membership as well as awarding certification in eight forensic fields, but has been denounced as a diploma mill. Robert O'Block is the founder of ACFEI, which he established in Branson, Missouri in 1992, after being rejected for membership by a credentialing organization for forensic handwriting experts. The standards for certification originally were nonexistent, only requiring payment. Currently, to become a "certified forensic examiner," one only needs to pay the $165 membership fee in addition to $495 to take a one-hour online course and 100 question multiple choice exam. The questions on the multiple-choice exam have been described as mostly common sense, requiring no advanced knowledge. There is no prior education requirement nor do they appear to verify references or qualifications, as evident by the published anecdote that they certified a house cat, Zoe D. Katz. The ACFEI is ultimately engaged in deceitful practice that provides false credentials to unqualified and/or otherwise qualified gullible individuals and muddies the water for legitimate organizations such as the American Board of Criminalistics which has a rigorous forensic certification program. It warrants comment that certification alone, even from legitimate organizations, does not make one a forensic expert (Bartos 2012).

14.2 Publications and Other Information Sources

There are numerous resources available for those interested in the practice of bloodstain pattern analysis, including published journals, newsletters, books, as well as internet websites. Each of these will be discussed in more detail below.

14.2.1 Journals

There are a number of journals that publish articles related to bloodstain pattern analysis. These vary greatly with respect to quality, impact, and reliability. Ideally, journals that are scientific and peer-reviewed are of the highest value; however, there exist differences in stringency of the peer-review process that contribute to noticeable quality control issues for some ostensibly scientific journals. Journals such as Forensic Science International, the Journal of Forensic Sciences, Science and Justice, and the Journal of Forensic Identification have a rigorous peer-review process and thus often produce articles of high scientific significance.

In addition, there exist journals that are scientific but not peer-reviewed, nonscientific but peer-reviewed, and those that are neither scientific nor peer-reviewed. In recent years, the number of online journals has increased dramatically, many of questionable quality. Another recent development has been a "pay-to-publish" model, where authors are funding the distribution of their research. This has serious issues as to the substance of the published research. The quality of some profits from a rigorous peer-review process, while others will publish anything the authors are willing to pay. There are some advantages to these online journals such as their short timelines for publication, potentially free access to interested readers, and the worldwide reach of the internet thus enabling wide and rapid dissemination of results. However, caution must be taken when reading references in these journals as the experimental results and conclusions that are reported may be the product of poor design and/or faulty analysis, or the information they report may not be accurate.

14.2.2 Newsletters

Most of the previously mentioned organizations publish periodic newsletters that can include valuable information on bloodstain pattern analysis, both on the state of events in the field and current research. In addition to communication of newsworthy information for the membership, newsletters can serve to share laboratory research as well as important philosophical views and opinions. Many of these newsletters are available both in print and online. Newsletters are more informal but share similar attributes with online journals in that they enable a rapid sharing of information.

14.2.3 Books

There are a number of available books that are either dedicated to the subject or have significant space devoted to bloodstain pattern analysis. The value of books is as teaching tools rather than for reference, thus providing basic information to a reader. Books endeavor to contain an objective and comprehensive discussion of a topic. Unlike journal articles, books, including the present one, are not peer-reviewed in the strict sense (i.e., double blind and prior to publication); thus, they can be written by persons of varying levels of knowledge and competence. Consequently, the quality of information contained within may promote common myths and misconceptions rather than provide uniformly high quality of knowledge of the science of bloodstain patterns. Still, many of these books are reviewed subsequent to publication in journal reviews and on the internet, and thus there are methods for an interested reader to evaluate a book prior to purchase.

14.2.4 Internet Resources

The internet has many uses, but websites should not be uncritically relied on as a quality resource of bloodstain pattern analysis information. Its primary use should be as a searching mechanism for finding more reliable, published, and peer-reviewed resources.

14.3 Training and Education

There exist numerous forensic science education and training opportunities. Universities and Colleges with forensic science programs offer the best source of educational resources, while training in addition to that provided by the employer is often available in the form of workshops presented by private companies or professional associations. A thorough discussion of the nature of and distinction between training and education is presented in Chapter 8.

14.3.1 Continuing Education

The availability of continuing education opportunities can take many forms and includes attendance at courses (both in-person and online), workshops, lectures, seminars, webinars, and attendance at professional society meetings (i.e., AAFS, IAI, IABPA, etc.). Continuing education in the field of crime scene investigation, crime scene reconstruction, and bloodstain pattern analysis offers the attendee the opportunity to remain current in these fields, providing information on new scientific research and technologies.

14.4 Proficiency Tests

Proficiency tests are an important part of quality control in several areas of forensic science. Along with peer review of reports, both internal and external proficiency tests are methods for assessing an analyst's competency. Although the intention of proficiency tests in the area of bloodstain pattern analysis is good, the tests themselves are too simplistic and focus on a single measurement or identification of a category of pattern rather than a reconstruction. Consequently, implementation of these tests does not appropriately assess an analyst's competency in performing a meaningful bloodstain pattern reconstruction, which is the primary goal of this analysis, and unfortunately only tests the rote skills of a technician. Designing an appropriate proficiency test is not trivial, and other than using real crime scene evidence photographs with contextual information, the authors are not convinced that it can be accomplished.

In recent years, results from proficiency tests have been used as being indicative of error rates for specific forensic science disciplines. With respect to bloodstain pattern interpretations, this is inappropriate for many reasons, not limited to the overly simplistic design of the tests, the absence of blind testing, and the total lack of crime scene context. The ultimate purpose of bloodstain pattern analysis is crime scene reconstruction, and proficiency tests fall well short of providing a realistic challenge.

References

AAFS Standards Board (2019a). *Standard for a quality assurance program in bloodstain pattern analysis (ANSI/ASB 030-2019).* http://www.asbstandardsboard.org/wp-content/uploads/2019/12/030_Std_e1.pdf (accessed 1 October 2020).

AAFS Standards Board (2019b). *Standard for the validation of procedures in bloodstain pattern analysis, first edition (ANSI/ASB 072-2019).* http://www.asbstandardsboard.org/wp-content/uploads/2019/05/072_Std_e1.pdf (accessed 1 October 2020).

AAFS Standards Board (2020a). *Standard for report writing in bloodstain pattern analysis (ANSI/ASB 031-2020).* http://www.asbstandardsboard.org/wp-content/uploads/2020/07/031_Std_e1.pdf (accessed 1 October 2020).

AAFS Standards Board (2020b). *Standards for a bloodstain pattern analyst's training program (ANSI/ASB 032-2020).* http://www.asbstandardsboard.org/wp-content/uploads/2020/05/032_Std_e1.pdf (accessed 1 October 2020).

Bartos, L. (2012). *No forensic background no problem?* https://www.propublica.org/article/no-forensic-background-no-problem (accessed 1 October 2020).

De Forest, P.R. (2018). *Physical aspects of blood traces as a tool in crime scene investigation. CAC News*, pp. 30–33.

Morrison, G.S., Neumann, C., and Geoghegan, P.H. (2020). Vacuous standards – subversion of the OSAC standards - development process. *Forensic Science International: Synergy* 2: 206–209.

National Research Council of the National Academies. Committee on Identifying the Needs of the Forensic Science Community. Committee on Science, Technology, and Law Policy and Global Affairs. Committee on Applied and Theoretical Statistics Division on Engineering and Physical Sciences (2009). Strengthening Forensic Science in the United States: A Path Forward. Washington, DC: The National Academies Press.

SWGSTAIN (2010). *Scientific working group on bloodstain pattern analysis: SWGSTAIN history.* http://www.swgstain.org/documents/SWGSTAIN%20HISTORY.pdf (accessed 14 January 2015).

SWGSTAIN (2015) *Home.* http://www.swgstain.org/Home (accessed 14 January 2015).

CHAPTER 15

Concluding Remarks and Looking to the Future

As a conclusion to this book, we would like to take the opportunity to emphasize some of the more salient issues previously discussed. In addition, we will explore some additional concepts that we believe have had a negative impact on the physical "analysis" and interpretation of blood traces and will need to be addressed in the future.

15.1 Importance of Science on the Front End

Several authors have articulated the fundamental concept that the scientific enterprise does not commence at the threshold of the laboratory (De Forest 1999; Almog 2006; San Pietro and Kammrath 2018). At least it should not. It seems as if it is assumed by many that either science is not needed at the scene or it is being routinely utilized as it should be. They cannot or refuse to fathom the difference between a technician and a scientist. This problem is dovetailed with viewing the forensic laboratory as a mere testing facility. The problems are interlocked. No legitimate reason exists for allowing non-scientist investigators to circumscribe and limit the application of science for physical evidence problem-solving. Scientific inquiry and skepticism must be utilized in situ where the evidence originally resides. Every crime is unique, and its scene is a partially superimposed record of layers of what happened prior to the onset of the incident under investigation, a record of what happened during the incident, and a record

Blood Traces: Interpretation of Deposition and Distribution, First Edition. Peter R. De Forest, Peter A. Pizzola, and Brooke W. Kammrath.

of post-event occurrences. This amalgam of muddled layers makes it extraordinarily difficult to decipher and recognize physical evidence in many crime scenes.

Unrecognized evidence is just that – it remains undetected and therefore remains at the scene never to be submitted to the laboratory for analysis. Lacking a scientific-based selection process, relevant evidence may not be recovered and, perhaps compounding the problem, irrelevant material will be collected and submitted to the laboratory for analysis. Crime scenes can be exceedingly complex, with investigators faced with the difficult task of discerning the relevant from the irrelevant traces (i.e., the "signal" from the "noise"). Since each case is different and it is very difficult to sort the relevant physical evidence from the irrelevant background, scientific expertise is necessary to ensure that the criminal justice system is being adequately served. Without a scientific assessment on the front end of an investigation, the system is inviting a risk of wrongful convictions and a failure to prosecute the appropriate individuals.

Evidence recognition is a daunting challenge. For the physical evidence investigation to proceed methodically, flexibility is required to meet the needs of each case. Neither a random or aimless approach nor a rigid or strict one will achieve our ultimate goal. Adherence to the scientific method, as discussed in Chapter 2, is the solution that simultaneously provides a systematic approach coupled with the freedom for innovation and adaptability. An integral component of the scientific method is hypothesis testing. This is the necessity to challenge, question, and attempt to destroy one's own hypothesis. If the physical evidence does not fit with the original hypothesis, the initial one must be modified or a new hypothesis is formulated and subjected to the same degree of testing. This is a cyclic process. Rigorous hypothesis testing is one of the most difficult and counter-intuitive concepts to implement and faithfully carry out.

As previously stated, it is imperative to employ this process at crime scenes and thus avoid relying on untested or untestable assumptions which can lead an investigation down the wrong path. This is the core of the scientific method and the essence of science. Effective use of the scientific method must be employed by someone skilled in this process. Serious science students are educated with this philosophy over a period of years at the university. Hypotheses that are formed and refined in this way can successfully guide an investigation and provide a customized framework to fit the specific unique characteristics of a case. Hence, the perhaps seemingly unattainable goal that appeared to be self-contradictory, i.e., one that combines the positive attributes of both a rigid approach and an adaptable one, is achieved. The resulting custom-tailored approach to a given case is both flexible and systematic. The failure to appreciate and understand the concept of hypothesis testing is one of the most profound issues in the field of criminalistics and affects all aspects of its practice, from evidence recognition to crime scene reconstruction.

15.2 The Integration of Physical Evidence with Police Investigations

One must avoid the natural instinct to justify what is ostensibly a scientific position with conventional police investigation information. For example, non-scientific information that could bias a scientific investigation could include a suspect's identification card at the scene, a cell phone found on the floor alongside victim, statements by persons of interest, etc. This falls under the subject of bias, which is discussed in Chapter 7. Rigid adherence to the scientific method is the best tool to prevent bias from corrupting a scientific investigation. It is commonplace for lecturers to discuss these non-science aspects in the middle of supposed scientific discussion without explaining to the audience that scientific investigators must not allow themselves to be influenced by police investigative materials that tend to suggest conclusions.

Ideally, the scientific findings regarding the physical evidence should inform the police investigation in a timely fashion. Although it is common for DNA evidence to lead police officers to potential suspects, rarely are traces used to create investigative leads. This is a missed opportunity.

15.3 Troubling Developments and Perceptions

Despite the considerable sums of money being spent by the federal government in response to the National Research Council of the National Academy of Sciences' (2009) report, the central problems in bloodstain pattern analysis and forensic science are not being addressed by those organizations that were established for this purpose. For example, the Organization of Scientific Area Committee (OSAC) subcommittee for bloodstain pattern analysis directly adopted many documents with little modification with respect to the nomenclature and the philosophy from its predecessor organization established by the FBI, the Scientific Working Group on Bloodstain Pattern Analysis.

The bloodstain pattern OSAC subcommittee has demonstrated a narrow focus on taxonomical classification of patterns and as one example the determination of impact origins in lieu of the overarching and more challenging problem of interpretation and reconstruction of the overall event. There is a failure to recognize that a substantial number of configurations simply cannot be identified as originating in a particular manner coupled with significant uncertainty regarding sequence of deposition. There is a misguided emphasis in the scientific literature on the topic of bloodstain patterns written by physicists and forensic scientists with a focus on determining a precise origin in space. It is incorrectly assumed by these authors that there is a significant improvement in the science by experiments which address the question of reducing the uncertainty of the locus in three-dimensional space. Many researchers and practitioners continue to fail to recognize the futility of improvements in the precision of the area of origin determinations considering the inherent uncertainty and lack of value of increased precision for a reconstruction.

There are several cases discussed in Chapters 10–12 which demonstrate the inability to prove what action produced a blood trace configuration, and the timing or sequence of deposition may remain unknown. Additionally, there is the question of relevance for determining this information for every blood trace at a crime scene. The following example scenario illustrates part of this problem. An individual is fatally stabbed in the femoral artery; however, he is able to run about 50 feet before collapsing and expiring. At the suspected location where the initial stabbing took place, no detectable spatter is evident. There is however a tremendous amount of bloodspatter at the position of his collapse and ultimate death, in addition to a trail of blood traces between the two locations. The arterial spatter configuration observed at the end of the trail is of no value to understanding where the stabbing took place. Although the trail itself is important in helping to locate the site of the initial stabbing, the exact location may remain indeterminate. This entire process may have taken 30 seconds to transpire, but it may be the first seconds or less that is the most informative regarding the reconstruction of the event. This is fundamentally an example of several of the Blood Trace Principles (specifically #s 2 and 4). It is unfortunate that members of the bloodstain pattern community often do not recognize or appreciate these limitations and others enunciated in the Blood Trace Principles.

15.4 Testing Facilities & the Creeping Inversion

A serious overarching problem that the authors have discussed in various science forums is that systems for the delivery of forensic science services are being viewed and utilized primarily as "Testing Facilities" (De Forest et al. 2010a, De Forest et al. 2010b; De Forest and Kammrath 2014). Stated somewhat differently, this conceptualization focuses on *source attribution* and is directed narrowly to question "**who**" while ignoring the potential of other interrogatives in extracting information from the traces that comprise the physical evidence record. It is critically important for criminal justice stakeholders, users and even criminalists themselves, to re-conceptualize a forensic service as more than a mere testing facility. A full-service forensic science system that provides scientific problem-solving expertise across the physical evidence continuum, from the crime scene to case resolution, is essential for optimum utilization of the physical evidence record in case solutions.

There has been a decades long drift of criminalistics laboratories evolving toward becoming more narrowly focused testing laboratories, mimicking the clinical laboratory model, as opposed to facilities for providing scientific problem-solving services. The forensic laboratory should not mimic the clinical model since a clinical laboratory deals with a narrow range of tests from a fixed menu to be performed on a limited set of samples. The clinical model entails conducting tests to provide data to the medical doctor, who in addition to interpreting the results also orders the laboratory requests. The medical doctor possesses considerable expertise in framing scientific questions to be answered by the laboratory, in contrast to his or her non-scientist criminal justice counterpart. It is unfortunate that the necessity for scientific expertise in the front end of a criminal investigation is not appreciated in the same manner as it is in a medical investigation. Recalling that no two crime scenes are the same, an infinite number of critical questions can be formulated from the traces present at a crime scene which are not captured in the typical laboratory request form; the abductive reasoning of the criminalist is pivotal at this juncture in the physical evidence continuum.

As a consequence of the testing facility paradigm, one could fairly state that there has been a "creeping inversion" involving the respective roles of criminalists and law enforcement investigators (De Forest and Kammrath 2014). The ongoing and unfortunate trend is that forensic scientists are increasingly becoming more specialized, and consequently, and perhaps unwittingly, taking on the roles of reactive technicians. Simultaneously, law enforcement investigators are being cast in the roles that should be filled by scientists: interpreting what happened at a crime scene, developing hypotheses, conducting ad hoc experimentation for casework. Typically, law enforcement investigators are not scientists and are generally ill-equipped for conducting research in these areas, including the subject of this book. The knee-jerk reaction of many forensic scientists to this sentiment might be that they themselves are not glorified technicians since they use complex instrumentation on a daily basis. However, they are confusing technology with science. This is the "flip side" of the problem associated with the law enforcement investigators. Many are led to believe that they are the equivalent of scientists since they use advanced crime scene equipment. This is reinforced by many training programs and workshops. As a consequence, the need for scientific expertise at the scene goes unrecognized and unmet. These misguided beliefs are understandable, and often law enforcement personnel cannot be held responsible, as there are few actual criminalists with meaningful crime scene expertise to fill the role as the scientist at the crime scene. Both scientists and investigators have a lot to learn about the value of collaborative teamwork at the crime scene. In the experience of the authors, the scientist and the technical investigator are a model that can work very effectively together; mutual respect is necessary in addition to an appreciation for each other's strengths and weaknesses. Unfortunately, the potential value of the role of the scientist at the crime scene has been usurped over the past several decades.

15.5 The Pernicious Effects and Fallout from Bloodstain Workshops

As previously discussed in Chapter 9, training and education are often terms that are confused with one another; they are not synonyms, and for professional purposes, they must be distinguished from one another. It is regrettable when bloodstain pattern organizations and committees conflate the two terms, which they unfortunately do too often. Education is received at the university level with an emphasis on understanding, whereas training is a form of professional development with a focus on facilitating an individual's performance and is obtained in workshops, during scientific presentations at science meetings, and "on the job" training.

In our opinion, most of the problems seen currently with the interpretation and testimony in the area of bloodstain pattern analysis can be laid at the feet of the history of the 40-hour course. There are three major pernicious consequences stemming from these in the current practice of blood trace interpretations:

> Consequence #1: From the outset, the 40-hour workshops gave attendees misplaced confidence that they had become experts in the area as a result of attendance. This created a large number of instant experts. We are unaware of any attendees or "students" that failed to pass the course. These workshops would have been useful solely to raise awareness of the potential value of blood trace configurations at crime scenes for qualified criminalists and instead went beyond this laudable purpose and produced many unqualified "instant experts". Incredibly, with only

the training from a 40-hour workshop, some individuals even hung out a shingle and began offering courses of their own.

Consequence #2: There were no academic criteria for admission to these workshops. Many, or perhaps most, attendees had not completed any university course work let alone earned 4-year undergraduate or graduate science degrees. This has exacerbated the lack of appreciation for science and scientific problem solving in the field of blood trace configuration interpretations.

Consequence #3: These 40-hour courses have created the indelible impression that blood trace configuration interpretation is a standalone activity rather than being an integral part of crime scene investigation. There are both theoretical and practical reasons why the investigation of blood trace configurations must be seamlessly integrated with the overall crime scene investigation, as emphasized throughout this book.

The development of expertise in all areas of crime scene investigation, including the interpretation of blood trace configurations, must start with a sound scientific education. The actual major may be unimportant as long as the degree is a traditional science degree. Science degrees are hierarchical. Second level courses have pre-requisites. Third-level courses have sophomore courses as prerequisites. For example, a junior-level course such as physical chemistry cannot be taken without having previously enrolled in and passed full-year course sequences in general chemistry, organic chemistry, calculus, and physics. The Forensic Science Education Programs Accreditation Commission (FEPAC) guidelines for forensic science programs could be used as a minimum criterion for evaluating science degrees. It needs to be appreciated that the substitution of a 4-year degree with some science courses is not a meaningful equivalent of a science degree.

15.6 Future Directions

Future progress necessitates a sharp departure from the extant paradigm and practice. The following describes the most urgent of these issues.

1. **There needs to be a much more sophisticated understanding, appreciation, and utilization of physical traces of blood.**
   Throughout this book, we have advocated for a holistic view in developing an understanding of physical evidence, including blood trace configurations. The interpretation of blood traces has a long and checkered history, only some of which is grounded in scientific experimentation and analysis. Somewhere along the way, the field of bloodstain pattern interpretation has gone off course. In both bloodstain pattern committees and in the literature, there is a misguided focus on classification as opposed to the holistic view of the physical aspects of blood traces to aid in reconstruction. Pattern classifications alone do not illuminate or solve cases. There is a giant leap between pattern classification and reconstruction, and it is the latter that provides the richest and most useful information. In addition, as previously discussed, there is a lack of understanding regarding the necessity for truly scientific interpretations in blood traces.

The added benefit of a holistic approach to casework is that information from the different traces may be inextricably interwoven and thus be mutually supportive. As demonstrated in the "Vertical Crime Scene" case (Chapter 10), the blood evidence combined with the other traces and trajectories provided a nuanced coherent reconstruction.

2. **Increased and assured funding for forensic science research and laboratory operations, with a focus on scientific integrity.**
Forensic science research activity and laboratories are perennially underfunded and receive money as responses from crisis to crisis rather than a constant influx of appropriate funds. There has been exponential growth in forensic science over the last few decades, but it has not had a purposeful direction. There is increasing emphasis on the practice of forensic science narrowly focused on using the testing facility model (as explained earlier in this chapter), which is inadequate, inappropriate, and incomplete, as it only addresses part of the problem. There needs to be a meaningful and purposeful movement for providing sufficient funds for forensic science research and scientific laboratory operations.

In 2011, Mnookin et al. in the UCLA Law Review strongly advocated for the creation of a research culture that "...permeates the entire field of forensic science." It was their opinion that this research culture must be "central and foregrounded" whose core values are empiricism, transparency, and an on-going critical perspective. The research culture that these authors advocated could perhaps be more aptly thought of as a science culture. Regardless of what we refer to it as, it is clear that in many quarters, we need to shift from a police culture to a science culture in forensic science laboratories and in crime scene investigation units. The core value of transparency as presented by Mnookin et al. is a singularly critical one in our field that needs to be discussed in considerable depth.

The concept of a scientific culture of transparency should also be extended to the laboratory and crime scene functions of working with the defense community. If forensic science is to achieve the level of scientific integrity that is required by the criminal justice system for fundamental fairness, it is necessary that the defense community be treated in an entirely different way than in the recent past. Thus, the defense must be apprised of the same information on casework and given the same access to forensic scientists as is the prosecution. Of course, this must be done in a reasonable fashion so that active investigations are not undermined and prosecutions are not compromised. The defense's access to the evidence is commonly subsequent to the prosecution's, but ideally could be even contemporaneous. It should never be prior to a thorough documentation and analysis by the forensic laboratory. Unfortunately, too often the access to the analysis is blocked by the prosecution, and even necessitates a court order for the defense to obtain reports and case notes. It is a misguided view that the prosecution "owns" the laboratory access and the results of the forensic examinations. There must be a mechanism to provide communication between the forensic community and the defense. The laboratory management and accrediting bodies must take a much stronger stance on this particular unjustified practice of disparate access to information. It is our collective opinion that it is both unfortunate and injudicious that ASCLD/LAB's guiding principle number four (1994) states that professional forensic scientists should communicate with the defense "...when communications are permitted by law and agency practice." This same language was adopted by ANAB (2018) which absorbed ASCLD/LAB. Naturally, there is little recourse when the law is a limiting factor. However, in the absence of a legal obstacle, there is a serious issue. This seems to suggest that despite the laudatory spirit of the guiding principle, a real opportunity for information sharing can be undercut through a local antediluvian policy which thwarts the timely disclosure of findings to the defense. This cannot be what Mnookin et al. (2011) were advocating. The underlying philosophy of the forensic science services, whether it be relative to crime scene or laboratory services, must exhibit neutrality with respect to the access to scientific findings from the analysis and interpretation of physical evidence.

3. **Future progress in forensic science requires a greater attention to scientific issues throughout the physical evidence continuum, including educational requirements, problem definition at the scene, laboratory practices, and their ultimate interpretation and reporting of conclusions.**
   Earlier in this chapter, we discuss the need for more science on the front end, but this is only one area that would benefit from the injection of more stringent scientific requirements. The authors recommend that in moving forward, the emphasis should be for a minimum basic education of a baccalaureate with a major in a natural science be required for anyone engaging in blood trace reconstructions. It would be ideal if this was also applied to all aspects of physical trace-based crime scene investigations. In addition to the basic academic degree, structured training of the workshop variety is appropriate, in addition to extensive mentoring by someone with the academic credentials, skills, experience, etc.

We support both laboratory accreditation and individual certification. Nonetheless, there have been unintended consequences with respect to accreditation, namely, an unwillingness to conduct any examinations not included in the standard operating procedure (SOP) even though the examinations may be required to resolve issues in a given case. It is negligent to merely focus on doing the *job right*, the emphasis should be on doing the *right job*. In this context, one has to pose the right questions and do the most appropriate examination! It ceases to be a scientific enterprise if the scientist's envisioned modes of inquiry are not being pursued and examinations are not being conducted because they are not in the SOP. There is often no "standard" question or evidence trace. Therefore, the criminalist must have the scientific freedom to do both what is necessary and scientifically valid for an examination. Although following SOPs is appropriate for routine illicit drug testing and some questions of source attribution, the testing facility model is a poor fit with the rigorous scientific enterprise required for open-ended inquiries necessary for successful and accurate reconstructions. When cases present questions that cannot be answered by using extant SOPs, new procedures can be validated and implemented that contribute to the solution of a problem.

There is a need for forensic scientists to take more responsibility for how their work is being used by attorneys in legal proceedings and at the back end of the physical evidence continuum. As stated by San Pietro and Kammrath (2018), "we continue to allow the interpretation of the significance of evidence to be performed by non-scientist law enforcement personnel or attorneys who are free to impart their own bias onto its interpretation." Historically, concern for the misuse and misrepresentation of scientific results and testimonies has been considered off-limits by forensic scientists, as events in plea discussions and courtroom summations as many felt that it was outside our realm of control. However, we feel that it is imperative that the evidentiary significance be appropriately understood and conveyed by attorneys to the finders of fact. This can be facilitated by assuring that an expert's report is complete, including the "warts and wrinkles", and can stand on its own (e.g., contains a thoughtful conclusion and interpretation) and that expert testimony is comprehensive and representative of the complete evidence findings, understandable to a lay-person, and resistant to attorney manipulations.

It is hoped for the future of enhanced scientific contributions to criminal justice that there will be a paradigm shift in the policy, practice, and utilization of forensic sciences. As we have stressed throughout this book, there is a need for science across the entire physical evidence continuum. Blood traces are a potentially valuable type of physical evidence when they are appropriately recognized, interpreted, and integrated along with all of the relevant physical traces. It is also hoped that there will be fairness in access to all forensic science services and information developed. It is only through these actions that the latent contributions of science to justice can truly be realized.

References

Almog, J. (2006). Forensic science does not start in the lab: the concept of diagnostic field tests. Journal of Forensic Sciences 51 (6): 1228–1234.

ANAB (2018). *Guiding principles of professional responsibility for forensic service providers and forensic personnel, GD 3150.* https://anab.qualtraxcloud.com/ShowDocument.aspx?ID=6732 (1 October 2020).

ASCLD/LAB (1994). *Guiding principles of professional responsibility for crime laboratories and forensic scientists* (31 October 2014).

De Forest, P.R. (1999). Recapturing the essence of criminalistics. Science & Justice: Journal of the Forensic Science Society 39 (3): 196.

De Forest, P.R. and Kammrath, B.W. (2014). *The creeping inversion. Proceedings of the Annual Meeting of the Northeastern Association of Forensic Scientists*, Hershey, PA (3–6 November 2014).

De Forest, P.R., Mattheson, G., and Pizzola, P.A. (2010a). *Panel discussion: "testing facility paradigm and privatization of forensic science". American Society of Crime Lab Directors Annual Meeting,* Baltimore, MD (16 September 2010).

De Forest, P.R., Pizzola, P.A., Valentin, P., and Matheson, G. (2010b). *The criminalistics laboratory as a testing facility. Proceedings of the Joint Plenary Session of the Annual Meeting of the Northeastern Association of Forensic Scientists and the New England Division of IAI*, Manchester, VT (November 2010).

Mnookin, J.L., Cole, S.A., Dror, I.E. et al. (2011). The need for a research culture in the forensic sciences. UCLA Law Review. 58: 725.

National Research Council of the National Academies. Committee on Identifying the Needs of the Forensic Science Community. Committee on Science, Technology, and Law Policy and Global Affairs. Committee on Applied and Theoretical Statistics Division on Engineering and Physical Sciences (2009). Strengthening Forensic Science in the United States: A Path Forward. Washington, DC: The National Academies Press.

San Pietro, D. and Kammrath, B.W. (2018). Forensic science: a forensic scientist's perspective. The SciTech Lawyer 15 (1): 34–37.

Bibliography

AAFS Standards Board, ANSI/ASB 030-2019. 1st ed. (2019). *Standard for a quality assurance program in bloodstain pattern analysis.* http//www.abstandardsboard.org/published-documents/bloodstain-pattern-analysis-published documents (accessed 1 October 2020).

AAFS Standards Board, ANSI/ASB 031-2020, 1st ed. (2020). *Standard for report writing in bloodstain pattern analysis.* http//www.abstandardsboard.org/published-documents/bloodstain-pattern-analysis-published documents (accessed 1 October 2020).

AAFS Standards Board, ANSI/ASB 032-2020, 1st ed. (2020). *Standards for a bloodstain pattern analyst's training program.* http//www.abstandardsboard.org/published-documents/bloodstain-pattern-analysis-published documents (accessed 1 October 2020).

AAFS Standards Board, ANSI/ASB 072-2019. 1st ed. (2019). *Standard for the validation of procedures in bloodstain pattern analysis.* http//www.abstandardsboard.org/published-documents/bloodstain-pattern-analysis-published documents (accessed 1 October 2020).

Adam, C.D. (2012). Fundamental studies of bloodstain formation and characteristics. *Forensic Science International* 219: 76–87.

Adam, J.R., Lindblad, N.R., and Hendricks, C.D. (1968). The collision, coalescence, and disruption of water droplets. *Journal of Applied Physics* 39 (11): 5173–5180.

Adamson, A. (1967). Physical Chemistry of Surfaces, 2e, 21–41. New York: Interscience, Wiley.

Adler, W.F. (1977). Liquid drop collisions on deformable media. *Journal of Materials Science* 12 (6): 1253–1271.

Alcock, J. (1995). The belief engine. *The Skeptical Inquirer* 19 (3): 14–18.

Almog, J. (2006). Forensic science does not start in the lab: the concept of diagnostic field tests. *Journal of Forensic Sciences* 51 (6): 1228–1234.

Alter, M. (2013). *Camm trial in review.* http://www.wlky.com/news/local-news/david-camm/camm-trial-in-review/22429352 (accessed 1 October 2020).

ANSI National Accreditation Board (2018). *ANAB guiding principles of professional responsibility for forensic service providers and forensic personnel, GD 3150.* https://anab.qualtraxcloud.com/ShowDocument.aspx?ID=6732 (accessed 1 October 2020).

Aplin, S., Reynolds, M., Mead, R.J., and Speers, S.J. (2019). The influence of hematocrit value on area of origin estimations for blood source in bloodstain pattern analysis. *Journal of Forensic Identification* 69 (2): 163–175.

Arthur, R.M., Cockerton, S.L., de Bruin, K.G., and Taylor, M.C. (2015). A novel, element-based approach for the objective classification of bloodstain patterns. *Forensic Science International* 257: 220–228.

Arthur, R.M., Hoogenboom, J., Baiker, M.C., and de Briuin, K.G. (2018). An automated approach to the classification of impact spatter and cast-off bloodstain patterns. *Forensic Science International* 289: 310–319.

Asano, M., Oya, M., and Hayakawa, M. (1971). Identification of menstrual blood stains by the electrophoretic pattern of lactic dehydrogenase isozymes. *Nihon hoigaku zasshi=The Japanese Journal of Legal Medicine* 25 (2): 148–152.

ASCLD/LAB (1994). *ASCLD/LAB guiding principles of professional responsibility for crime laboratories and forensic scientists* (accessed 31 October 2014).

Attinger, D., Moor, C., Donaldson, A. et al. (2013). Fluid dynamics topics in bloodstain pattern analysis: comparative review and research opportunities. *Forensic Science International* 231: 375–396.

Blood Traces: Interpretation of Deposition and Distribution, First Edition. Peter R. De Forest, Peter A. Pizzola, and Brooke W. Kammrath.

Attinger, D., Liu, Y., Bybee, T., and De Brabanter, K. (2018). A data set of bloodstain patterns for teaching and research in bloodstain pattern analysis: impact beating spatters. *Forensic Science International* 18: 648–654.

Attinger, D., Comiskey, P.M., Yarin, A.L., and De Brabanter, K. (2019). Determining the region of origin of blood spatter patterns considering fluid dynamics and statistical uncertainties. *Forensic Science International* 298: 323–331.

Balko, R. and Carrington, T. (2018). *Bad science puts innocent people in jail – and keeps them there: how discredited experts and fields of forensics keep sneaking into the courtroom. Washington Post* (March 21). https://www.washingtonpost.com/outlook/bad-science-puts-innocent-people-in-jail--and-keeps-them-there/2018/03/20/f1fffd08-263e-11e8-b79d-f3d931db7f68_story.html (accessed 1 October 2020).

Balthazard, V., Piédelièvre, R., Desoille, H., and Dérobert, L. (1939). Étude des Gouttes de Sang Projeté. *Annales de Médecine Légale de Criminologie, Police Scientifique, Médecine Sociale, et Toxicologie* 19: 265–323.

Bartos, L. (2012). *No forensic background? No problem.* https://www.propublica.org/article/no-forensic-background-no-problem (accessed 2 August 2020).

Batchelor, C.K. and Batchelor, G.K. (2000). An Introduction to Fluid Dynamics. New York: Cambridge University Press.

Bauer, M. and Patzelt, D. (2002). Evaluation of mRNA markers for the identification of menstrual blood. *Journal of Forensic Sciences* 47 (6): 1278–1282. https://doi.org/10.1520/JFS15560J.

Beard, K.V. and Pruppacher, H.R. (1969). A determination of the terminal velocity and drag of small water drops by means of a wind tunnel. *Journal of the Atmospheric Sciences* 26: 1066–1072.

Beard, K.V., Ochs, H.T., and Kubesh, R.J. (1989). Natural oscillations of small raindrops. *Nature* 342: 408–410.

Betz, P., Peschel, O., Stiefel, D., and Eisenmenger, W. (1995). Frequency of blood spatters on the shooting hand and of conjunctival petechiae following suicidal gunshot wounds to the head. *Forensic Science International* 76: 47–53.

Bevel, T. (1983). Geometric bloodstain interpretation. *FBI Law Enforcement Bulletin* 52 (5): 7–10.

Bevel, T. (2001). Applying the scientific method to crime scene reconstruction. *J of Forensic Identification* 51: 150–162.

Bevel, T. and Gardner, R. (2008). Bloodstain Pattern Analysis, 3e. Boca Raton, FL: CRC Press, Taylor & Francis Group.

Beyerstein, B.L. (1996). Distinguishing Science from Pseudoscience. Simon Fraser University. http://www.dcscience.net/beyerstein_science_vs_pseudoscience.pdf (accessed 1 October 2020).

Bird, J.C., Tsai, S.S., and Stone, H.A. (2009). Inclined to splash: triggering and inhibiting a splash with tangential velocity. *New Journal of Physics* 11 (6): 063017.

Blanchard, D.C. (1950). The behaviour of water drops at terminal velocity in air. *Transactions - American Geophysical Union* 31 (6): 836–842.

Bodziak, W.J. (1996). Use of leuco crystal violet to enhance shoe prints in blood. *Forensic Science International* 82: 45–52.

Boonkhong, K., Karnjandecha, M., and Alyarak, P. (2010). Impact angle analysis of bloodstains using a simple image processing technique, sonklanakarin. *Journal of Science and Technology* 32: 169.

Boyd, S., Bertino, M.F., and Seashols, S.J. (2011). Raman spectroscopy of blood samples for forensic applications. *Forensic Science International* 208 (1–3): 124–128.

Brown, K.C. and Watkins, M.D. (2016). *Ultraviolet and infrared photographs. Evidence Technology Magazine.* http://www.evidencemagazine.com/index.php?option=com_content&task=view&id=2280.

Browne, N.B. and Keeley, S.M. (2007). Asking the Right Questions: A Guide to Critical Thinking, 8e. Upper Saddle River, NJ: Pearson Prentice Hall.

Bruning, A. and Wiethold, F. (1934). Die Untersuchung von SelbstmorderschuBwaffen. *Deutsche Zeitschrift für die gesamte gerichtliche Medizin* 23: 71–82.

Burnett, B.R. (1991). Detection of bone and bone-plus-bullet particles in backspatter from close-range shots to heads. *Journal of Forensic Sciences* 36 (6): 1745–1752.

Butler, J.M. (2005). Forensic DNA Typing: Biology, Technology, and Genetics of STR Markers, 2e. New York: Academic Press.

Camana, F. (2013). Determining the area of convergence in bloodstain pattern analysis: a probabilistic approach. *Forensic Science International* 231: 131–136.

Camm v. Faith (2019). 937 F.3d 1096.

Carroll, K. and Mesler, R. (1981). Splashing liquid drops form vortex rings and not jets at low Froude numbers. *Journal of Applied Physics* 52 (1): 507.

Carter, A. (1999). Bloodstain pattern analysis with a computer. In: Scientific and Legal Applications of Bloodstain Pattern Interpretation (ed. S.H. James), 20–21. Boca Raton, FL: CRC.

Carter, F. (2001). The directional analysis of bloodstain patterns: theory and experimental validation. *Canadian Society of Forensic Science Journal* 34: 173–189.

Carter, A. (2005). Bloodstain pattern analysis with a computer. In: Principles of Bloodstain Pattern Analysis: Theory and Practice (eds. S.H. James, P.E. Kish and T.P. Sutton), 241–261. Boca Raton, FL: CRC, Taylor & Francis.

Carter, A.L. and Podworny, E.J. (1991). Bloodstain pattern analysis with a scientific calculator. *Journal of the Canadian Society of Forensic Science* 24 (1): 37–42.

de Castro, T.C., Nickson, T., Carr, D., and Knock, C. (2013). Interpreting the formation of bloodstains on selected apparel fabrics. *International Journal of Legal Medicine* 127: 251–258.

de Castro, T.C., Taylor, M.C., Kieser, J.A., Carr, D.J., Duncan, W. (2015). *Systematic investigation of drip stains on apparel fabrics: the effects of prior-laundering, fibre content and fabric structure on final stain appearance. Forensic Sci. Int.* 250, 98–109.

de Castro, T.C., Carr, D., Taylor, M.C. et al. (2016). Drip bloodstains appearance on inclined apparel fabrics: effect of prior-laundering, fibre content and fabric structure. *Forensic Science International* 266: 488–501.

Chandrasekar, V., Cooper, W.A., and Bringi, V.N. (1988). Axis ratios and oscillations of raindrops. *Journal of the Atmospheric Sciences* 45 (8): 1323–1332.

Ching, B., Golay, M.W., and Johnson, T.J. (1984). Droplet impacts upon liquid surfaces. *Science* 226: 535–537.

Chisum, W.J. (1998, 4th Quarter). *Pitfalls in bloodstain pattern interpretation. CAC (California Association of Criminalists) News*, pp. 14–17.

Chisum, J. (2007). Reconstructions using bloodstain evidence. In: Crime Reconstruction (eds. J. Chisum and B. Turvey), 313–359. Waltham, MA: Academic Press.

Cho, Y., Springer, F., Tulleners, F., and Ristenpart, W. (2015). Quantitative bloodstain analysis: differentiation of contact transfer patterns versus spatter patterns on fabric via microscopic inspection. *Forensic Science International* 249: 233–240.

Cleland, C. (2002). Methodological and epistemic differences between historical science and experimental science. *Philosophy of Science* 69: 474–496.

Collaborative Testing Service – Forensic Testing Program. *Test No. 19-5601: Bloodstain Pattern Analysis, 113.*

Colloff, P. (2018a). *Blood will tell: part I.* ProPublica. https://features.propublica.org/blood-spatter/mickey-bryan-murder-blood-spatter-forensic-evidence/ (accessed 1 October 2020).

Colloff, P. (2018b). *Blood will tell: part II.* ProPublica. https://features.propublica.org/blood-spatter/joe-bryan-conviction-blood-spatter-forensic-evidence/ (accessed 1 October 2020).

Comiskey, P.M., Yarin, A.L., Kim, S., and Attinger, D. (2016). Prediction of blood back spatter from a gunshot in bloodstain pattern analysis. *Physical Review Fluids* 1 (4): 043201.

Comiskey, P.M., Yarin, A.L., and Attinger, D. (2019). Implications of two backward blood spatter models based on fluid dynamics for bloodstain pattern analysis. *Forensic Science International* 301: 299–305.

Committee on Science, Engineering, and Public Policy (2009). On Being a Scientist: A Guide to Responsible Conduct in Research, 3e, 5. Washington, DC: National Academy of Sciences,

National Academy of Engineering, and Institute of Medicine of the National Academies. The National Academies Press.

Connolly, C., Illes, M., and Fraser, J. (2012). Affect of impact angle variations on area of origin determination in bloodstain pattern analysis. *Forensic Science International* 223: 233–240.

Courts, C., Madea, B., and Schyma, C. (2012). Persistence of biological traces in gun barrels—an approach to an experimental model. *International Journal of Legal Medicine* 126 (3): 391–397.

Dale, W.M. and Becker, W.S. (2007). The Crime Scene: How Forensic Science Works. New York: Kaplan Trade, Inc.

Das, R., Collins, A., Verma, A. et al. (2015). Evaluating simulant materials for understanding cranial backspatter from a ballistic projectile. *Journal of Forensic Sciences* 60 (3): 627–637.

Daubert v. Merrell Dow Pharmaceuticals, Inc. (1993). 509 U.S. 579, 589.

Davidson, P.L., Taylor, M.C., Wilson, S.J. et al. (2012). Physical components of soft-tissue ballistic wounding and their involvement in the generation of blood backspatter. *Journal of Forensic Sciences* 57: 1339–1342.

Davies, J.T. and Rideal, E.K. (1963). Interfacial Phenomena, 2e. New York: Academic Press.

De Forest, P.R. (1990). A review of interpretation of bloodstain evidence at crime scenes. *Journal of Forensic Science* 35 (6): 1491–1495.

De Forest, P.R. (1991). *Trace evidence: a holistic view and approach. Proceedings of the International Symposium on the Forensic Aspects of Trace Evidence, FBI Academy, Quantico, USGPO,* Quantico, VA (June 1991), 9–15.

De Forest, P.R. (1998). Proactive forensic science. *Science & Justice* 38 (1): 1–2.

De Forest, P.R. (1999). Recapturing the essence of criminalistics. *Science & Justice: Journal of the Forensic Science Society* 39 (3): 196.

De Forest, P.R. (2001). What is trace evidence? In: Forensic Examination of Glass and Paint (ed. B. Caddy), 1–25. New York: Taylor & Francis.

De Forest, P.R. (2005). Crime scene investigation. In: Crime Scene Investigation, Encyclopedia of Law Enforcement (ed. L. Sullivan), 111–116. Thousand Oaks, CA: Sage Publications.

De Forest, P.R. (2018). *Physical aspects of blood traces as a tool in crime scene investigation. CAC News*, pp. 30–33.

De Forest, P.R. and Haag, L.C. (1990). *Trajectory reconstructions at crime scenes using low-power lasers and positioning stages. Proceedings of the International Symposium on the Forensic Aspects of Mass Disasters and Crime Scene Reconstruction*, FBI Academy, Quantico, VA, USGPO (23–29 June 1990), 275–76.

De Forest, P.R. and Kammrath, B.W. (2014). *The creeping inversion. Proceedings of the Annual Meeting of the Northeastern Association of Forensic Scientists*, Hershey, PA (3–6 November 2014), 35–36.

De Forest, P.R. and Lentini, J.J. (2003). *Reducing the prevalence of pseudoscientific interpretations in complex physical evidence casework. Proceedings of the 55th Annual Meeting of the American Academy of Forensic Sciences*, Chicago, IL (17–22 February 2003).

De Forest, P.R. and Pizzola, P.A. (1995). Letter to the editor. *Journal of Forensic Science*, Response to Letter to the Editor by Herbert MacDonell, November 1995 40 (6): 928–931.

De Forest, P.R., Gaensslen, R.E., and Lee, H.C. (1983). Forensic Science: An Introduction to Criminalistics. New York: McGraw-Hill Book Company.

De Forest, P.R., Bucht, R., Kammerman, F. et al. (2009). Blood on Black-Enhanced Visualization of Bloodstains on Dark Surfaces. Washington, DC: US Dept of Justice. NIJ, Award 2006-DN-BX-K026.

De Forest, P.R., Mattheson, G., and Pizzola, P.A. (2010). *Panel Discussion: "Testing Facility Paradigm and Privatization of Forensic Science" at American Society of Crime Lab Directors Annual Meeting, Baltimore, MD (16 September 2010).*

De Forest, P.R., Matheson, G.B., and Pizzola, P.A. (2010a). *The testing facility paradigm and the privatization of forensic science. American Society of Crime Lab Directors Annual Meeting*, Baltimore, MD (16 September 2010).

De Forest, P.R., Pizzola, P.A., Valentin, P., and Matheson, G. (2010b). *The criminalistics laboratory as a testing facility. Proceedings of the Joint Plenary Session of the Annual Meeting of the Northeastern Association of Forensic Scientists and the New England Division of IAI*, Manchester, VT (8–12 November 2010).

De Forest, P., Bucht, R., Buzzini, P. et al. (2015). *The making of the criminalistics maestro: on the knowledge, skills, and abilities to oversee and coordinate the work on non-routine and complex cases. Proceedings of the 67th Meeting of the American Academy of Forensic Sciences*, Orlando, FL (February 2015).

De Forest, P.R., Pizzola, P.A., Ristenbatt III, R.R., and Diaczuk, P.R. (2015). *Bloodstain patterns associated with gunshot wounds – misconceptions. Proceedings of the 67th Annual Meeting of the American Academy of Forensic* Sciences, Orlando, FL (February 2015).

DeLougherly, T.G. (2015). Tests of hemostasis and thrombosis. In: Hemostasis and Thrombosis, 3e (ed. T.G. Lougherly). Switzerland: Springer.

Denison, D., Porter, A., Mills, M., and Schroter, R.C. (2011). Forensic implications of respiratory derived blood spatter distributions. *Forensic Science International* 204 (1–3): 144–155.

Diaczuk, P.J., Herschman, Z., Pizzola, P.A., and De Forest, P.R. (2002). *A new experimental model for evaluating mechanisms of gunshot spatter. Proceedings of the 99th Semiannual Seminar of the California Association of Criminalists*, San Francisco, CA (7–11 May 2002).

Diaczuk, P.J., Hieptas, J., Pizzola, P.A., and De Forest, P.R. (2003). *A new experimental model for evaluating mechanisms of gunshot spatter. Proceedings of the 55th Annual Meeting of the American Academy of Forensic Sciences*, Chicago, IL (17–22 February 2003).

Donaldson, A.E., Walker, N.K., Lamont, I.L. et al. (2011). Characterising the dynamics of expirated bloodstain pattern formation using high-speed digital video imaging. *International Journal of Legal Medicine* 125: 757–762.

Doty, K.C. (2015). *A Raman 'spectroscopic clock' for bloodstain age determination: the first week after deposition*. Presentation at 41st Annual Meeting of the Northeastern Association of Forensic Scientists. Criminalistics/Crime Scene Scientific Session, Hyannis, MA: University at Albany, State University of New York (12 October 2015).

Doyle, J.M. (2010). Learning from error in American criminal justice. *The Journal of Criminal Law and Criminology* 100: 109–148.

Dror, I.E. (2020). Cognitive and human factors in expert decision making: six fallacies and the eight sources of bias. *Analytical Chemistry* 92 (12): 7998–8004.

DuBey, I. S. (2019). *A study of the impact of the physical properties of blood on the interpretation of bloodstain patterns in forensic investigations.*

Dubyk, M. and Liscio, E. (2016). Using a 3D laser scanner to determine the area of origin of an impact pattern. *Journal of Forensic Identification* 66 (3): 259–272.

Duggar, A.S. (2000). *Identification of initial and secondary stains in overlapping bloodspatter patterns.* Master's thesis. John Jay College of Criminal Justice, CUNY, New York.

Eastman Kodak Company (1985). Kodak Filters for Scientific and Technical Uses (B-3), 3e. Rochester, NY: Eastman Kodak Company.

Eckert, W.G. and James, S.H. (1987). Blood Evidence in Crime Scene Investigation. Wichita, KS: INFORM Publications.

Eckert, W.G. and James, S.H. (1989). Interpretation of Bloodstain Evidence at Crime Scenes. New York: Elsevier Science Publishing Company.

Eco, U. and Sebeok, T. (eds.) (1983). The Sign of Three: Dupin, Holmes, Peirce. Bloomington, IA: Indiana University Press.

Elliott, T.A. and Ford, D.M. (1972). Dynamic contact angles. Part 7. Impact spreading of water drops in air and aqueous solutions of surface active agents in vapour on smooth paraffin wax surfaces. *Journal of the Chemical Society, Faraday Transactions 1: Physical Chemistry in Condensed Phases* 68: 1814–1823.

EngagedScholarship @ Cleveland State University (n.d.). *The sam sheppard case: 1954–2000.* https://engagedscholarship.csuohio.edu/sheppard/#browse (accessed 1 October 2020).

Everett, D.H. (1972). Manual of symbols and terminology for physicochemical quantities and units, appendix II: definitions, terminology and symbols in colloid and surface chemistry. *Pure and Applied Chemistry* 31 (4): 577–638.
Federal Bureau of Investigation (FBI) (2020). *Rapid DNA. General information: laboratory services.* https://www.fbi.gov/services/laboratory/biometric-analysis/codis/rapid-dna (accessed 4 October 2020).
Flippence, T. and Little, C. (2011). Calculating the area of origin of spattered blood on curved surfaces. *Journal of Bloodstain Pattern Analysis* 27: 3–16.
Flower, W.D. (1928). The terminal velocity of drops. *Proceedings of the Physical Society of the London* 40: 167–176.
Foote, G.B. (1975). The water drop rebound problem: dynamics of collision. *Journal of the Atmospheric Sciences* 32: 390–402.
Ford, D.M. (1974). *Impact spreading of aqueous drops.* Doctoral dissertation. University of Nottingham.
Ford, D.M. and Elliott, T.A. (1974). Dynamic contact angles. *Journal of the Chemical Society* 70: 423–430.
Fukai, J., Shiiba, Y., Yamamoto, T. et al. (1995). Wetting effects on the spreading of a liquid droplet colliding with a flat surface: experiment and modeling. *The Physics of Fluids* 7 (2): 236–247.
Gaensslen, R.E. (1983). Sourcebook in Forensic Serology, Immunology, and Biochemistry. Washington, DC: US Department of Justice, National Institute of Justice.
Gardner, R.M. (1995). Letter to the editor in response to MacDonell letter. *IABPA Newsletter* 11 (2): 4.
Gardner, R.M. (2002). Directionality in swipe patterns. *Journal of Forensic Identification* 52: 57–93.
Geoghegan, P.H., Spence, C.J.T., Wilhelm, J. et al. (2016). Experimental and computational investigation of the trajectories of blood drops ejected from the nose. *International Journal of Legal Medicine* 130 (2): 563–568.
Gillies, D.A. (1956). Scientific evidence in the Sheppard case. *Police Science Legal Abstracts and Notes* 4: 136–142.
Glynn, C. and Ambers, A. (2021). Rapid DNA analysis – need, technology, and applications. In: Portable Spectroscopy and Spectrometry, vol. 2: Applications (eds. R. Crocombe, P.E. Leary and B.W. Kammrath). Hoboken, NJ: Wiley.
Goetz, B. (2017). Bloodstain pattern analysis. In: Forensic Science Reform (eds. W.J. Koen and C.M. Bowers), 279–298. Academic Press https://doi.org/10.1016/B978-0-12-802719-6.00009-1.
Gonor, A.L. and Yakovlev, V.Y. (1977). Impact of a drop on a solid surface. *Fluid Dynamics* 12 (5): 767–771.
Gray, D., Frascione, N., and Daniel, B. (2012). Development of an immunoassay for the differentiation of menstrual blood from peripheral blood. *Forensic Science International* 220 (1–3): 12–18. https://doi.org/10.1016/j.forsciint.2012.01.020.
Green, A.W. (1975). An approximation for the shape of large raindrops. *Journal of Applied Meteorology* 14: 1578–1583.
Greenberg, J. (2016). *How many police departments are in the United States?*, Politifact. https://www.politifact.com/punditfact/statements/2016/jul/10/charles-ramsey/how-many-police-departments-are-us/ (accessed 1 October 2020).
Gross, H. (1924). Criminal Investigation: A Practical Textbook for Magistrates, Police Officers and Lawyers. Adapted from the System Der Kriminalistik. London: Sweet & Maxwell, Limited.
Gunn, R. and Kinzer, G.D. (1949). The terminal velocity of fall for water droplets in stagnant air. *Journal of Meteorology* 6: 243–248.
Hakim, N. and Liscio, E. (2015). Calculating point of origin of blood spatter using laser scanning technology. *Journal of Forensic Sciences* 60 (2): 409–417.
Hall, J.E. (2015). Hemostasis and blood coagulation. In J.E. Hall (Ed.) Guyton and Hall Textbook of Medical Physiology. 13 (pp 483–494). Elsevier, Philadelphia, PA. https://vitalebooks.store/product/guyton-and-hall-textbook-of-medical-physiology-e-book-13th-edition/?msclkid=24e1ebbf430e16afa1c5b4c874aa22fc (accessed 1 October 2020).

Haung, Y.C., Hammit, F.G., and Yang, W.J. (1973). Hydrodynamic phenomena during highspeed collision between liquid droplet and rigid plane. *Journal of Fluids Engineering* 95 (2): 276–294.

Hesselink, W.F. (1931). Blutspuren in der kriminalistischen Praxis. *Angewandte Chemie* 44 (31): 653–655.

Hobbs, R.L. (1981). The impression to drop size ratio for the raindrop foil impactor. *Journal of Applied Meteorology* 20 (3): 301–308.

Hobbs, P.V. and Kezweeny, A.J. (1967). Splashing of a water drop. *Science* 155 (3): 1112–1114.

Hockensmith, R. (2020). *Forensic Program Coordinator, Collaborative Testing Services, Inc.*, Personal communique.

Holbrook, M. (2010). Evaluation of blood deposition on fabric: distinguishing spatter and transfer stains. *IABPA News* 26: 3–10.

Huffman, S.W., Lukasiewicz, K.B., and Brown, C.W. (2003). FTIR hyperspectral images of microscopic droplets of splattered blood. *Microscopy Today* 11 (3): 10–15.

Hulse-Smith, L. and Illes, M. (2007). A blind trial evaluation of crime scene methodology for deducing impact velocity and droplet size from circular bloodstains. *Journal of Forensic Sciences* 52: 65–69.

Hulse-Smith, L., Mehdizadeh, N., and Chandra, S. (2005). Deducing Drop Size and Impact Velocity from Circular Bloodstains. *Journal of Forensic Sciences* 50 (1): 54–63. doi. https://doi.org/10.1520/JFS2003224.

Hurley, N.M. and Pex, J.O. (1990). Sequencing of bloody shoe impressions by blood spatter and blood droplet drying times. *International Association of Bloodstain Pattern Analysts Newsletter* 6: 1–8.

Illes, M. and Boué, M. (2011). Investigation of a model for stain selection in bloodstain pattern analysis. *Canadian Society of Forensic Science Journal* 44: 1–12.

Innocence Project (2020). *Jeff Deskovic.* https://www.innocenceproject.org/cases/jeff-deskovic/ (accessed 1 October 2020).

International Association for Identification (IAI) (2020). *Bloodstain certification requirements.* http://www.theiai.org/certifications/bloodstain/requirements.php (accessed 1 October 2020).

IUPAC (1997). Online version (2019-) created by S. J. Chalk. ISBN 0-9678550-9-8. doi). Compendium of chemical terminology. In: The "Gold Book", 2e (eds. A.D. McNaught and A. Wilkinson). Oxford: Blackwell Scientific Publications https://doi.org/10.1351/goldbook.

Jackowski, C., Thali, M., Aghayev, E. et al. (2006). Postmortem imaging of blood and its characteristics using MSCT and MRI. *International Journal of Legal Medicine* 120: 233–240.

James, S.H., Kish, P.E., and Sutton, T.P. (2005). Principles of Bloodstain Pattern Analysis: Theory and Practice, 45–48. Boca Raton, FL: CRC Press.

Jermy, M. and Taylor, M. (2012). The mechanics of bloodstain formation. In: Forensic Biomechanics (eds. J. Kieser, M. Taylor and D. Carr), 100–136. West Sussex, UK: Wiley.

Joris, P., Develter, W., Jenar, E. et al. (2014). Calculation of bloodstain impact angles using an active bloodstain shape model. *Journal of Forensic Radiology and Imaging* 2: 188–198.

Joris, P., Develter, W., Jenar, E. et al. (2015). Hemovision: an automated and virtual approach to bloodstain pattern analysis. *Forensic Science International* 251: 116–123.

Jussila, J. (2004). Preparing ballistic gelatine–review and proposal for a standard method. *Forensic Science International* 141 (2–3): 91–98.

Jussila, J., Leppäniemi, A., Paronen, M., and Kulomäki, E. (2005). Ballistic skin simulant. *Forensic Science International* 150 (1): 63–71.

Kabaliuk, N., Jermy, M.C., Morison, K. et al. (2013). Blood drop size in passive dripping from weapons. *Forensic Science International* 228: 75–82.

Kabaliuk, N., Jermy, M.C., Williams, E. et al. (2014). Experimental validation of a numerical model for predicting the trajectory of blood drops in typical crime scene conditions, including droplet deformation and breakup, with a study of the effect of indoor air currents and wind on typical spatter drop trajectories. *Forensic Science International* 245: 107–120.

Kane, J.W. and Sternheim, M.M. (1978). Life Science Physics, 250–255. New York: Wiley.

Karger, B. and Brinkmann, B. (1997). Multiple gunshot suicides: potential for physical activity and medico-legal aspects. *International Journal of Legal Medicine* 110 (4): 188–192.

Karger, B., Nüsse, R., Schroeder, G. et al. (1996). Backspatter from experimental close-range shots to the head I. Macrobackspatter. *International Journal of Legal Medicine* 109: 66–74.

Karger, B., Nüsse, R., Tröger, H.D., and Brinkmann, B. (1997). Backspatter from experimental close-range shots to the head II. Microbackspatter and the morphology of bloodstains. *International Journal of Legal Medicine* 110: 27–30.

Karger, B., Rand, S.P., and Brinkmann, B. (1997). Experimental bloodstains on fabric from contact and from droplets. *International Journal of Legal Medicine* 111 (1): 17–21.

Karger, B., Nüsse, R., and Bajanowski, T. (2002). Backspatter on the firearm and hand in experimental close-range gunshots to the head. *The American Journal of Forensic Medicine and Pathology* 23 (3): 211–213.

Kassin, S.M., Dror, I.E., and Kukuck, J. (2013). The forensic confirmation bias: Problems, perspectives, and proposed solutions. *Journal of Applied Research in Memory and Cognition* 2: 42–52.

Kenner, T., Leopold, H., and Hinghofer-Szalkay, H. (1977). The continuous high-precision measurement of the density of flowing blood. *Pflügers Archiv European Journal of Physiology* 370: 25–29.

Kinnel, P.I.A. (1972). The acoustic measurement of water drop impacts. *Journal of Applied Meteorology* 11 (4): 691–694.

Kirk, P.L. (1953). Crime Investigation: Physical Evidence and the Police Laboratory. New York: Interscience Publishers, Inc.

Kirk, P.L. (1955). *Affidavit of Paul Leland Kirk in STATE OF OHIO vs. SAMUEL H. SHEPPARD.* State of Ohio, Cuyahoga County, Court of Common Pleas, Criminal Branch, No. 64571, 1–27.

Kirk, P.L. (1967). Blood-a neglected criminalistics research area. *Law Enforcement Science and Technology* 1: 267–272.

Kleiber, M., Stiller, D., and Weigand, P. (2001). Assessment of shooting distance on the basis of bloodstain analysis and histological examinations. *Forensic Science International* 119: 260–262.

Kneubuehl, B.P. (2012). Maximum flight velocity of blood drops in analysing blood traces. *Forensic Science International* 219: 205–207.

Kodet-Sherwin, L., Pizzola, P.A., and De Forest, P.R. (1988). *A critical assessment of the phenomenon of secondary bloodspatter*. Presentation at the 40th Annual Meeting of American Academy of Forensic Science, Philadelphia.

Koen, W.J. (2017). Bloodstain pattern analysis: case study: David Camm. In: Forensic Science Reform (eds. W.J. Koen and C.M. Bowers), 272–278. Academic Press https://doi.org/10.1016/B978-0-12-802719-6.00009-1.

Kruger, J. and Dunning, D. (1999). Unskilled and unaware of it: how difficulties in recognizing one's own incompetence lead to inflated self-assessments. *Journal of Personality and Social Psychology* 77: 1121–1133.

Kunz, S.N., Brandtner, H., and Meyer, H.J. (2015). Characteristics of backspatter on the firearm and shooting hand—an experimental analysis of close-range gunshots. *Journal of Forensic Sciences* 60 (1): 166–170.

Kunz, S.N., Adamec, J., and Grove, C. (2017). Analyzing the dynamics and morphology of cast-off pattern at different speed levels using high-speed digital video imaging. *Journal of Forensic Sciences* 62: 428–434.

Laan, N., De Bruin, K.G., Slenter, D. et al. (2015). Bloodstain pattern analysis: implementation of a fluid dynamic model for position determination of victims. *Scientific Reports* 5 (1): 1–8.

Laber, T.L. (1985). Diameter of bloodstains as a function of origin, distance fallen, and volume of drop. *IABPA Newsletter* 2 (1): 12–16.

Laber, T.L. and Epstein, B.P. (1983). Bloodstain Pattern Analysis. Minneapolis, MN: Callen Publishing, Inc.

Laber, T.L. and Epstein, B.P. (1994). Experiments and Practical Exercises in Bloodstain Pattern Analysis, 4e. Minneapolis, MN: Callen Publishing, Inc.

Laber, T.L. and Epstein, B.P. (2001). Substrate effects on the clotting time of human blood. *Canadian Society of Forensic Science Journal* 34: 209–214.

Laber, T.L., Epstein, B.P., and Taylor, M.C. (2007). *High Speed Digital Video Analysis of Bloodstain Pattern Formation from Common Bloodletting Mechanisms Project Report. MFRC Project No. 06-S-02.* http://www.ameslab.gov/mfrc/bloodstain_pattern_formation (accessed 1 October 2020).

Lee, H.C. (1982). Identification and grouping of bloodstains. In: Forensic Science Handbook (ed. R. Saferstein), 267–337. Englewood Cliffs, NJ: Prentice-Hall, Inc.

Lee, H.C., Gaensslen, R.E., and Pagliaro, E.M. (1986). Bloodstain volume estimation. *IABPA News* 3 (2): 47–54.

Lee, V.K., Lanzi, A.M., Ngo, H. et al. (2014). Generation of multi-scale vascular network system within 3D hydrogel using 3D bio-printing technology. *Cellular and Molecular Bioengineering* 7 (3): 460–472.

Lentini, J.J. (2015). It's time to lead, follow, or get out of the way – criminalistics section. *AAFS Academy News.* 45 (5).

Li, J., Li, X., and Michielsen, S. (2016). Alternative method for determining the original drop volume of bloodstains on knit fabrics. *Forensic Science International* 263: 194–203.

Liscio, E., Bortot, S., Frankcom, J. et al. (2015). A preliminary validation of the use of 3D scanning for bloodstain pattern analysis. *J. Bloodstain Pattern Analysts* 31: 3–10.

List, R. and Hand, M.J. (1971). Wakes of freely falling water drops. *The Physics of Fluids* 14 (8): 1648–1655.

List, R., Low, T.B., and McTaggart-Cowan, J.D. (1974). Collisions of raindrops with chaff. *Journal of Applied Meteorology* 13: 796–799.

Liu, Y., Attinger, D., and De Brabanter, K. (2020). Automatic classification of bloodstain patterns caused by gunshot and blunt impact at various distances. *Journal of Forensic Sciences* 65: 729–743.

Lochte, T. and Ziempke, E. (1914). *Die Untersuchnung von Blutspuren.* Translated as "The Investigation of Blood Tracks", "Gerichtsärztliche und polizeiärztliche Technik: eine Handbuch für Studierende, Ärzte, Medizinalbeamte und Juristen". 152–166.

Lock, J.A. (1982). The physics of air resistance. *The Physics Teacher* 20 (3): 158–160.

Lowe, G.D.O. (1987). Blood rheology in vitro and in vivo. *Baillière's Clinical Haematology* 1: 597–636.

Lytle, L.T. and Hegecock, D.G. (1978). Chemiluminescence in the visualisation of forensic bloodstains. *Journal of Forensic Sciences* 23: 550–562.

MacDonell, H.L. (1971). Interpretation of bloodstains: physical considerations. In: Legal Medicine Annual (ed. C. Wecht), 91–136. New York: Appleton-Century Crofts.

MacDonell, H. L. (1972). Flight Characteristics and Stain Patterns of Human Blood (Vol. 71, No. 4). National Institute of Law Enforcement and Criminal Justice, Washington, DC.

MacDonell, H.L. (1981). Criminalistics: blood stain examination. In: Forensic Sciences, vol. 3 (ed. C. Wecht), 37-4–37-17. New York: Matthew Bender & Company, Inc.

MacDonell, H.L. (1982). Bloodstain Pattern Interpretation. Corning, NY: Laboratory of Forensic Science.

MacDonell, H.L. (1993). Bloodstain Patterns. Elmira Heights, NY: Golos Printing, Inc.

MacDonell, H.L. (2011). Credit where it's due. *Journal of Forensic Identification* 61: 210–221.

MacDonell, H.L. and Bialousz, L.F. (1971). Flight Characteristics and Stain Patterns of Human Blood. Washington, DC: U.S. Department of Justice, Law Enforcement Assistance Administration.

MacDonell, H.L. and Bialousz, L.F. (1973). Laboratory Manual on the Geometric Interpretation of Human Bloodstain Evidence, vol. 61. Painted Post, NY: Laboratory of Forensic Science.

MacDonell, H.L. and Brooks, B.A. (1977). Detection and significance of blood in firearms. In: Legal Medicine Annual 1977 (ed. C. Wecht), 185–199. New York: Appleton-Century-Crofts.

MacDonell, H.L. and Copp Organization, Inc, & United States of America (1977). Preserving bloodstain evidence at crime scenes. *Law Order* 25: 66–69.

MacDonell, H.L. and De Lige, K. (1990). On measuring the volume of very small drops of fluid blood and correlation of this relationship to bloodstain diameter. *IABPA News* 7: 14–25.

MacDonell, H.L. and Panchou, C.G. (1979). Bloodstain pattern interpretation. *Identification News* 29: 3–5.

MacDonell, H.L. and Panchou, C.G. (1979). Bloodstain patterns on human skin. *Journal of the Canadian Society of Forensic Science* 12: 134–141.

Magarvey, R.H. and Taylor, B.W. (1956). Free-fall breakup of large drops. *Journal of Applied Physics* 27 (10): 1129–1135.

Magono, C. (1954). On the shape of water drops falling in stagnant air. *Journal of Meteorology* 11: 77–79.

Marin, N., Buszka, J., and Pizzola, P.A. (2010). *Scientific Analysis of Bloodstain Patterns: Laboratory Workbook*. Special Investigations Unit (SIU), NYC Office of Chief Medical Examiner, Department of Forensic Pathology, Forensic Sciences Training Program. National Institute of Justice (NIJ) Grant 2009-DN_BX-K205.

Marin, N., Buszka, J., and Pizzola, P.A. (2011). *Measurement of the spread of blood droplets on impact*. Unpublished experiments conducted at NY OCME, Special Investigations Unit.

Markert, C.L. and Ursprung, H. (1962). The ontogeny of isozyme patterns of lactate dehydrogenase in the mouse. *Developmental Biology* 5 (3): 363–381. https://doi.org/10.1016/0012-1606(62)90019-2.

McDonald, J.E. (1954). The shape and aerodynamics of large raindrops. *Journal of Meteorology* 11: 478–494.

McTaggart-Cowan, J.D. and List, R. (1975). Collision and break-up of water drops at terminal velocity. *Journal of the Atmospheric Sciences* 32: 1401–1411.

Mehdizadeh, N.Z., Chandra, S., and Mostaghim, J. (2004). Formation of fingers around the edges of a drop hitting a metal plate with high velocity. *Journal of Fluid Mechanics* 510: 353–373.

Meneely, K. (2011). Alternative resources in bloodstain pattern analysis. *Journal of Bloodstain Pattern Analysis* 27: 3–8.

Merrill, E.W. (1969). Rheology of blood. *Physiological Reviews* 49: 863–888.

Miskelly, G.M. and Wagner, J.H. (2005). Using spectral information in forensic imaging. *Forensic Science International* 155: 112–118.

Mnookin, J.L., Cole, S.A., Dror, I.E. et al. (2011). The need for a research culture in the forensic sciences. *UCLA Law Review* 58: 725.

Mole, R.H. (1948). Fibrinolysin and the fluidity of the blood post-mortem. *The Journal of Pathology and Bacteriology* 60: 413–427.

Moore, C.C. (2003). Demonstrative aid for bloodstain pattern examiners. *Journal of Forensic Identification* 53 (6): 639–646.

Moran, T.A. and Viele, C.S. (2005). Normal clotting. *Seminars in Oncology Nursing* 21: 1–11.

Morawitz, P. (1906). Uber einige post-mortale blutveranderunguen. *Beitr z chem Physiol u Path* 8: 1–10.

Morrison, G.S., Neumann, C., and Geoghegan, P.H. (2020). Vacuous standards – subversion of the OSAC standards - development process. *Forensic Science International: Synergy* 2: 206–209.

National Academy of Sciences, N. A (2009). On Being a Scientist: A Guide to Responsible Conduct in Research. Washington, DC: National Academies Press.

National Institute of Forensic Science (NIFS), Australia & New Zealand (2019). *A Multi-Disciplinary Approach to Crime Scene Management, Australia New Zealand Policing Agency (ANZPAA)*, State of Victoria ©, version 1.0.

National Research Council (U.S.) (2004). Committee on Scientific Assessment of Bullet Lead Elemental Composition Comparison. Forensic Analysis: Weighing Bullet Lead Evidence. Washington, DC: National Academy Press.

National Research Council of the National Academies. Committee on Identifying the Needs of the Forensic Science Community. Committee on Science, Technology, and Law Policy and Global Affairs. Committee on Applied and Theoretical Statistics Division on Engineering and

Physical Sciences (2009). Strengthening Forensic Science in the United States: A Path Forward. Washington, DC: The National Academies Press.

Neitzel, G.P. and Smith, M. (2017). The Fluid Dynamics of Droplet Impact on Inclined Surfaces with Application to Forensic Blood Spatter Analysis. Washington, DC: Department of Justice (NCJRS). Award # 2013-DN-BX-K003, Document # 251439, 4.

Nour-Eldin, F. and Wilkinson, J.F. (1957). The blood clotting factors in human saliva. *The Journal of Physiology* 136: 324–332.

Nowack, L., Collins, R., Li, G. et al. (2011). Computer analysis of bloodstain patterns on angled surfaces. *Journal of Bloodstain Pattern Analysis.* 27: 17–28.

Park, R. (2000a). Voodoo Science: The Road from Foolishness to Fraud, 38. New York: Oxford University Press.

Park, R.L. (2000b). Voodoo science and the belief gene. *The Skeptical Inquirer* 24 (5): 26–27.

People v. Castro (1989). 545 N.Y.S.2d 985 (Sup. Ct. 1989).

Perkins, M. (2005). The application of infrared photography in bloodstain pattern documentation of clothing. *Journal of Forensic Identification* 55 (1): 1–9.

Peschel, O., Kunz, S.N., Rothchild, M.A., and Mutzel, E. (2011). Blood stain pattern analysis. *Forensic Science, Medicine and Pathology* 7: 257–270.

Peterson, A.W., Caldwell, D.J., Rioja, A.Y. et al. (2014). Vasculogenesis and angiogenesis in modular collagen–fibrin microtissues. *Biomaterials Science* 2 (10): 1497–1508.

Petrucci, R.H. (1982). General Chemistry: Principles and Modern Applications, 3e. New York: Macmillan Publishing Co., Inc.

Pex, J.O. (2009). The identification and significance of hemospheres in crime scene investigation. *IABPA News* 25 (1): 8–18.

Pex, J.O. and Vaughan, C.H. (1987). Observations of high velocity bloodspatter on adjacent objects. *Journal of Forensic Sciences* 32: 1587–1594.

Pigolkin, I., Leonova, E.N., Dubrovin, I.A., and Nagornov, M.N. (2014). The new working classification of blood stain patterns. *Sudebno-Meditsinskaia Ekspertiza* 57 (1): 11–15.

Piotrowski, E. (1895). Über Entstehung, Form, Richtung u. Ausbreitung der Blutspuren nach Hiebwunden des Kopfes. Vienna, Austria: Slomski.

Pizzola, P.A. (1984). *Blood droplet dynamics and their implication for bloodstain pattern interpretation at crime scenes.* M.S. Thesis. John Jay College of Criminal Justice/City University of New York.

Pizzola, P.A. (1998). *Improvements in the detection of gunshot residue and considerations affecting its interpretation.* Doctoral dissertation. City University of New York, 90.

Pizzola, P.A., Roth, S., and De Forest, P.R. (1986). Blood droplet dynamics – I. *Journal of Forensic Sciences* 31 (1): 36–49.

Pizzola, P.A., Roth, S., and De Forest, P.R. (1986). Blood droplet dynamics – II. *Journal of Forensic Sciences* 31 (1): 50–64.

Pizzola, P.A., Sherwin, L.K., Perkins, J.C., and De Forest, P.R. (1988). *A critical assessment of the phenomenon of gunshot backspatter. Proceedings of the 40th Annual Meeting of the American Academy of Forensic Sciences*, Philadelphia, PA (February 15–20).

Pizzola, P., De Forest, P.R., Martir, K. et al. (1993). *Blood droplet dynamics III. Presented at the Annual Meeting of the American Academy of Forensic Science*, Boston, MA (February 1993).

Pizzola, P.A., Buszka, J.M., Marin, N. et al. (2012). Commentary on "3D bloodstain pattern analysis: ballistic reconstruction of the trajectories of blood drops and determination of the centres of origin of the bloodstains" by Buck et al.[Forensic Sci. Int. 206 (2011) 22–28]. *Forensic Science International* 220 (1–3): e39–e41.

Poe, E.A. (1927). Collected Works of Edgar Allan Poe. New York: Walter J. Black.

Radford, G.E., Taylor, M.C., Kieser, J.A. et al. (2015). Simulating backspatter of blood from cranial gunshot wounds using pig models. *International Journal of Legal Medicine* 130 (4): 985–994. https://doi.org/10.1007/s00414-015-1219-s.

Raymond, M.A. and Hall, R.L. (1986). An interesting application of infrared reflection photography to blood splash pattern interpretation. *Forensic Science International* 31: 189–194.

Raymond, M.A., Smith, E.R., and Liesegang, J. (1996). The physical properties of blood - forensic considerations. *Science & Justice* 36: 153–160.

Raymond, M.A., Smith, E.R., and Liesegang, J. (1996). Oscillating blood droplets - implications for crime scene reconstruction. *Science & Justice* 36 (3): 161–171.

Rein, M. (1993). Phenomena of liquid drop impact on solid and liquid surfaces. *Fluid Dynamics Research* 12 (2): 61–93.

Reynolds, M. and Raymond, M.A. (2008). New bloodstain measurement process using Microsoft Office Excel 2003 autoshapes. *Journal of Forensic Identification* 58: 453–468.

Reynolds, M. and Silenieks, E. (2016). Considerations for the assessment of bloodstains on fabrics. *Journal of Bloodstain Pattern Analysis* 32: 15–20.

Reynolds, M., Franklin, D., Raymond, M.A., and Dadour, I. (2008). Bloodstain measurement using computer-fitted theoretical ellipses: a study in accuracy and precision. *Journal of Forensic Identification* 58: 469–484.

Reynolds, M.E., Raymond, M.A., and Dadour, I. (2009). The use of small bloodstains in blood source area of origin determinations. *Canadian Society Forensic Science Journal* 42: 133–146.

Riddel, J.P., Aouizerat, B.E., Miaskowski, C., and Lillicrap, D.P. (2007). Theories of blood coagulation. *Journal of Pedicatric Oncology Nursing* 24: 123–131.

Ristenbatt, R.R. (2008). Review of:). Bloodstain Pattern Analysis with an Introduction to Crime Scene Reconstruction, 3e. Boca Raton: CRC Press (Taylor & Francis Group), 402p. J For. Sci. 54, p 234. DOI: https://doi.org/10.1111/j.1556-4029.2008.00932.x.

Ristenbatt, R.R. and Shaler, R. (1995). Letter to the editor in response to MacDonell letter. *IABPA Newsletter* 11 (2): 5–10.

Rizer, C. (1955). Police Mathematics: A Textbook in Applied Mathematics for Police, Police Science Series. Springfield, IL: Charles C. Thomas.

Robbins, K.S. (1996). Suggested IABPA terminology list. *IABPA Newsletter* 12 (4): 15–17.

Rogers, N. (2009). *Hematocrit implications for bloodstain pattern analysis*. MS thesis. University of Western Australia, Perth.

Rossi, C., Holbrook, M., James, S.H., and Mabel, D. (2012). Medical and forensic aspects of blood clot formation in the presence of saliva – a preliminary study. *Journal of Bloodstain Pattern Analysis* 28: 3–12.

Rossi, C., Herold, L.D., Bevel, T. et al. (2018). Cranial backspatter pattern production utilizing human cadavers. *Journal of Forensic Sciences* 63 (5): 1526–1532.

Rubio, A., Esperança, P., and Martrille, L. (2014). Backspatter simulation: comparison of a basic sponge and a complex model. *Journal of Forensic Identification* 64 (3): 285.

Ryan, R.T. (1976). The behaviour of large, low surface tension water drops falling at terminal velocity in air. *Journal of Applied Meteorology* 15: 157–165.

San Pietro, D. and Kammrath, B.W. (2018). Forensic science: a forensic scientist's perspective. *The SciTech Lawyer* 15 (1): 34–37.

San Pietro, D., Kammrath, B.W., and De Forest, P.R. (2019). Is forensic science in danger of extinction? *Science & Justice* 59: 199–202.

Schyma, C., Lux, C., Madea, B., and Courts, C. (2015). The 'triple contrast'method in experimental wound ballistics and backspatter analysis. *International Journal of Legal Medicine* 129 (5): 1027–1033.

Seiyama, A., Suzuki, Y., and Maeda, N. (1993). Increased viscosity of erythrocyte suspension upon hemolysis. *Colloid & Polymer Science* 271: 63–69.

Shaler, R. (2012). Crime Scene Forensics: A Scientific Method Approach. Boca Raton, FL: CRC Press, Taylor & Francis Group.

Shar, S. and Staretz, M. (2015). *Estimating the age of bloodstains by comparing oxidized and reduced hemoglobin absorption spectra*. Presentation at 41st Annual Meeting of the Northeastern

Association of Forensic Scientists. Criminalistics/Crime Scene Scientific Session. Hyannis, MA: Cedar Crest University.

Shen, A.R., Brostow, G.J., and Cipolla, R. (2006). *Toward automatic blood spatter analysis in crime scenes. Proceedings of the Institution of Engineering and Technology Conference on Crime and Security*, London, UK (13–14 June 2006), 378–383.

Šikalo, Š. and Ganić, E.N. (2006). Phenomena of droplet–surface interactions. *Experimental Thermal and Fluid Science* 31 (2): 97–110.

Slemko, J.A. (2003). Bloodstains on fabric: the effects of droplet velocity and fabric composition. *IABPA Newsletter* 19 (3): 3–11.

Spengler, J.D. and Gokhale, N.R. (1973). Drop impactions. *Journal of Applied Meteorology* 12: 316–321.

Spilhaus, A.F. (1948). Raindrop size, shape and falling speed. *Journal of Meteorology* 5: 108–110.

Stephens, B.G. and Allen, T.B. (1983). Back spatter of blood from gunshot wounds—observations and experimental simulation. *Journal of Forensic Science* 28 (2): 437–439.

Stone, I.C. (1982). Characteristics of firearms and gunshot wounds as markers of suicide. *American Journal of Forensic Medicine and Pathology*. 13: 275–280.

Stone, I.C. (1987). Observations and statistics relating to suicide weapons. *Journal of Forensic Sciences* 32: 711–716.

Stotesbury, T., Illes, M., Wilson, P., and Vreugdenhil, A. (2015). A commentary on synthetic blood substitute research and development. *Journal of Bloodstain Pattern Analysis*. 31: 3–6.

Stotesbury, T., Bruce, C., Illes, M., and Hanley-Dafoe, R. (2016). Design considerations for the implementation of artificial fluids as blood substitutes for educational and training use in the forensic sciences. *Forensic Science Policy and Management* 7: 81–86.

Stotesbury, T., Illes, M., Jermy, M. et al. (2016). Three physical factors that affect crown growth of the impact mechanism and its implication for bloodstain pattern analysis. *Forensic Science International* 266: 254–262.

Stotesbury, T., Taylor, M.C., and Jermy, M.C. (2017). Drip stain formation dynamics of blood onto hard surfaces and comparison with simple fluids for blood substitute development and assessment. *Journal of Forensic Sciences* 62: 74–82.

Stow, C.D. and Hadfield, M.G. (1981). An experimental investigation of fluid flow resulting from the impact of a water drop with an unyielding dry surface. *Proceedings of the Royal Society of London* 373: 419–441.

Stufflebeam, R. (2017). Popper: Conjectures and Refutations. Department of Philosophy, University of New Orleans. https://www.youtube.com/watch?v=q3qUzLvOKJo (accessed 1 October 2020).

Sweet, M.J. (1993). Velocity measurements of projected bloodstains from a medium velocity impact source. *Canadian Society of Forensic Science Jouranl* 26 (3): 103–110.

SWGSTAIN (2002). *Scientific working group on bloodstain pattern analysis. Minutes of fall meeting*. Quantico (VA). World Wide Web. http://www.swgstain.org/documents/Fall2002Minutes.pdf (accessed 28 January 2015).

SWGSTAIN (2009). *Scientific working group on bloodstain pattern analysis. Recommended terminology*. World Wide Web. http://www.swgstain.org/documents/SWGSTAIN%20Terminology.pdf (accessed 19 May 2015).

Takeichi, S., Wakasugi, C., and Shikata, I. (1984). Fluidity of cadaveric blood after sudden death. Part I. Post mortem fibrinolysis and plasma catecholamine level. *The American Journal of Forensic Medicine and Pathology* 5: 223–228.

Takeichi, S., Wakasugi, C., and Shikata, I. (1985). Fluidity of cadaveric blood after sudden death. Part II. Mechansim of release of plasminogen activator from blood vessels. *The American Journal of Forensic Medicine and Pathology* 6: 25–29.

Takeichi, S., Tokunaga, I., Hayakumo, K., and Maeiwa, M. (1986). Fluidity of cadaveric blood after sudden death: part III. Acid-base balance and fibrinolysis. *The American Journal of Forensic Medicine and Pathology* 7: 35–38.

Tanton, R.L. (1979). Jury preconceptions and their effect on expert scientific testimony. *Journal of Forensic Sciences* 24 (3): 681–691.

Taylor, M.C., Laber, T.L., Epstein, B.P. et al. (2011). The effect of firearm muzzle gases on the backspatter of blood. *International Journal of Legal Medicine* 125 (5): 617–628.

Taylor, M.C., Laber, T.L., Kish, P.E. et al. (2016). The reliability of pattern classification in bloodstain pattern analysis, part 1: bloodstain patterns on rigid non-absorbent surfaces. *Journal of Forensic Sciences* 61: 922–927.

Taylor, M.C., Laber, T.L., Kish, P.E. et al. (2016). The reliability of pattern classification in bloodstain pattern analysis-part 2: bloodstain patterns on fabric surfaces. *Journal of Forensic Sciences* 61: 1461–1466.

Templeman, H. (1990). Errors in blood droplet impact angle reconstruction using a protractor. *Journal of Forensic Identification* 40 (1): 15–22.

Thali, M.J., Kneubuehl, B.P., Zollinger, U., and Dirnhofer, R. (2002). The “skin–skull–brain model”: a new instrument for the study of gunshot effects. *Forensic Science International* 125 (2–3): 178–189.

Thali, M.J., Kneubuehl, B.P., Zollinger, U., and Dirnhofer, R. (2002). A study of the morphology of gunshot entrance wounds, in connection with their dynamic creation, utilizing the “skin–skull–brain model”. *Forensic Science International* 125 (2–3): 190–194.

Thorwald, J. (1966). Crime and Science: The New Frontier in Criminology. New York: Harcourt, Brace & World, Inc.

Thurston, G.B. and Henderson, N.M. (2006). Effects of flow geometry on blood viscoelasticity. *Biorheology* 43 (6): 729–746.

United States v. Davis (1996). 103 F.3d 660 (8th Cir.).

Verhoff, M.A. and Karger, B. (2003). Atypical gunshot entrance wound and extensive backspatter. *International Journal of Legal Medicine* 117: 229–231.

Virchow, R. (1871). Die Cellularpathologie (Vol. 1). 4th ed, p. 194, Berlin.

Virkler, K. and Lednev, I.K. (2010). Raman spectroscopic signature of blood and its potential application to forensic body fluid identification. *Analytical and Bioanalytical Chemistry* 396 (1): 525–534.

Wallace, E.W., Cunningham, M.J., and Boggiano, D. (2016). Crime Scene Unit Management: A Path Forward. New York: Routledge: Taylor and Francis Group.

Walraven, J. (1995a). *The Simpson trial transcripts: Dr. Fredrich Reiders 7/24 and 8/14.* https://www.simpson.walraven.org/ (accessed 1 October 2020).

Walraven, J. (1995b). *The Simpson trial transcripts: SA Martz 7/25 and 7/26.* https://www.simpson.walraven.org/ (accessed 1 October 2020).

Weimann, W. (1931). Uber das Verspritzen von Gewebeteilen aus EinschuBoffnungen und seine kriminalistische Bedeutung. *Deutsche Zeitschrift für die gesamte gerichtliche Medizin* 17: 92–105.

Weinstock, R., Leong, G.B., and Silva, J.A. (2000). Ethics. In: Encyclopedia of Forensic Sciences (eds. J.A. Siegel, G.C. Knupfer and P.J. Saukko), 706–712. San Diego, CA: Academic Press.

Wells, J.K. (2006). *Investigation of factors affecting the region of origin estimate in bloodstain pattern analysis.* Master’s thesis. University of Canterbury. http://ir.canterbury.ac.nz/bitstream/handle/10092/1419/thesis_fulltext.pdf?sequence=2&isAllowed=y.

White, B. (1986). Bloodstain patterns on fabrics: the effect of drop volume, dropping height and impact angle. *Journal of the Canadian Society of Forensic Science* 19 (1): 3–36.

Whitehead, P.H. and Divall, G.B. (1973). Assay of “soluble fibrinogen” in bloodstain extracts as an aid to identification of menstrual blood in forensic science: preliminary findings. *Clinical Chemistry* 19 (7): 762–765. https://doi.org/10.1093/clinchem/19.7.762.

Wiedeman, M.P. (1963). Dimensions of blood vessels from distributing artery to collecting vein. *Circulation Research* 12 (4): 375–378.

Williams, E., Neumann, E., and Taylor, M. (2012). The development of a passive, closed-system pig blood collection apparatus for bloodstain pattern analysis research and crime scene reconstruction. *Journal of Bloodstain Pattern Analysis* 28: 11–18.

Williams, E., Graham, E.S., Jermy, M.C. et al. (2019). The dynamics of blood drop release from swinging objects in the creation of cast-off bloodstain patterns. *Journal of Forensic Sciences* 64: 413–421.

Willis, C., Piranian, A.K., Donaggio, J.R. et al. (2001). Errors in the estimation of the distance of fall and angles of impact blood drops. *Forensic Science International* 123: 1–4.

Wobus, H., Murray, F.W., and Koenig, L.R. (1971). Calculation of the terminal velocity of water drops. *Journal of Applied Meterology* 10 (4): 751–754.

Wolson, T.L. (1995). Documentation of bloodstain pattern evidence. *Journal of Forensic Identification* 45 (4): 396–408.

Wonder, A.Y. (2003). Fact or fiction in bloodstain pattern evidence. *Science & Justice* 43 (3): 166–168.

Wonder, A.Y. and Yezzo, G.M. (2015). Bloodstain Patterns: Identification, Interpretation, and Application, vol. 20–21, 145. Amsterdam: Elsevier Academic Press.

Worthington, A.M. (1963). A Study of Splashes. With an Introduction and Notes by Irwin. New York: K.G. The McMillan Co.

Yen, K., Thali, M.J., Kneubuehl, B.P. et al. (2003). Blood-spatter patterns: hands hold clues for the forensic reconstruction of the sequence of events. *The American Journal ofForensic Medicin. and Pathology* 24: 132–140.

Yuen, S., Taylor, M.C., Owen, G., and Elliot, D.A. (2017). The reliability of swipe/wipe classification and directionality determination methods in bloodstain pattern analysis. *Journal of Forensic Sciences* 62: 1037–1042.

Zajac, R., Osborne, N., Singley, L., and Taylor, M. (2015). Contextual bias: what bloodstain pattern analysts need to know. *J. Bloodstain Pattern Analysis* 31: 7–16.

Ziempke, E. (1914). Die Untersuchnung von Blutspuren. Translated: “The investigation of blood tracks”. In: Gerichtsarziliche und polizieartzliche Technik (ed. T. Lochte), 152–166. Wiesbaden: J.F. Bergmann.

Zwingli, M. (1941). Uber Spuren an der SchieBhand nach Schub mit Faustfeuerwaffen. *Archiv für Kriminologie* 108: 1–26.

Appendix 1

Fundamentals Revisited

Unfortunately, past experiences dictate that some of the basic tenets of scientific crime scene investigation should be reviewed here.

- Physical evidence cannot be wrong. Only the human failure to recognize and preserve it, the human failure to properly interpret it, or human corruption can lead to an incorrect conclusion.
- One should make a conscious effort to utilize the scientific method. This must be developed as a habit.
- Any hypothesis is provisional. One must consider that additional facts may come to light at any time. A working hypothesis should have predictive value, and it should serve as a custom-made, adaptive framework for a scientific investigation.
- One can uncover facts or physical evidence. One can overlook facts or physical evidence. One cannot change facts – physical evidence can be altered, contaminated, or destroyed.
- Always consider, evaluate, and assign proper weight to alternate hypotheses. This is essential in developing valid hypotheses. As a side benefit, this is useful in anticipating opposing party approaches.
- One should openly invite criticism of hypotheses. Often it is helpful to work with a qualified independently minded partner as a "sounding board." This is also useful with an honest opposing expert.
- Never approach a case with preconceived notions. Approach the case with an analytical "open" mind, not necessarily a "blank" one. One should never let the emotional content of a case, or the expectations of others, influence one's objectivity.
- Never succumb to the desire to make facts conform to theories. Don't try to change the facts. Don't identify with a hypothesis. Don't become "married" to it, no matter how aesthetically appealing it may be. Be willing to change the hypothesis whenever necessary. This often runs counter to "human nature," but it is at the core of the scientific method.
- Widen one's perspective. One should put oneself in the place of an "opposing expert." Make an honest effort to prove that the assumptions of those you are "working for" in the case are wrong. One should be mindful that a scientist does not care about the particular outcome of the legal case.

Blood Traces: Interpretation of Deposition and Distribution, First Edition. Peter R. De Forest, Peter A. Pizzola, and Brooke W. Kammrath.

- Nothing replaces firsthand scene experience. When forced to work exclusively with secondary information, appropriate caution must be applied.
- Although often overlooked, ceilings must be carefully examined for the presence of physical evidence.
- The effect of scene experience is synergistic with laboratory work – either generally or specifically. Criminalistics functions are handicapped where the scientific problem to be explored is defined and limited by a non-scientifically trained investigator or similarly circumscribed by naïve or uninformed evidence submissions. In such a case the, laboratory can only be reactive.
- Be mindful of one's own limitations and the limitations of the science. Do not let one's ego get in the way. Never feel obligated to always have an answer. In some cases, the available information is limited, and a detailed reconstruction may not be possible. In others, a detailed reconstruction may be possible, but it may not be directly relevant to the issues in the case. It may not address competing scenarios.
- There is no such thing as a perfect crime scene investigation. "Monday morning quarterbacking" by scientists and others will always be possible. Although perfection will never be achieved, in this or any other human endeavor, it should be sought.

Index

Blood Traces: Interpretation of Deposition and Distribution, First Edition. Peter R. De Forest, Peter A. Pizzola, and Brooke W. Kammrath.

W